Integrated Applications of Geo-Information in Environmental Monitoring

Integrated Applications of Geo-Information in Environmental Monitoring

Editors

Weicheng Wu
Yalan Liu

MDPI • Basel • Beijing • Wuhan • Barcelona • Belgrade • Manchester • Tokyo • Cluj • Tianjin

Editors
Weicheng Wu
East China University of Technology
China

Yalan Liu
Chinese Academy of Sciences
China

Editorial Office
MDPI
St. Alban-Anlage 66
4052 Basel, Switzerland

This is a reprint of articles from the Special Issue published online in the open access journal *Remote Sensing* (ISSN 2072-4292) (available at: https://www.mdpi.com/journal/remotesensing/special_issues/Geo-Information_technology_applications).

For citation purposes, cite each article independently as indicated on the article page online and as indicated below:

LastName, A.A.; LastName, B.B.; LastName, C.C. Article Title. *Journal Name* **Year**, *Volume Number*, Page Range.

ISBN 978-3-0365-5755-7 (Hbk)
ISBN 978-3-0365-5756-4 (PDF)

Contents

About the Editors

Weicheng Wu

Weicheng Wu holds a PhD in Geography from the University of Paris I (Pantheon-Sorbonne), is full-time Professor and Team Leader at the Key Lab of Digital Land and Resources, and PhD supervisor at the Faculty of Earth Sciences of the East China University of Technology (ECUT). Dr Wu used to be an international scientist and remote sensing specialist with the University of Sassari in Italy and the International Center for Agricultural Research in the Dry Areas (ICARDA). He had led and/or participated in more than 20 international cooperation projects focused in Central and Western Asia and Northern Africa (CWANA) and the European Mediterranean region, funded by different international consortia such as ESA, EU, IFAD, USAID, AusAID, CGIAR, etc. After joining ECUT in 2018, he has managed several projects on assessments of the sustainable utilization of land resources and of the impacts of natural disasters on food security and society using geo-information technology and artificial intelligence techniques, supported by ECUT and the Jiangxi Government. He was the Chair of the International Conference on Geo-information Technology and its Applications (ICGITA 2019), and is an Associate Editor of the *International Journal of Remote Sensing* and Guest Editor of *Remote Sensing and ISPRS International Journal of Geo-Information.* His research interests include analyses of human–nature interactions, carbon sequestration/emission, natural disaster risk zoning and food security. He has authored more than 100 scientific publications, and in 2018, Dr Wu was granted the title of "Ten Leading Chinese Talents on Science and Technology in Europe" and selected by the "Talent Program of Jiangxi Government".

Yalan Liu

Yalan Liu holds a PhD in Cartography and Remote Sensing from the University of Chinese Academy of Sciences, achieved in 2004, is a full-time Professor at the Aerospace Information Research Institute under the Chinese Academy of Sciences (AIR/CAS), and part-time Professor in the Regional Centre for Space Science and Technology Education in Asia and the Pacific (China) (affiliated with the United Nations). She is also the Director of the Lab of Spatial Information Integration Technology at AIR/CAS, specializing in remote sensing and geographic information systems (GIS), including technology, theory, and applications, since 1996. Her research interests include intelligent information extraction, the integration of spatial information and their applications in environment monitoring, disaster mitigation, smart cities, and decision support for sustainable development. She has been leading three projects funded by the National Natural Science Foundation, National Key Research and Development Program, and China High-resolution Earth Observation System Program, and has participated in two projects supported by the National Science and Technology Support Program as well as two international cooperation projects. She served as Guest Editor and reviewer for several international journals, such as *ISPRS International Journal of Geo-Information and Remote Sensing*, and has published more than 100 scientific papers. Dr Liu has received several prizes of Scientific and Technological Progress Awards (STPA), including one First Class Prize of STPA of the Chinese Academy of Sciences, one Third Class Prize of STPA of the Beijing Government, and one Second Class Prize of the State STPA of China.

Preface to "Integrated Applications of Geo-Information in Environmental Monitoring"

The aims of this Special Issue (SI) of Remote Sensing are to demonstrate the state of the art in the development and application of geo-information technology, including optical and radar remote sensing, GIS, GPS, and computing systems in various environmental fields. Motivated by the high-quality presentations at the International Conference on Geo-information Technology and its Applications (ICGITA 2019), this book covers the up-to-date research in land cover change and desertification analyses, geo-disaster risk and damage evaluations, mining area restoration assessments, the improvement and development of algorithms, and coastal environmental monitoring and object targeting. This book can be used as a reference for graduates and scientists in environmental sciences, disaster managers, urban planners, and big data miners. The Guest Editors are grateful to the authors for their contributions of high-quality studies, and, specifically, to the MDPI team for their effective assistance and cooperation. Without their devotion and support, it would have been impossible to publish this book.

Weicheng Wu and Yalan Liu

Editors

remote sensing

MDPI

Editorial

Editorial for the Special Issue: "Integrated Applications of Geo-Information in Environmental Monitoring"

Weicheng Wu [1,2,*] and Yalan Liu [3]

1 Key Laboratory of Digital Land and Resources, East China University of Technology, Nanchang 330013, China
2 Faculty of Earth Sciences, East China University of Technology, Nanchang 330013, China
3 Aerospace Information Research Institute, Chinese Academy of Sciences, Beijing 100101, China
* Correspondence: wuwch@ecut.edu.cn

Abstract: Geo-information technology has been playing an increasingly important role in environmental monitoring in recent decades. With the continuous improvement in the spatial resolution of remote sensing images, the diversification of sensors and the development of processing packages, applications of a variety of geo-information, in particular, multi-resolution remote sensing and geographical data, have become momentous in environmental research, including land cover change detection and modeling, land degradation assessment, geohazard mapping and disaster damage assessment, mining and restoration monitoring, etc. In addition, machine learning algorithms such as Random Forests (RF) and Convolutional Neural Networks (CNN) have improved and deepened the applications of geo-information technology in environmental monitoring and assessment. The purpose of this Special Issue is to provide a platform for communication of high-quality research in the world in the domain of comprehensive application of geo-information technology. It contains 10 high-level scientific papers on the following topics such as desertification monitoring, governance of mining areas, identification of marine dynamic targets, extraction of buildings, and so on.

Citation: Wu, W.; Liu, Y. Editorial for the Special Issue: "Integrated Applications of Geo-Information in Environmental Monitoring". *Remote Sens.* **2022**, *14*, 4251. https://doi.org/10.3390/rs14174251

Received: 11 June 2022
Accepted: 11 July 2022
Published: 29 August 2022

1. Desertification and Sand-Control Monitoring and Assessment

Soil degradation and even desertification are a serious environmental problem that affects the production activities and quality of life of local residents. Kim et al. developed a method fusing optical vegetation index and time-series phase coherence Synthetic Aperture Radar (SAR) data to monitor aeolian erosion sequences in desert areas using the Interferometric SAR (InSAR) technique. Additionally, the surface changes before and after desertification control activities in the Kubuqi Desert were monitored and analyzed [1], and the research demonstrates the effectiveness of InSAR techniques when applied for monitoring desertification and sand control. Li et al. [2] comprehensively evaluated sand-control effectiveness based on multitemporal Landsat images from 1990 to 2020 in the Mu Us Desert, China [2], by linking sand-control with socioeconomic and environmental data via multiple linear and logistic regression models. At the same time, the driving forces of desertification were also analyzed. This study shows that from 1991 to 2020, 8712 km^2 or 63% of the desert has been converted into pastures and shrublands with a greenness increase of 0.3509 in GDVI; the effectiveness of sand-control is favored by the rational agropastoral activities and policies; though desertification occurs locally, it is associated with both climatic and socioeconomic factors, such as wind speed, precipitation, water availability, distance to roads and animal husbandry. Globally, the rationalization of agricultural and animal husbandry activities through policies and other means is favorable for sand-control activities [2].

2. Geohazard Mapping and Restoration Assessment in Mining Areas

Since industrialization, mining activities have been increased significantly worldwide, causing a series of geo-environmental problems, especially geodisasters, for example, landslides, collapses and subsidence. Hence, high-resolution geohazard prediction and mapping are essential but challenging due to limitations in the acquisition of high-resolution data of the mining areas. In response to this problem, Qin et al. [3] combined geological prospecting data, topographic data and Gaofen-1 satellite data to construct a multi-source dataset and used machine learning technology to model the geological hazard in the Liaojiaping Orefield in Central China. The Random Forests (RF)-based mining-induced geohazard mapping (MGM) model they constructed shows excellent predictive performance, which provides a new method for geological hazard mapping [3]. Rare earth elements (REE) have a wide range of usages in the world, and the local enterprises and governments have heavily exploited these REE resources through open-pit or chemical leaching approaches in southern Jiangxi, China. Such REE mining has caused serious damage to the environment. Xie et al. [4] developed a new set of remote sensing indicators, i.e., Mining and Restoration Assessment Indicators (MRAIs), to assess land degradation caused by mining and the effectiveness of remediation. Compared with the single vegetation index, these new indicators have stronger sensitivity and wider dynamic range than the usual indices such as NDVI, GDVI, EVI, SAVI and ARVI. They are hence more suitable for evaluating the mining and recovery of REE mines [4]. This study not only provides more efficient remote sensing indicators for mining and restoration evaluation, but also makes up for the shortcomings of only relying on the known vegetation indices for evaluation of the restoration effectiveness.

3. Improvement of Algorithms for Classification

The rapid development of urbanization has led to a significant increase in the density of buildings. While providing convenience for living, problems such as illegal construction and conflicts between people and land are also prominent. It greatly increases the difficulty of supervision for governments. Therefore, remote sensing and geographic information technologies to accurately extract buildings and monitor them dynamically are one of the important applications of geo-information technology in environmental monitoring. Ma et al. [5] proposed an improved Convolutional Neural Network (CNN) Inception V3 architecture to evaluate the degree of damage to buildings after an earthquake in Yushu, Qinghai, China. This method improved the accuracy of classification and performed better than the traditional machine learning classifiers and also avoided the disadvantages of the traditional CNN, such as difficulty in function selection and image segmentation. Additionally, it provides a new attempt to evaluate the damage degree of buildings after the earthquake through remote sensing images. Wu et al. [6] proposed an improved anchor-free instance segmentation method based on CenterMask with spatial and channel attention-guided mechanisms for accurate extraction of buildings in high-resolution remote sensing images. In comparison with methods of Mask R-CNN, Mask Scoring R-CNN and CenterMask, their method is able to achieve the state-of-the-art performance at real-time speed, which makes it possible to extract buildings accurately in real time [6].

As a basic content of environmental monitoring, land use/land cover change (LUCC) monitoring has always been a research hotspot. Yu et al. [7] effectuated monitoring research on LUCC in the Xiong'an New Area based on new bands (purple, yellow, and red edges) of the GF-6 Wide Field of View (WFV) images (16 m in resolution) using the Double-Constrained Change Detection approach. The accuracy of monitoring through the two red-edge bands of GF-6WFV is higher than that of GF-1WFV. This result provides theoretical support for the application of GF-6WFV in LUCC study [7].

4. Coastal and Ocean Applications

Another major content included in this Special Issue is the application of optical and radar remote sensing to the coastal and oceanic research, for example, extraction and identification of coastlines and tracking the dynamic targets on the sea surface. The ocean occupies about 71% of the Earth's surface; it is rich in minerals and biological and chemical resources and has great development prospects. At the same time, socioeconomic development in the world has also brought about environmental problems in coastal areas and ocean such as sea level rise, marine debris and ocean garbage patches. However, limited by sensors and resolution, the monitoring of the marine environment has certain limitations. In recent years, with the rapid development of geo-information technology, many scholars have shifted the target of environmental monitoring from the traditional land surface to the ocean. Nazeer et al. [8] tracked the positional changes of the coastline by extracting historical coastline positions from topographic maps, medium-resolution Landsat (30 m) and high-resolution (3 m) imagery, respectively. Their research shows that human activities have a significant impact on the movement of the shoreline of Karachi. Additionally, the difference in human activities between the east and west coasts makes the coastline in the eastern region very fragile, while the western region is more stable. Therefore, management of the coast by the government and prevention of illegal marine encroachment are crucial for coastal protection [8]. Jiang et al. used the wideband echo simulation method based on a frequency domain to analyze the influence of ocean wave motion on the synthetic aperture radar (SAR) image of a target. Furthermore, they introduced a rectangular wave-beam-based geometrical optics and physical optics (GO/PO) method to calculate the electromagnetic (EM) scattering properties for the identification of complex targets [9]. The methods they proposed are capable of simulating the SAR image of the target on the sea surface, which is an important advancement in identifying dynamic targets on the sea surface. Massarelli et al. used high-resolution airborne data to identify and map mussel farming in the first and second inlet of Mar Piccolo, Italy. On this basis, factors that could harm the environmental status of the Mar Piccolo ecosystem were assessed. Their map made it possible to determine anthropogenic pressure on the Mar Piccolo of Taranto and the necessary actions for better management of the area [10].

Author Contributions: Conceptualization, W.W. and Y.L.; writing—original draft preparation, review, and editing, W.W. and Y.L. All authors have read and agreed to the published version of the manuscript.

Funding: This research received no external funding.

Acknowledgments: The Guest Editors of this Special Issue would like to thank all authors who have contributed to this volume for sharing their scientific results and for their excellent collaboration. Special gratitude will go to the community of distinguished reviewers for their constructive inputs. The *Remote Sensing* editorial team is acknowledged for its support during all phases related to the successful completion of this issue.

Conflicts of Interest: The author declares no conflict of interest.

References

1. Kim, J.R.; Lin, C.W.; Lin, S.Y. The Use of InSAR Phase Coherence Analyses for the Monitoring of Aeolian Erosion. *Remote Sens.* **2021**, *13*, 2240. [CrossRef]
2. Li, J.; Wu, W.; Fu, X.; Jiang, J.; Liu, Y.; Zhang, M.; Zhou, X.; Ke, X.; He, Y.; Li, W.; et al. Assessment of the Effectiveness of Sand-Control and Desertification in the Mu Us Desert, China. *Remote Sens.* **2022**, *14*, 837. [CrossRef]
3. Qin, Y.; Cao, L.; Boloorani, A.D.; Wu, W. High-Resolution Mining-Induced Geo-Hazard Mapping Using Random Forest: A Case Study of Liaojiaping Orefield, Central China. *Remote Sens.* **2021**, *13*, 3638. [CrossRef]
4. Xie, L.; Wu, W.; Huang, X.; Ou, P.; Lin, Z.; Wang, Z.; Song, Y.; Lang, T.; Huangfu, W.; Zhang, Y.; et al. Mining and Restoration Monitoring of Rare Earth Element (REE) Exploitation by New Remote Sensing Indicators in Southern Jiangxi, China. *Remote Sens.* **2020**, *12*, 3558. [CrossRef]
5. Ma, H.; Liu, Y.; Ren, Y.; Wang, D.; Yu, L.; Yu, J. Improved CNN Classification Method for Groups of Buildings Damaged by Earthquake, Based on High Resolution Remote Sensing Images. *Remote Sens.* **2020**, *12*, 260. [CrossRef]
6. Wu, T.; Hu, Y.; Peng, L.; Chen, R. Improved Anchor-Free Instance Segmentation for Building Extraction from High-Resolution Remote Sensing Images. *Remote Sens.* **2020**, *12*, 2910. [CrossRef]
7. Yu, J.; Liu, Y.; Ren, Y.; Ma, H.; Wang, D.; Jing, Y.; Yu, L. Application Study on Double-Constrained Change Detection for Land Use/Land Cover Based on GF-6 WFV Imageries. *Remote Sens.* **2020**, *12*, 2943. [CrossRef]
8. Nazeer, M.; Waqas, M.; Shahzad, M.I.; Zia, I.; Wu, W. Coastline Vulnerability Assessment through Landsat and Cubesats in a Coastal Mega City. *Remote Sens.* **2020**, *12*, 749. [CrossRef]

9. Jiang, W.Q.; Wang, L.Y.; Li, X.Z.; Liu, G.; Zhang, M. Simulation of a Wideband Radar Echo of a Target on a Dynamic Sea Surface. *Remote Sens.* **2021**, *13*, 3186. [CrossRef]
10. Massarelli, C.; Galeone, C.; Savino, I.; Campanale, C.; Uricchio, V.F. Towards Sustainable Management of Mussel Farming through High-Resolution Images and Open Source Software—The Taranto Case Study. *Remote Sens.* **2021**, *13*, 2985. [CrossRef]

remote sensing

MDPI

Article

The Use of InSAR Phase Coherence Analyses for the Monitoring of Aeolian Erosion

Jung-Rack Kim [1,*], Cheng-Wei Lin [2] and Shih-Yuan Lin [3]

[1] Department of Geoinformatics, University of Seoul, Seoulsiripdaero 163, Dongdaemum-gu, Seoul 02504, Korea
[2] Sinotech Engineering Consultants, Geotechnical Engineering Research Center, Taipei 114, Taiwan; cwlin@sinotech.org.tw
[3] Department of Land Economics, National Chengchi University, No.64, Sec. 2, Zhinan Rd., Wenshan Dist., Taipei 116, Taiwan; syl@mail2.nccu.tw
* Correspondence: kjrr001@uos.ac.kr; Tel.: +82-02-649-2880

Abstract: Aeolian erosion occurring in sand deserts causes significant socio-economical threats over extensive areas through mineral dust storm generation and soil degradation. To monitor a sequence of aeolian erosion in a sand desert area, we developed an approach fusing a set of remote sensing data. Vegetation index and Interferometric Synthetic Aperture Radar (InSAR) phase coherence derived from space-borne optical/SAR remote sensing data were used. This scheme was applied to Kubuqi Desert in Inner Mongolia where the effects of activity to combat desertification could be used to verify the outcome of the approach. We first established time series phase coherence and conducted a functional operation based on principal component analysis (PCA) to remove uncorrelated noise. Then, through decomposition of vegetation effect, where a regression model together with the Enhanced Vegetation Index (EVI) was employed, we estimated surface migration caused by aeolian interaction, that is, the aeolian erosion rate (AER). AER metrics were normalized and validated by additional satellite and ground data. As a result, the spatiotemporal migration of the target environment, which certainly induced dust storm generation, was traced and analyzed based on the correlations among surface characteristics. It was revealed that the derived AER successfully monitored the surface changes that occurred before and after the activities to combat desertification in the target area. Employing the established observation scheme, we expect a better understanding of the aeolian process in sand deserts with enhanced spatio-temporal resolution. In addition, the scheme will be beneficial for the evaluation of combating desertification activities and early warning of dust storm generations.

Keywords: aeolian process; desertification; multi-sensor fusion; interferometric SAR; time-series analysis

Citation: Kim, J.-R.; Lin, C.-W.; Lin, S.-Y. The Use of InSAR Phase Coherence Analyses for the Monitoring of Aeolian Erosion. *Remote Sens.* **2021**, *13*, 2240. https://doi.org/10.3390/rs13122240

Academic Editors: Weicheng Wu and Yalan Liu

Received: 3 May 2021
Accepted: 6 June 2021
Published: 8 June 2021

1. Introduction

Dust storms are now emerging as a major threat affecting public health and social activities worldwide, including northeastern Asian countries. Since most dust originates from the arid desert, desert expansion caused by climate change results in increased dust levels. For instance, Tegen and Fung (1995) [1] estimated that 30–50% of aerosol in the global atmosphere originates from regolith over arid areas. From a regional point of view, a dust storm from Inner Mongolia and the Tarim Basin, where arid climate conditions interact with the surrounding environment, is now regarded as a disastrous environmental hazard in northeastern Asia [2–5]. It imposes significant ill effects on surrounding areas; for example, the occurrence of dust storms in South Korea has increased from four times between 1985 and 1989 to 26 times between 2000 and 2004 and caused significant socio-economic problems [6]. Therefore, dust source monitoring has attracted lots of research interest. However, there is still not yet an effective and accurate monitoring method of the aeolian activity over arid desert because of the difficult accessibility to such regions and the time requirements for the full range of data collection. Remote sensing, especially for space-borne data, can be a powerful solution for such difficulties. A few studies that employ medium resolution satellite products such as Total Ozone Mapping Spectrometer (TOMS) [7,8] and Moderate Resolution Imaging Spectroradiometer (MODIS) [9,10] have been conducted to tackle the technical issues of the dust source mapping using aerosol optical depth (AOD). However, the most significant aspects of the aeolian process in the arid desert, including the propagation of the transition region between the desert and soil-rich areas, and the aeolian surface migration speed and tendency, cannot be observed properly using the typical medium resolution AOD products. This is because the primary interest if we focus on the environment in the dust source area of an arid desert, is in the change in the topographic conditions rather than the dust aerosol trail. The generation of dust, especially over the sandy desert, was induced mainly by a process called saltation. The particles caused by saltation are suspended into the air and transported. It is why sandy deserts are considered as a dust source by the United Nations Convention to Combat Desertification (UNCCD) [11]. Hence, it is necessary to define the degree of surface change consequence into dust generation interacting with the wind. In further discussions, we will use "aeolian erosion" as a term encompassing all topographic migration processes such as transportation and deposition inducing saltation and subsequent removal of minerals. The rate of aeolian erosion in the following is defined as the aeolian erosion rate (AER), which can be measured by observation methods including the proposed phase coherence time-series analysis used in this study. It should be noted that there is no universal approach to measure the strength of aeolian erosion. Therefore, we use AER as an index to identify the relative (dimensionless) strength of aeolian erosion within an independent observation system. Furthermore, if we compare the AERs derived from different observations, it is necessary to adjust their data range into cross-comparable quantities. Such adjusted AERs are referred to as normalized AERs (n-AER) in this study. Once n-AERs are converted into physically meaningful values, for instance, a changed weight per unit time and area with a certain level of calibration, they can be treated as a quantified AER. Except in very particular cases where high-resolution volumetric tracing is used, the quantified AER is not directly measurable by a remote sensing approach.

Recently, very high-resolution satellite image products enable detection of a subtle surface change caused by dust generation—especially sand dune migration with 2D projected migration. In spite of such merit, the very high-resolution satellite observation has been rarely used for desertification monitoring because of their technical problems, for example, the dependence on the climate factors, the revisiting time, and the inability to trace continuous topographic conditions. The technical barriers are more obvious, as the task requires the observation of external environmental factors governing the aeolian surface migration as well as the migration itself. Hence, a sophisticated monitoring system that applies remote sensing techniques is required to monitor the origin and consequences of the AER on arid deserts using various recent satellite images. Considering that the most significant identifier of aeolian processes over the arid desert is aeolian topographic erosion such as sand dune migration inducing saltation and suspension of particles, the full aeolian interactions that induce dust storms can be reconstructed by observing topographic changes in the arid desert and surrounding areas affected by surface erosion. However, conventional image analysis by visual inspection cannot directly measure the aeolian process and its governing factors.

To tackle such problems, we developed a monitoring system to observe and model the progress of desertification and the aeolian activity using a synergistic fusion of remotely sensed data, including the Synthetic Aperture Radar (SAR) and medium resolution optical image. The constructed approach was tested over the Kubuqi Desert located in Northeast China for two reasons. Firstly, it is considered a major contributor to dust storms over the surrounding areas in eastern Asian countries [12,13]. Secondly, the test area has been managed by well-organized international organizations through a series of combating desertification activities for a decade. Hence, the environmental evolution was expected to be observed using satellite images, verifiable by the on-ground observations over a long time period, and to be converted into quantified AERs. Moreover, it was expected that the governing parameters of aeolian interaction and consequent AERs in the target area were extracted within the observation capabilities of the proposed scheme.

The target area is described in Section 2. The processing methodologies for Interferometric Synthetic Aperture Radar (InSAR) and auxiliary data sets are given in Section 3, and the results are demonstrated in Section 4. A discussion on the detected aeolian migration and the inter-comparison work validating the InSAR observations is presented in Section 5.

2. Test Site

In past decades, the Chinese Loess Plateau and arid Gobi Deserts of Inner Mongolia have been observed as a source of mineral dust [14]. However, recent research (Zhang et al., 2008) [15] identified few deserts in northern China, such as Badain Jaran, Kubuqi, and Taklamakan, as the more prominent sources of mineral dust. Although Wang et al. (2007, 2009) [16,17] reported that dune activities in deserts in China ceased after the 1990s, northern China still suffers from undergoing desertification. Therefore, the observation of aeolian processes, especially dune migration in the deserts of northern China, is key to reconstructing the dust storm generation process, as the sand dune activity is the most obvious consequence and the index of aeolian activities over the sand desert. The deserts in northern China, one of the largest arid desert areas on Earth, extend from the Tarim Basin to Inner Mongolia as shown in Figure 1a. Among them, the Kubuqi Desert's expanse threatens the highly densified residential areas in Inner Mongolia and Shanxi province [18]. In addition, considering the land cover types, Kubuqi Desert has a highly peculiar pattern in which the barren fields, moving sand dunes, forests, and built-up urban center complicatedly coexist. Such a landscape provides an optimal location to conduct comparative observations of erosion over different surface characteristics; thus, this area was selected as the test site of our monitoring system. Liu et al. (2005) [19] once tried to identify the influence of wind factor over sand dune movement mainly using ground survey data on Kubuqi Desert. Our purposes are quite similar to that, but this study was performed to build a comprehensive and universal monitoring scheme to identify the effects of wind-topographic interactions using remote sensing data fusion.

Kubuqi Desert is well confined by the riverside and Ordos plateau as shown in Figure 1b and the weakly connected Southern Mu-Us Desert. Thus, the environments between the arid desert and background vegetated fields are well distinguished, compared to the background Ordos plateau. It should be noted that the area's contribution as a dust source is significant. Analysis conducted using MODIS satellite image and atmospheric circulatory simulation confirmed that a large amount of sand dust (66% probability of generated dust storm) in the Kubuqi Desert region migrated to the Korean peninsula (Yun et al., 2011) [20]. This value is higher than that in the Gobi Desert (55%), a known source of dust, which indicates that it is demanding to conduct detailed monitoring of the Kubuqi Desert region to understand the sand dust erosion process in Northern China. Additionally, due to the severe threats that occurred in this region, in the last decade, NGOs and Chinese local governments have conducted activities to combat desertification in the Kubuqi Desert and the surrounding areas. Since those efforts have dramatically changed the desert environment in Kubuqi (Figure 1c,d), the region became an ideal test site to develop the monitoring system. Furthermore, it provided a study case where comprehensive validation by in-field observation was feasible, allowing all outcomes of this study to be validated using long-term on-site observations. The origin and morphological characteristics of the Kubuqi Desert were reported in Yang et al. (2016) [21].

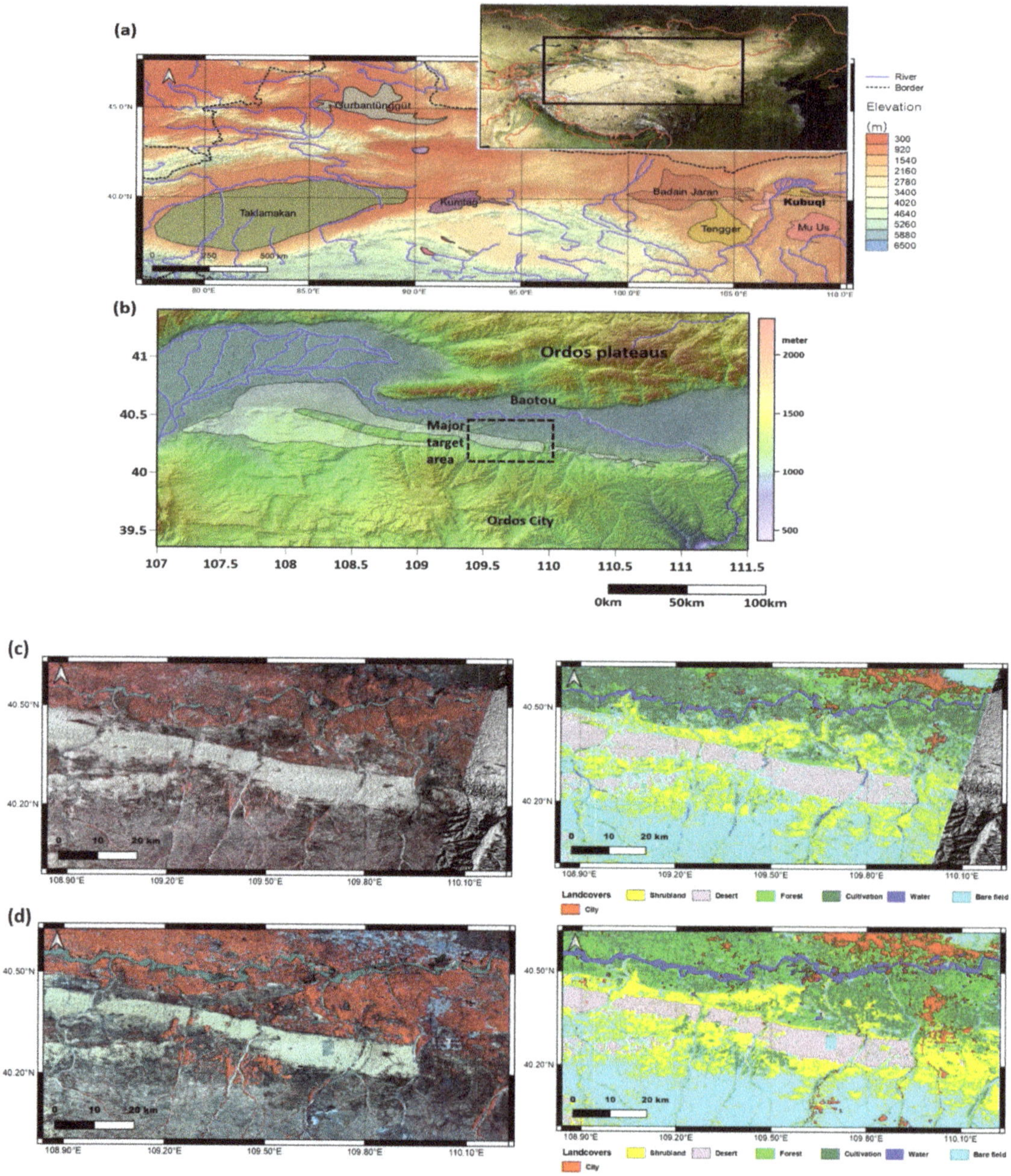

Figure 1. (**a**) North-Eastern deserts in China. Kubuqi is located on the eastern side. (**b**) Location and geological context of the target area over Kubuqi. (**c**) 17 July 2007 (Near-IR-G-B) LANDSAT-5 image and land cover classification. (**d**) 12 September 2019 (Near-IR-G-B) Sentinel-2 multispectral image and the associated land cover classification. Land cover classification was performed by a maximum likelihood algorithm. The changes in land cover around desert areas were noticed. The land cover classification results were employed in the aeolian erosion rate (AER) extraction process (see Section 3).

Regarding climate conditions over the target area, it should be noted that the annual precipitation in Kubuqi is 300 mm/year and concentrates on the period from July to August according to the local government information (http://www.baotou.gov.cn/ (accessed on 1 January 2021). Thus, dust generation usually occurs from March to May before the seasonal precipitation. Such seasonal concentration of the aeolian process is a reason why we adopted the long-term time-series analysis of phase coherence.

3. Method and Data Sets

To achieve the objectives of this study, satellite data sets covering sufficient temporal and spatial domains, including InSAR and auxiliary data sets, were extensively employed. Then, we were able to evaluate both the progress of aeolian erosion and the effects of anti-desertification activities. InSAR phase coherence was chosen as a prime observation. Meanwhile, a series of InSAR observations were necessary for measuring surface changes. We first established the time-series approach on the target domain and then developed a further processing scheme to decompose the components which were not involved in the surface changes. Consequently, the progress of desertification and the effects of activities to combat desertification were properly measured. Then, the driving factors of surface changes and their consequences could be inferred. The data indices and processing approaches, which are capable of measuring the cause and consequence of aeolian erosion, were based on the backgrounds as follows.

3.1. Erosion Measurement with InSAR Phase Decorrelation

Since the target area is an arid desert, it was presumed that dune mobility is the most obvious representative of aeolian surface erosion. Monitoring of dune mobility on certain spatial domains is feasible by the following methods:

(1) 2D transition vector derivation by high-resolution satellite/Unmanned Aerial Vehicle (UAV) orthographic image comparison [22–24]: This approach has the advantage of obtaining accurate dune mobility only by acquiring a pair of satellite images. However, it is limited to only measuring 2D projected dune mobility on ortho-image space.

(2) Observation of 3D volumetric change based on high-resolution Laser Detection and Ranging (LIDAR)/UAV continuous stereo observations [25–27]: Digital Elevation Models (DEMs) generated from precisely registered stereo/LIDAR observations at two observation times allows the calculation of volumetric change. However, the possibilities of consecutive 3D observations are even more restricted.

Our target area covered a 200 by 150 km spatial domain and more than a decadal time period. In terms of data availability and accuracy of remote sensing measurements, the employment of 2D/3D dune mobility tracing is practically impossible. Hence, the application of InSAR techniques was proposed as an alternative way to cover the spatio-temporal domain of the target area. Although InSAR was developed for the measurement of topographic deformation, phase angle analyses, which is a conventional approach of InSAR, are not adequate to trace severely ongoing topographic changes such as sand dune migrations. This is because the large change of base topography by saltation processes on desert dune fields causes a temporal decay of phase coherence (Liu et al., 2001) [28], called decorrelation. However, it also represents that phase coherence is strongly correlated with the strength of dune migrations, as shown in studies such as Wegmuller et al. (2000) [29]. Phase coherence represents cross-correlation between two SAR images derived from the InSAR technique and can be described in the following relationship.

$$\text{coh} = \sum_{i=1}^{N} |< S_{mi} S_{si}^{*} >| / \sqrt{< S_{mi} S_{mi}^{*} >< S_{si} S_{si}^{*} >} \quad (1)$$

where S_{mi} and S^{*}_{si} are the complex conjugated signals of master and slave SAR images, N is the total number of InSAR pixels within the estimated window, and <> assigns an average operation.

Nowadays, some studies have established methods for monitoring sand dune dynamics and aeolian migrations by the InSAR phase coherence footprint as shown by Gomez et al. (2018) [30], Havivi et al. (2018) [31], and Ullmann et al. (2019) [32]. However, decorrelation is not only a function of dune migration rate, it is also involved with multiple other components such as vegetation [33–35], InSAR geometry [33,36], and climate effects [37,38]. Thus, in order to extend phase coherence monitoring on the extensive desert environment, it is essential to decompose the decorrelation part contributed by the other factors.

The overall InSAR phase coherence can be expressed by the aggregation of the following components [39].

$$Coh = Coh_{thermal}\, Coh_{spatial}\, Coh_{temporal} \tag{2}$$

where thermal coherence $coh_{thermal}$ can be expressed as follows [40]:

$$coh_{thermal} = 1/\sqrt{\left(1 + SNR^{-1}_{m\_image}\right)\left(1 + SNR^{-1}_{s\_image}\right)} \tag{3}$$

where SNR_{m_image} is the signal to noise of the master SAR image and SNR_{s_image} is the signal to noise of the slave SAR image. Thermal phase coherence can be negligible due to the high signal to noise values of the target InSAR systems in this study. Tamm et al. (2016) [41] proved it using range-dependent noise parameters in SAR metadata.

Spatial coherence $coh_{spatial}$ can be expressed as follows [42]:

$$Coh_{spatial} = 1 - \frac{2|Bp|R_y cos^2(\theta - \alpha)}{\lambda\rho} \tag{4}$$

where Bp is the perpendicular baseline, Ry is the range resolution, θ is the incidence angle, λ is the wavelength range of SAR, and α is the local surface slope and ρ is the distance between the sensor and the object [36]. To understand the characteristics of spatial coherence, we herein introduced the SAR data used in this study. We employed two different InSAR data sets, including the Advanced Land Observing Satellite (ALOS) and the Phased Array type L-band Synthetic Aperture Radar (PALSAR) 1/2 [43,44] over a 10-year period by the combined observations by PALSAR 1 and 2, and C-band Sentinel-1 [45,46], collected from 2016 to 2020 (see Tables 1 and 2). Spatial coherence, especially with short perpendicular baselines such as Sentinel-1 and PALSAR1/2, was very close to 1.0 as shown in Tables 1 and 2. Moreover, the effects of local surface slope on spatial coherence, as shown in Equation (4), was tiny. This was because the terrain slopes in the test area were only a few degrees or less; thus the variations of spatial coherence were not high (refer to *Spatial coh1* and 2 values in Tables 1 and 2 whose maximum differences were only 0.03 to 0.05 in the cases of PALSAR-1/2 and Sentinel-1) and were alternatively managed by the consequent functional analysis. In addition, the window size for phase coherence calculations was chosen to optimize the minimize noise and to prevent over-sampling of phase coherence. The empirical or typical optimal window sizes are 5 in Sentinel-1, 5 in PALSAR-1, and 3 in PALSAR-2.

Table 1. Employed PALSAR 1 (left part)/2 (right part) InSAR pairs and their acquisition conditions together with spatial phase coherences (*Bp*: perpendicular baseline (m), *Bt*: temporal baseline (day), *Spatial coh1*: spatial phase coherences in mean topographic slope angle (3.34 °). *Spatial coh2*: spatial phase coherences in mean topographic slope angle (28.3 °). The samples of topographies to calculate *Spatial coh1* and 2 were taken from a low sloped plain and high-relief dune field.

Master Date	Slave Date	*Bp*	*Bt*	*Spatial Coh1*	*Spatial Coh2*	Master Date	Slave Date	*Bp*	*Bt*	*Spatial Coh1*	*Spatial Coh2*
11 July 2007	26 August 2007	251	46	0.955	0.967	18 July 2015	23 September 2015	85	70	0.985	0.987
26 August 2007	11 October 2007	389	46	0.931	0.948	15 July 2015	02 December 2015	−1.6	140	1.000	1.000
11 October 2007	26 February 2007	1209	138	0.785	0.839	23 September 2015	02 December 2015	−83	70	0.985	0.988
26 February 2008	12 April 2008	437	46	0.922	0.942	10 February 2016	13 July 2016	95	154	0.983	0.986
26 February 2008	28 May 2008	274	92	0.951	0.964	10 February 2016	23 September 2015	25	−140	0.995	0.996
12 April 2008	28 May 2008	−163	46	0.971	0.978	13 July 2016	10 February 2016	−94.9	−154	0.983	0.986
28 August 2008	13 October 2008	955	46	0.830	0.873	13 July 2016	21 September 2016	53.9	70	0.990	0.992
-	-	-	-	-	-	13 July 2016	30 November 2016	−141.8	140	0.974	0.979
-	-	-	-	-	-	21 September 2016	30 November 2016	−88	70	0.984	0.987
-	-	-	-	-	-	21 September 2016	08 February 2017	69.8	140	0.987	0.990
-	-	-	-	-	-	30 November 2016	08 February 2017	157.8	70	0.971	0.976

Table 2. Employed Sentinel-1 pairs and their acquisition conditions together with spatial phase coherences (*Bp*: perpendicular baseline (m), *Bt*: temporal baseline (day), *Spatial coh1*: spatial phase coherences in mean topographic slope angle (3.34 °), *Spatial coh2*: spatial phase coherences in mean topographic slope angle (28.3 °)).

2016/2017											
Master Date	Slave Date	*Bp*	*Bt*	*Spatial Coh1*	*Spatial Coh2*	Master Date	Slave Date	*Bp*	*Bt*	*Spatial Coh1*	*Spatial Coh2*
16 May 2016	09 June 2016	17	25	0.988	0.992	12 March 2017	05 April 2017	63	24	0.954	0.968
09 June 2016	03 July 2016	7	25	0.995	0.997	05 April 2017	11 May 2017	12	36	0.991	0.994
03 July 2016	20 August 2016	3	47	0.998	0.999	11 May 2017	04 June 2017	38	24	0.972	0.981
20 August 2016	07 October 2016	49	48	0.965	0.978	04 June 2017	10 July 2017	110	36	0.919	0.944
25 September 2016	07 October 2016	75	13	0.947	0.967	10 July 2017	03 August 2017	74	24	0.945	0.963
07 October 2016	12 November 2016	111	37	0.921	0.951	03 August 2017	08 September 2017	81	37	0.940	0.959
12 November 2016	06 December 2016	35	25	0.975	0.984	08 September 2017	02 October 2017	78	24	0.943	0.961
06 December 2016	11 January 2017	179	37	0.873	0.920	02 October 2017	07 November 2017	5	48	0.996	0.997
-	-	-	-	-	-	07 November 2017	01 December 2017	52	24	0.962	0.974
2018/2019											
Master Date	Slave Date	*Bp*	*Bt*	*Spatial Coh1*	*Spatial Coh2*	Master Date	Slave Date	*Bp*	*Bt*	*Spatial Coh1*	*Spatial Coh2*
06 January 2018	11 February 2018	13	36	0.990	0.993	02 December 2018	01 January 2019	26	24	0.981	0.987
11 February 2018	07 March 2018	11	24	0.992	0.994	01 January 2019	06 February 2019	5	48	0.996	0.997
07 March 2018	12 April 2018	52	36	0.962	0.974	06 February 2019	02 March 2019	143	24	0.895	0.928
12 April 2018	06 May 2018	49	24	0.964	0.975	02 March 2019	07 April 2019	28	36	0.979	0.986
06 May 2018	11 June 2018	3	37	0.998	0.998	07 April 2019	01 May 2019	64	24	0.953	0.968
11 June 2018	05 July 2018	41	48	0.970	0.979	01 May 2019	06 June 2019	83	36	0.939	0.958

Table 2. *Cont.*

2018/2019											
Master Date	**Slave Date**	*Bp*	*Bt*	*Spatial Coh1*	*Spatial Coh2*	**Master Date**	**Slave Date**	*Bp*	*Bt*	*Spatial Coh1*	*Spatial Coh2*
05 July 2018	10 August 2018	83	93	0.939	0.958	06 June 2019	12 July 2019	27	36	0.980	0.986
10 August 2018	03 September 2018	173	24	0.873	0.913	12 July 2019	05 August 2019	34	24	0.975	0.983
03 September 2018	09 October 2018	174	36	0.872	0.912	05 August 2019	10 September 2019	45	48	0.967	0.977
09 October 2018	02 November 2018	94	24	0.931	0.953	10 September 2019	04 October 2019	104	24	0.923	0.948
-	-	-	-	-	-	04 October 2019	09 November 2019	123	36	0.909	0.938
-	-	-	-	-	-	09 November 2019	03 December 2019	10	24	0.993	0.995
-	-	-	-	-	-	03 December 2019	08 January 2020	79	36	0.942	0.960
2020											
Master Date	**Slave Date**	*Bp*	*Bt*	*Spatial Coh1*	*Spatial Coh2*	**Master Date**	**Slave Date**	*Bp*	*Bt*	*Spatial Coh1*	*Spatial Coh2*
08 January 2020	01 February 2020	71	4	0.948	0.964	18 July 2020	11 August 2020	116	24	0.915	0.941
01 February 2020	08 March 2020	5	36	0.996	0.997	11 August 2020	04 September 2020	148	24	0.891	0.925
08 March 2020	01 April 2020	54	24	0.960	0.973	04 September 2020	10 October 2020	99	28	0.927	0.950
01 April 2020	18 July 2020	102	108	0.925	0.949	-	-	-	-	-	-

Thermal and spatial coherences are related to the system characteristics or InSAR observation conditions, whereas temporal coherence is a major parameter to observe surface migration. Lee et al. (2012) [47] described the phase coherence in the time domain as shown below:

$$Coh_{temporal} = \exp\left(-C_t \Delta T\right) \tag{5}$$

where C_t is the decay constant of temporal coherence, which is determined based on the physical and geometric characteristics of the target surface, and ΔT is the time period of InSAR pair observations.

Once these have been addressed, it is possible to express the phase decorrelation, i.e., the decrease of phase coherence over a certain period, especially over the arid desert, as follows:

$$Total_{Decorr} = Decorr(AER) \times Decorr(Vege) \times Decorr(MO) \times Decorr(System) \tag{6}$$

where *Decorr(AER)* is the decorrelation induced by the aeolian erosion rate, *Decorr(Vege)* is the decorrelation induced by vegetation, *Decorr(MO)* is the decorrelation induced by soil moisture and other climatic factors, and *Decorr(System)* is the remnants of systematic effects.

However, considering the vegetation and its interaction with wind and moisture (Yun et al., 2019) [48], it is almost impossible to precisely model the decorrelation components indicated by the temporal baseline between InSAR pairs. As such, random and systematic effects are difficult to predict, and the application of a time series analysis of phase coherences was a potential approach to address the issues. Based on the phase coherence time-series results, the principal component analysis (PCA) was applied to extract the major component. To extract the trending component along the mean direction for further decomposition (see Gallagher et al., 2020 [49] for more details of centering effects in PCA), the centering of each phase coherence was not applied. The temporal outliers, e.g., temporal wind, moisture variations, and the remnants of systematic components as shown in Kim et al. (2020) [50], were expected to be excluded. Note that the decorrelation by the AER might have a directional trend. Thus, it will remain even in the first component of PCA. However, the decorrelation by vegetation is very strong and also has directional trends with seasonal change and growth of vegetation. To handle the decorrelation in the AER, a decomposition was further conducted through modeling.

3.2. Decomposition by Vegetation Decomposition and Overall Decomposition Procedure

Even though PCA analysis can effectively suppress the random change of phase coherence, we have to address the decorrelation by vegetation. Our approach was to model decorrelation with the vegetation indices. Based on Equation (5) and Santoro et al. (2009)'s temporal coherence approximation [51], we proposed phase coherence decomposition by introducing an external vegetation index:

$$Decorr(\text{Vege}) = \exp(Cm \times VI) \tag{7}$$

where Cm is the dependent constant on surface vegetation, and VI is the vegetation index.

Corresponding to the insensitivity of desert dune that occurred in the measurements, it was observed that the Normalized Vegetation Difference Index (NDVI) and other vegetation indices were not optimally presented in the temporal and spatial migrations. Particularly, NDVI had a problem presenting actual vegetation metrics [52,53]. Therefore, the Enhanced Vegetation Index (EVI) was applied in the analysis. We tackled the regression model by introducing LANDSAT-5/8 EVI to our target area. Two regression models between decorrelation and the vegetation index computed from the two SAR image acquisition times were proposed as follows.

$$Decorr(\text{Vege}) = \exp(Cm1 \times Avg(EVI1, EVI2) + Cm2 \times abs(EVI1 - EVI2)) \tag{8}$$

$$Decorr(\text{Vege}) = \exp(Cm1 \times Avg(EVI1, EVI2)) \tag{9}$$

We tested two models and found that it was difficult to establish two variable regression models as listed in Equation (8) and, therefore, employed Equation (9) as a default model instead. Noted that we used long-term (monthly or yearly) average EVIs of time InSAR image acquisition to avoid short-term variation of the vegetation index and void due to cloud coverage. Once the decorrelation component by vegetation was established, we applied it for decomposition either before or after PCA analysis based on time-series phase coherence. Decomposition is a simple division of decorrelation by vegetation from each phase coherence (Figure 2a) or component 1 of PCA analysis based on the phase coherence time-series as shown in Figure 2b. To distinguish the two processes, we named the procedure in Figure 2a as a prior processor and Figure 2b as a posterior processor.

It was worthwhile noting that the decorrelation model that employs EVI should be established in areas excluding desert as well as urban areas, artificial structures, and water surfaces. The land cover, which has more than a certain proportion of vegetation, would follow the decorrelation model as presented in Equation (9). Therefore, we introduced a land cover classification map for the decorrelation modeling and excluded urban areas, artificial structures, desert, and water surfaces from the EVI-decorrelation model as presented in both Figure 2a,b.

3.3. Inter-Comparison with Protrusion Coefficients (PC)

As demonstrated in Figure 2, we employed the protrusion coefficient (PC) as validation data sets. PC was introduced because the other data sets such as vegetation indices, which supposedly involved the aeolian process, had significant problems being directly compared to AER by InSAR analysis. In an arid desert environment, some rough objects such as leafless vegetation, shrub, and non-vegetation objects are largely effective in reducing the aeolian erosion process; however, unlike non-vegetation, vegetation indices do not reflect the strength of ground obstacles. Therefore, we needed quantity measuring roughness, which consequently represented resistance against aeolian erosion to validate the AER estimated by InSAR analysis. The background for the employment of PC for comparison with InSAR AER is as follows.

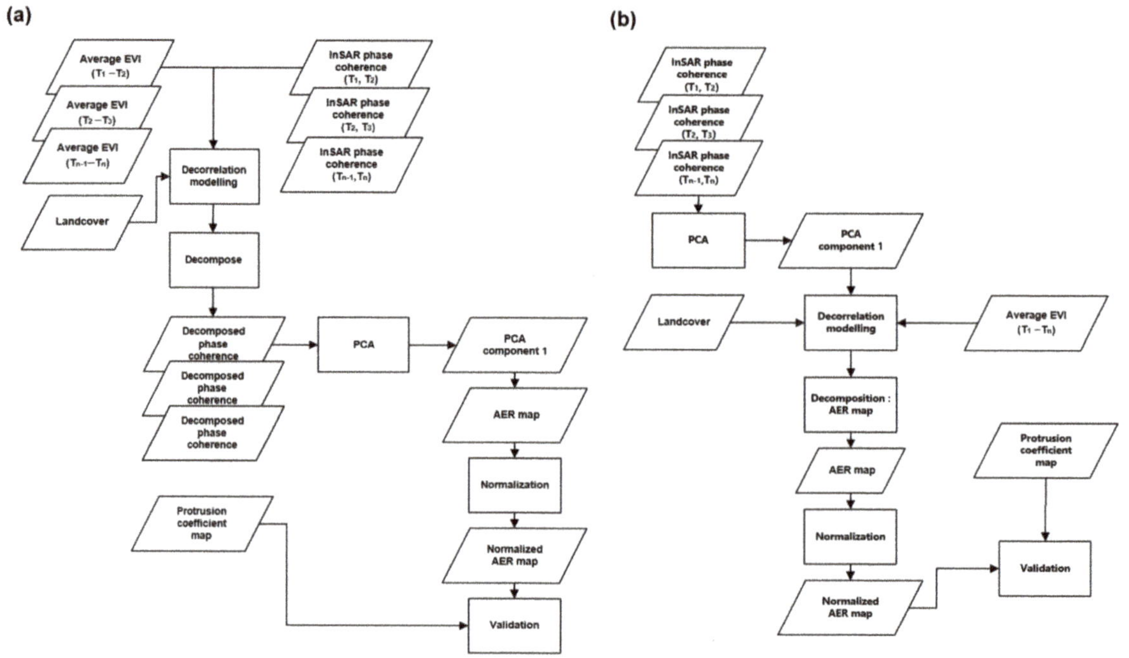

Figure 2. (**a**) Processing flow of phase coherence analyses for the establishment of the AER map using decomposed phase coherence, which is called a prior processor, (**b**) Processing flow of phase coherence analyses for the establishment of the AER map using the 1st principal component analysis (PCA) component and its decomposition, which is hereafter referred to as the posterior processor.

One of the crucial parameters governing aeolian erosion is aerodynamic roughness, also called Z_0. The smaller the Z_0 value (mostly covered desert, low vegetation area, etc.), the greater the probability of dust in the topsoil during the saltation process. Therefore, the essential information to identify the feasibility of erosion is the Z_0 of the target surface and its time-series, specifically the spatiotemporal migration of Z_0, expressed by the following function.

$$U(z) = \left(\frac{u*}{K}\right) \ln(z - d0)/Z_0 \tag{10}$$

where $U(z)$ is the vertical wind profile, Z_0 is the aerodynamic roughness length, $u*$ is the friction velocity, K is the Vol Karman constant (dimensionless) which defines the logarithmic law in a fluid flow, and $d0$ is the elevation of displacement plane.

The observation of Z_0 in an open field is available by the following methods. The first straightforward approach is to analyze high-resolution satellite and airborne DEMs and to identify aerodynamic obstacles such as man-made structures and vegetation on the DEM [54–56]. However, it should be noted that the spatio-temporal domains of the target area in this study could not be easily covered by the employment of high-resolution DEM products. The method which regressed Z_0 from the vegetation index [57,58] was not valid in many practical cases. Since the target area included large urban and artificial structures and some vegetation in the arid desert was not distinguished by the satellite vegetation index, therefore, this approach was not appropriate for our case.

An indirect derivation may be obtained through processing the low-medium resolution image of the Bidirectional Reflectance Distribution Function (BRDF). The BRDF can be obtained from a combination of multiple satellite images over one point [59] and measurements of the reflectivity function depending on the viewing and illuminating of geometry and surface roughness. By a kernel-based approach that has been commonly used for satellite products [60], the BRDF is defined as follows:

$$R_{surf} = k_0 + k_1 f_1 + k_2 f_2 \quad (11)$$

where R_{surf} is the BRDF of the surface, k_0 is an isotropic coefficient, k_1, f_1 is the volumetric coefficient and kernel while k_2, f_2 is the geometric coefficient and kernel. The volumetric and isotropic coefficients of the derived BRDFs are the dependent variable of PC and downward relationship with Z_o [61,62].

$$PC = \frac{k_1}{k_2} = a \times \log(Z_o) + b \quad (12)$$

It is known that the PC from the BRDF was involved with the strength of aeolian erosion on the condition that the other governing factors for aeolian erosion such as soil moisture are negligible. Therefore, the inter-comparison of the InSAR AER to PC from the BRDF had merit. Here, we employed MODIS BRDF products to extract PC maps as precedent researches have conducted [63,64].

4. Processing Results

The ALOS PALSAR time series were used for long-term monitoring, while Sentinel-1 imagery tested the availability of short-term AER measurements. Two L-band InSAR image series of PALSAR-1 and -2 were first employed to extract phase coherence maps. A minimum span tree (MST) [65] was formed in which the average of InSAR phase coherence between pairs could be the maximum value, and the InSAR phase coherence between the formed pair was extracted. Phase coherence maps generated by PALSAR-1 InSAR pairs during 2007–2008 represented the decorrelation trend before the activity to combat desertification as shown in Figure 3. On the contrary, Figure 4 generated from the PALSAR-2 InSAR time series in 2016–2017 demonstrated decorrelation status after combating desertification.

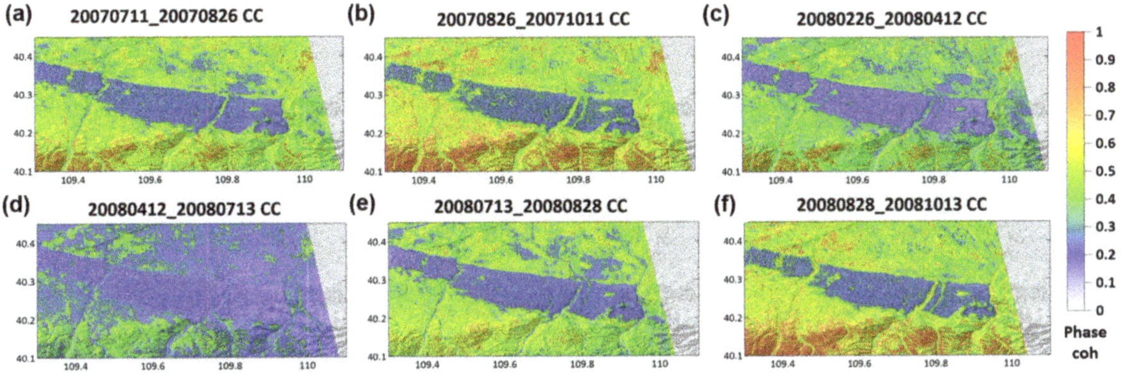

Figure 3. Interferometric Synthetic Aperture Radar (InSAR) phase coherence using PALSAR-1 during the 2007–2008 period, including (**a**) 11 July 2007–26 August 2007, (**b**) 26 August 2007–11 October 2007, (**c**) 26 February 2008–12 April 2008, (**d**) 12 April 2008–13 July 2008, (**e**) 13 July 2008–28 August 2008, and (**f**) 28 August 2008–13 October 2008. Note the dust storm season is usually from February to May, but mainly in March and April.

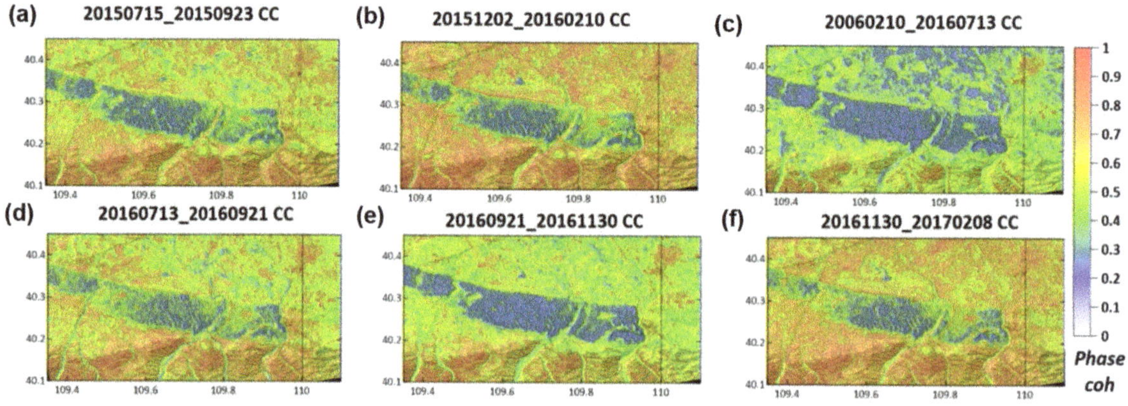

Figure 4. InSAR phase coherence using PALSAR-2, including pairs of (**a**) 15 July 2015–23 September 2015, (**b**) 2 December 2015–10 February 2016, (**c**) 10 February 2016–13 July 2016, (**d**) 13 July 2016–21 September 2016, (**e**) 21 September 2016–30 November 2016, and (**f**) 30 November 2016–8 February 2017.

Vegetation index EVI derived from the medium resolution (30–50 m/pixel) LANDSAT 5 and 8 data were extracted in the corresponding areas of InSAR observations using Google Earth Engine (GEE), as described later. The scattergrams in Figure 5 demonstrated extremely important results in the correlation between vegetation and phase coherence signals.

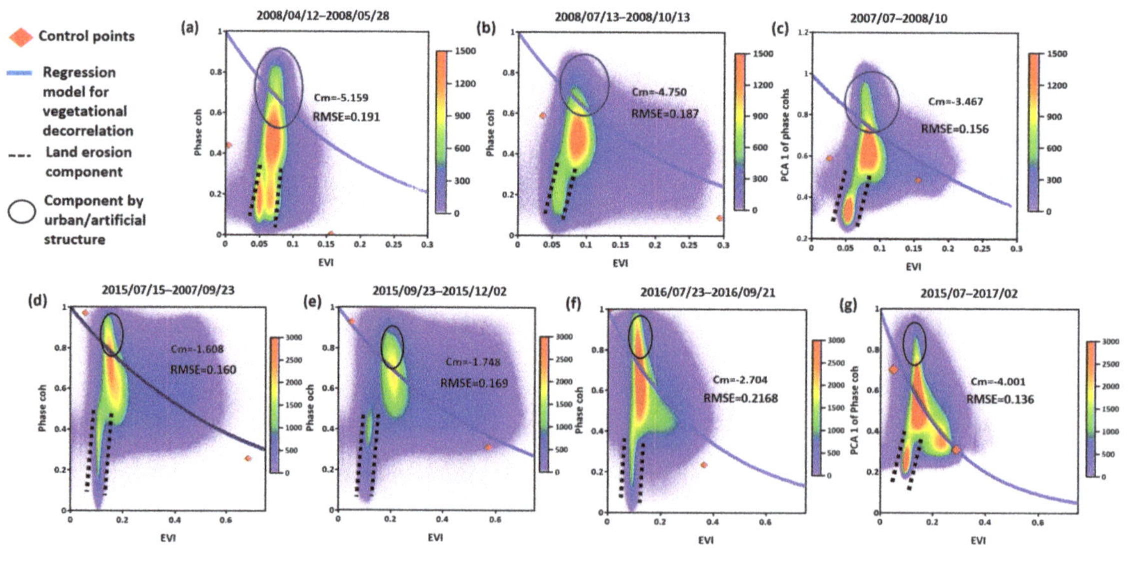

Figure 5. (**a**,**b**) The scattergram between the Enhanced Vegetation Index (EVI) and its corresponding single-phase coherence from the PALSAR-1 InSAR pair by a prior processor, (**c**) The scattergram between the EVI and PCA 1st component from the PALSAR-1 phase coherence time-series by a posterior processor, (**d**–**f**) The scattergram between the EVI and its corresponding single-phase coherence from the PALSAR-2 InSAR pair by a prior processor, (**g**) The scattergram between the EVI and PCA 1st component from the PALSAR-2 phase coherence time-series by a posterior processor. The corresponding regression models were presented as blues lines. Control points indicating high EVI, low phase coherence, and low EVI and high phase coherence were indicated by red points. The color represents the cumulative pixel number of the corresponding EVI-PCA bin. RMSEs were measured over the upper area of the land erosion component and demonstrated with the regression model constant *Cm* of Equation (1).

For all the scattergrams illustrated in Figure 5, three components were commonly observed: (1) the component of vegetation which is populated around the decorrelation model (blue lines); (2) elongated part into lower EVI and low phase coherence (black dotted lines); (3) some distribution with low EVI and high phase coherence presented in black ellipses. The components around the modeled decorrelation re clearly vegetated terrains, thus they are decomposed. Elongated components described in (2) with low EVI and low phase coherence assign aeolian erosion, referred to as the erosion component as low vegetation and low phase coherence can only be presented as the erosional surface by the decrease of Z_0 which induces more aeolian erosion as stated in Section 3.3. Components described in (3) are mainly artificial structures or bare terrain covered by very weak vegetation. The problem we observed here is that the decorrelation model of vegetation was frequently inclined in the component 3 direction when there was no consideration to exclude it from the modeling. Thus, we employed two ways to clarify the findings: (1) processing together with land cover maps, excluding from the modeling some specific land cover types such as water surface, desert, and urban area, which may distort the decorrelation model on vegetation; (2) further introducing two control points in regression modeling, one with low EVI and high phase coherence and the other with high EVI and low phase coherence. Both points were clearly located in natural vegetation, and the value was assigned within a sufficient surrounding buffer zone. These two control points were introduced in the regression with high weighting values. The effects of the control points may not have been significant, but they enabled us to observe the controllability of the regression model. We observed that the regression model of vegetation decorrelation through the posterior process was better established as it was not inclined towards artificial/urban components compared to other cases by a prior processor (refer to Figure 5c,g). Moreover, it is worthwhile highlighting that the erosion components assigned by black dotted lines, which was our main object to measure, were all better distinguished in the results by a posterior process, as shown in Figure 5c,g. The erosion components characterized the inverse relationship between the EVI and phase coherences, which were also observed in precedent studies regarding surface migration on arid sandy desert using phase coherence [28,29,31–33]. We measured and presented root mean square errors (RMSEs) between regression models by Equation (5) and observations. Those values demonstrated that the posterior processor was more effective to eradicate uncorrelated components of time-series phase coherence. Which corresponded to the precedent phase coherence analysis by Kim et al. (2021) [50]. The observations summarized above hence formed the basis of using a posterior process to extract AER as the default method.

In order to extract the decorrelation component by vegetation, we established a decorrelation map using an established regression model and EVI, as shown in Figure 6. Therefore, we were able to decompose the decorrelation map from phase coherence or the PCA 1st component of the phase coherence time-series, as demonstrated in Figure 7. It should be noted that the AER in Figure 7 is the relative strength of aeolian erosion between the maximum to minimum values after decomposition. Thus, the minimum AER value in an InSAR time-series analysis presents the minimum change of topography while the maximum AER value is the maximum change of topography in each time-series analysis. As we inferred from the Figure 5 cases, the results from a posterior processor showed the outcomes better distinguished the before (see Figure 7c) and after activities to combat desertification (Figure 7d), which were intensively conducted in the mid-2010s.

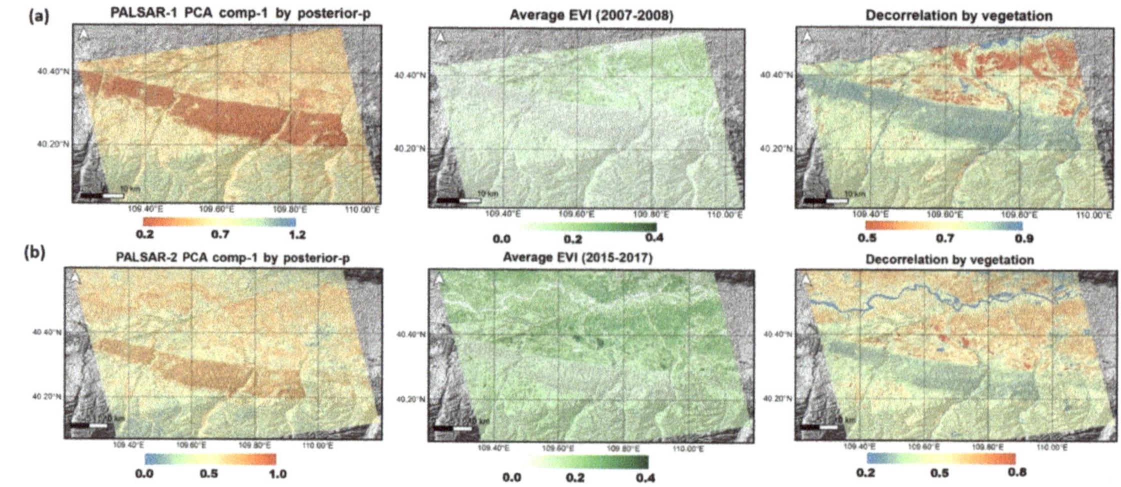

Figure 6. (**a**) Pre-decomposition stage in PALSAR-1 (2007–2008 period) by a posterior processor. (**b**) Pre-decomposition stage in PALSAR-2 (2015–2017 period) by a posterior processor with LANDSAT-8 EVI average. The left is the PCA component 1 of PALSAR-1/2 phase coherence, the middle corresponds to the LANDSAT-5/8 EVI average over the InSAR observation time period, and the right is the simulated decorrelation map by vegetation decorrelation model and the EVI map.

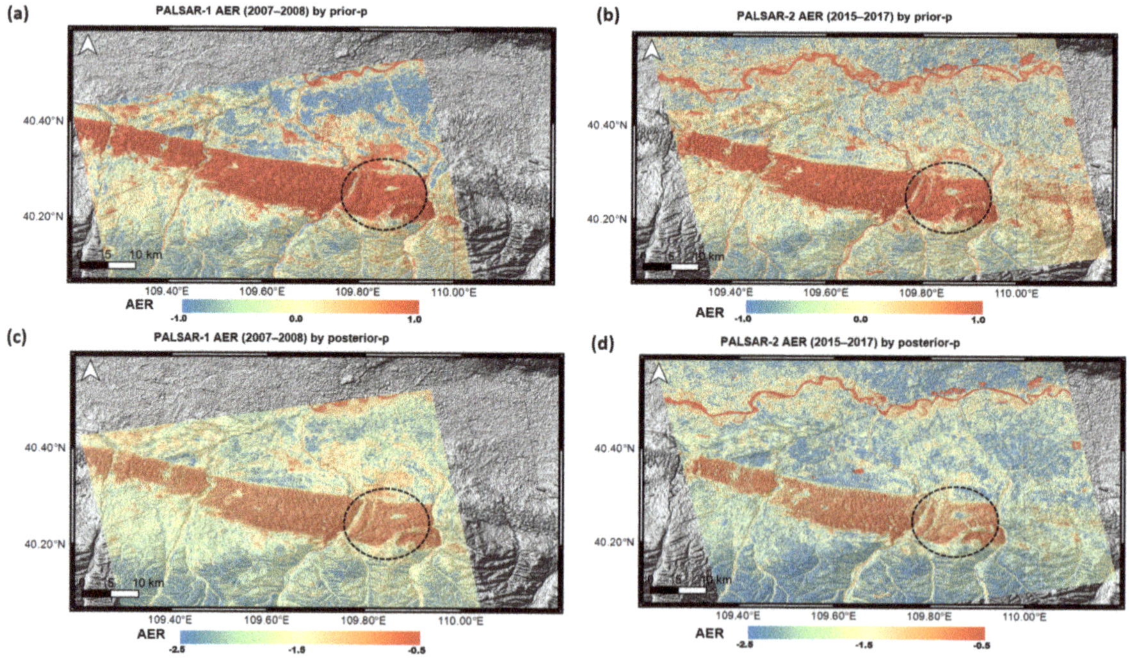

Figure 7. AER maps using ALOS PALSAR-1/2 during 2007–2008 and 2015–2017 periods using a prior processor in (**a**,**b**) and a posterior processor in (**c**,**d**). Noted that the effects of combating desertification activities are obvious in the posterior processor cases.

Sentinel-1 InSAR phase coherence time-series covering the 2016–2020 period were constructed in consecutive InSAR pairs. We established short-term time series for each year from 2016 to 2020. After the decomposition of vegetation, surface erosion components in each year were established (Figure 8). Since the erosion map created by a prior processor presented more noise because of temporal variation of the EVI and phase coherence, our default approach followed a posterior decomposition processor in all Sentinel-1 cases. The AER time-series analysis is for the inter-comparison of short-term aeolian erosion using the high temporal density of Sentinel-1 InSAR image acquisitions. Thus, a normalization procedure was necessary to convert all AER metrics in the same data range. The details of validation employing normalized Sentinel-1 AER are described in the next section.

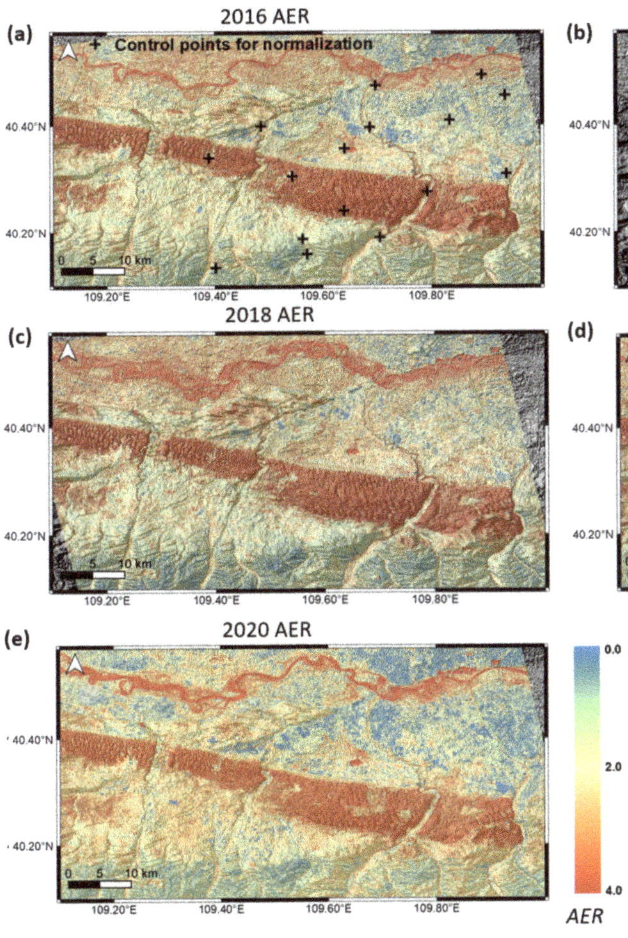

Figure 8. (**a–e**) Surface erosion map using Sentinel-1 during the 2016–2020 period. Noted that all AER maps were constructed by a posterior processor. The crosshairs in (**a**) indicate the location of ground control points for normalization. See the next section for the detailed normalization process. AERs of data sets are not controlled by each other's data ranges. See the next section for the detailed normalization process to assign controlled data ranges using ground control points.

5. Discussion

Since the decomposition procedure was conducted based on the vegetation index, it is not ideal to employ other vegetation indicators to assess extracted AER. Thus, we employed the PC for this purpose. The PC time-series representing the surface characteristics in the target area from 2000 to 2017 were reconstructed based on MODIS 43 BRDF products as shown in Figure 9 [66] in which the relatively clear discrimination between sand dune and other land cover types can be clearly observed. Thus, it is worth noting that the PC to InSAR erosion map comparison was effective for validating our approach.

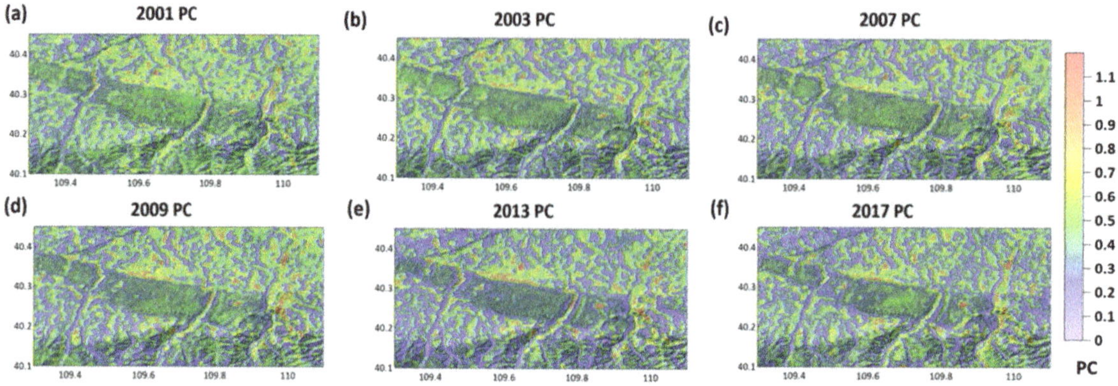

Figure 9. Time-series representations of protrusion coefficients (PCs) in 2001 (**a**), 2003 (**b**), 2007 (**c**), 2009 (**d**), 2013 (**e**), and 2017 (**f**) using MODIS BRDF MOD 42 products (the resolution is 500 m/pixel).

The scattergrams between the PCs and AERs extracted from Sentinel-1 in the 2016–2020 period, together with the established regression models, are shown in Figure 10 and Table 3. It is noted that there are commonly two data clusters (see blue and red ellipses in Figure 10a) together with the inverse proportional relationship in PC-AER distributions. The part encircled by the red ellipse is the aeolian land degradation component, and the part outlined in the blue ellipse represents the weak erosion components with low-to-high rough surfaces, such as vegetation and urban areas. The dispersions in PCs in both distributions mainly originated from resolution differences of the PC (500 m) and the AER (30 m) as aggregated land cover types in one PC pixel which induced a large dispersion. However, parts of the dispersion in the aeolian erosion ellipse can be attributed to the influences of other control factors of aeolian erosion, such as soil moisture, condensation, and clay ratio. The lower half part of the high roughness ellipse, i.e., the lower part in the blue ellipse, is certainly formed by the relatively flat but robust surface to aeolian erosion in rocky areas, concrete pavements, and the roadside; thus they have low PC and low AER values. Other upper parts of the low erosion ellipse can be explained as high roughness in surfaces found in forest and urban structures.

Table 3. The metrics of linear and LMA regression models between the PC and Sentinel-1 AER (R^2: coefficient of determination).

2016	**Model**	**R^2**	**RMSE**	**2017**	**Model**	**R^2**	**RMSE**	**2018**	**Model**	**R^2**	**RMSE**
Linear	$Y = -0.074 \times X - 0.822$	0.024	0.079	Linear	$Y = -0.087 \times X - 0.737$	0.023	0.072	Linear	$Y = -0.108 \times X - 0.668$	0.056	0.074
RMA	$Y = -0.469 \times X + 0.157$	0.372	0.137	RMA	$Y = -0.567 \times X + 0.426$	0.371	0.126	RMA	$Y = -0.458 \times X + 0.195$	0.396	0.12
2019	**Model**	**R^2**	**RMSE**	**2020**	**Model**	**R^2**	**RMSE**	-	-	-	-
Linear	$Y = -0.081 \times X - 0.796$	0.027	0.088	Linear	$Y = -0.213 \times X - 0.421$	0.152	0.084	-	-	-	-
RMA	$Y = -0.495 \times X + 0.231$	0.375	0.151	RMA	$Y = -0.547 \times X + 0.392$	0.45	0.671001	-	-	-	-

Therefore, the regression models of the AER and PC were not clear in both the least mean squares (LMA) and linear regression approaches as listed in Table 3. Considering the dispersion in low erosion and high erosion ellipses, we estimated that the PC-AER on the natural surface was similar to the linear regression model, which was actually the first-order polynomial presented in Table 3 rather than the high order polynomial cases. High data dispersion in the PC to AER regression showed the extra effects on surface changes such as soil moisture, wind strength, and human activities which induced aeolian erosion. The introduction of advanced biomass metrics that can be extracted from a sequence of new algorithms [67,68] or from new sensing systems [69,70] will be helpful for a precise validation of the AER in future studies. Thus, the developed AER tracing scheme will be useful for the activities to combat desertification as well as the early warning of dust storm generations after applying such precise validation processing.

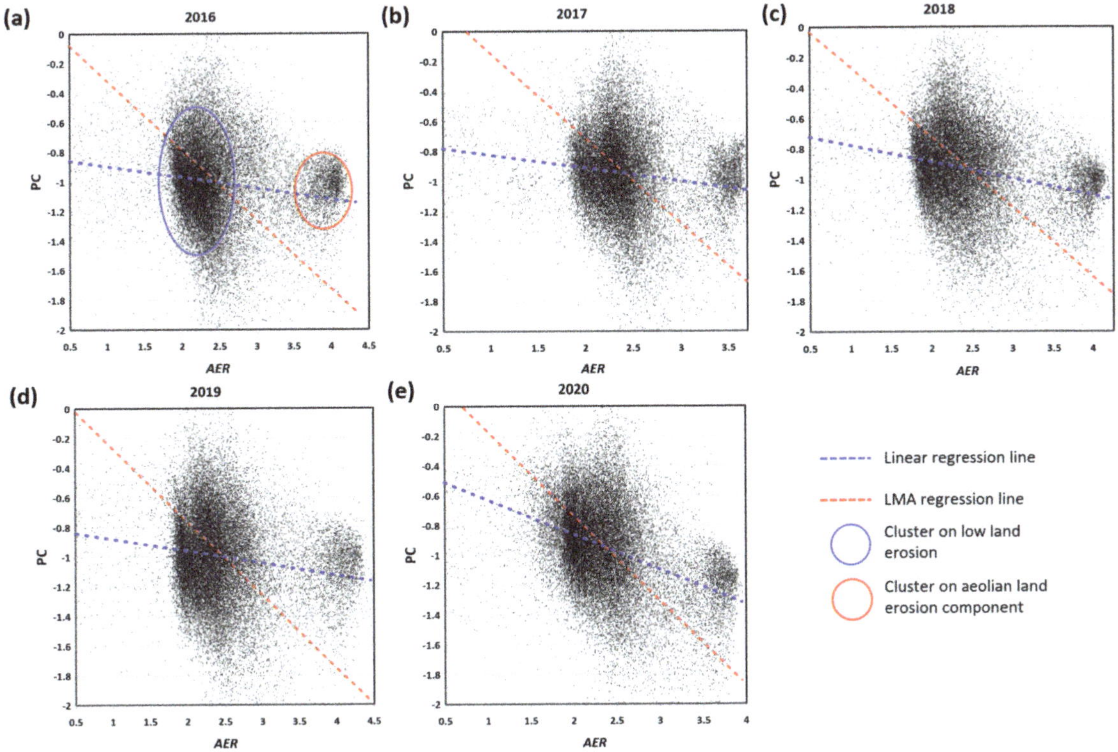

Figure 10. PC vs. InSAR AER in 2016 (**a**), 2017 (**b**), 2018 (**c**), 2019 (**d**), and 2020 (**e**). Note that the regression models between the PC and AER were built by an ordinal linear regression model (blue dotted lines) and a least mean squares (LMA) regression model (red dotted lines).

A further step applied was to normalize all derived AER measurements by different InSAR time-series analyses in the same standardized range. The processed outcomes can be comparable to each so it is referred to as n-AER as stated in Section 1. Regarding normalization, a linear regression model assuming a linear relationship between the values of the reference and subject images for points of the same land cover types as those reported in Sadeghi et al. (2017) [71] was applied in this study. Based on the dark set-bright set (DB) method developed by Hall et al. (1991) [72], a total of 16 erosional ground control points (EGCPs) appearing unchanged as settlement, river, and road were manually selected to derive the normalization coefficients (refer to Figure 8a to see the location of the selected points). Given the pattern of exponential decay in phase coherence presented in Equation (5), a linear relationship can be established:

$$n\text{-}AER_i = G_{iX} log(AER_i) + off\_set_i \quad (13)$$

where G is the gain value, *off* is offset to deploy logarithmic AER value into a normalized data range, and I is ith target InSAR time series observation. The G and *off_set* can be extracted from the linear system in Equation (13) and EGCPs were assigned to 0.0 with zero migration over settlement and road and 1.0 with full decorrelation over the water surface. The values of the potential EGCPs in the InSAR AER images over the whole observation period were recorded respectively and chosen carefully only in the case that there were no significant changes. Based on the coefficients, i.e., Gi and off_set_i derived through ordinary least squares estimation, the AER images were normalized through a linear transformation. The extracted n-AERs are demonstrated in Figure 11 for ALOS PALSAR 1/2 and in Figure 12 for Sentinel-1.

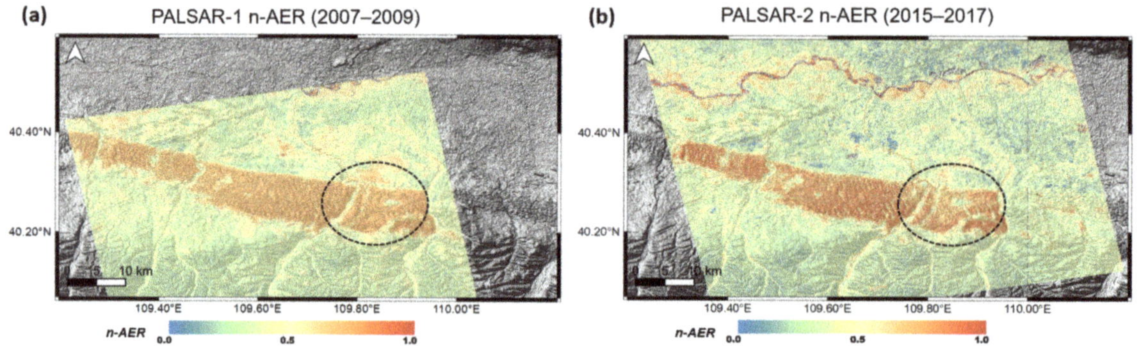

Figure 11. Normalized aeolian erosion (n-AER) maps using ALOS PALSAR-2 pairs of 2007–2009 (**a**) and 2015–2017 (**b**). Two n-AER maps now can be assessed against each other as the data ranges of the two maps used the process of employing control points (see Figure 8a). The effects of activities to combat desertification are evident as presented in circled areas.

Therefore the n-AER maps produced by independent phase coherence time-series analysis over the Kubuqi Desert demonstrated how the change in surface characteristics caused the mitigation of erosions. A further procedure for assessment of InSAR n-AER products was performed based on the inter-comparison to vegetation information and ground truth regarding activities to combat desertification. The n-AER covering a decadal period in comparison to the EVI of LANDSAT 5/8 and MODIS is presented in Table 4. It was observed that commonly in LANDSAT and MODIS EVIs, the low EVI area (<0.2, 0.15, and 0.1, respectively) was dramatically reduced in the 2016 observations compared to the 2007 observations. For instance, the area with an EVI lower than 0.1 changed from 77% in 2006 to 6.6% in 2016. When the MODIS EVIs were employed to address different sensor issues in the use of LANDSAT 5 and 8, low EVI areas were considerably reduced, especially in areas where the EVI was smaller than 0.1. More importantly, n-AER values in low EVI areas showed no differences, or worse, in the 2007 and 2016 observations. Thus, the values in Table 4 represented that the overall changes in n-AER were due to the increasing vegetation coverages in the entire Kubuqi area. However, natural conditions that induce aeolian land degradation were still severe. Conclusively, it has been proven that the main drivers of mitigating sand dune activities and aeolian land degradation for the last decade have been anthropogenic factors, including activities to combat desertification in this area.

These observations can be proved again by employing ground truth regarding activities to combat desertification. Figure 13a showed the extent of yearly planting areas brought about by efforts to combat desertification conducted by international NGOs and Chinese local government from 2006 to 2020. Figure 13b demonstrated the organized planting efforts that resulted in noticeable changes in both EVI and n-AER values as observed in PALSAR 1/2 time-series analyses covering the 10-year period. On the contrary, the short-term n-AERs by Sentinel-1 from 2016 to 2020 did not make significant changes. Obviously, old planting areas (2006–2010 and 2010) in Figure 13c,d were most successful in reducing aeolian erosion in the decadal period. Compared to the recent planting activities and their relatively weak anti-AER effects presented in Figure 13e–g, it is concluded that the activities to combat desertification during 2006–2010 successfully built a high aerodynamic surface roughness wall crossing Kubuqi Desert, which prevented aeolian saltation.

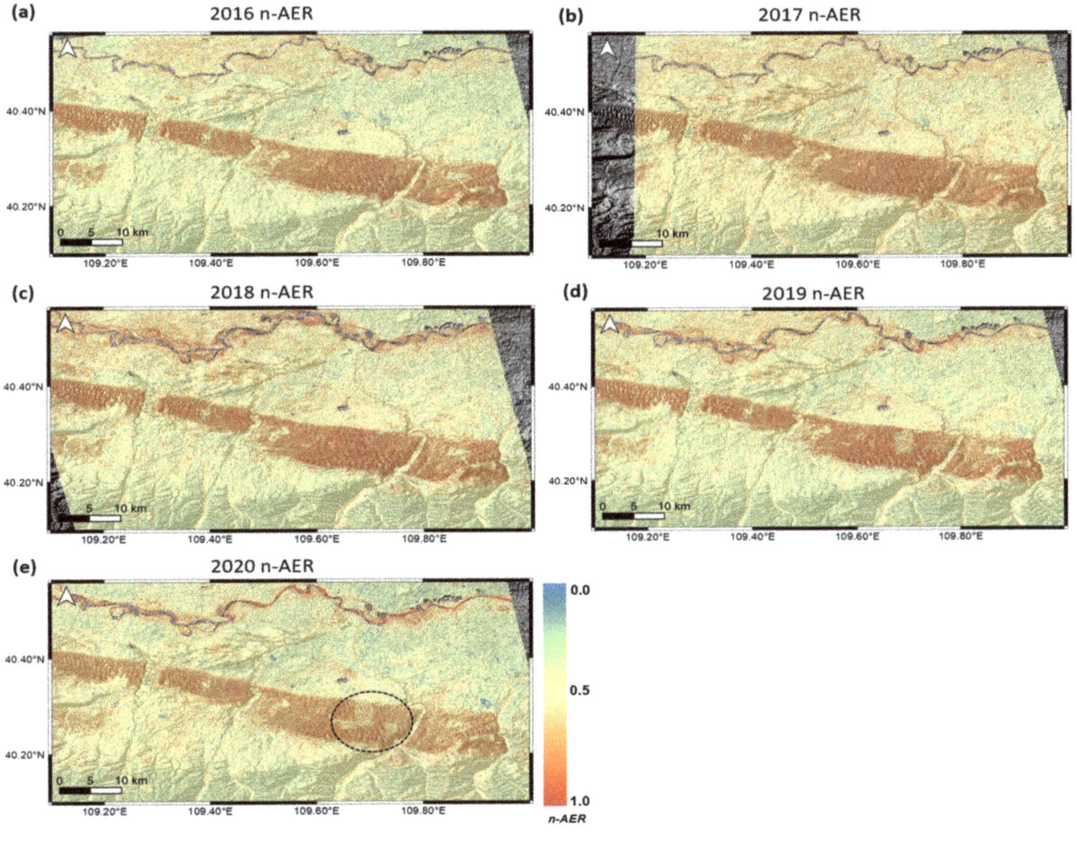

Figure 12. n-AER maps using Sentinel-1 in 2016 (**a**), 2017 (**b**), 2018 (**c**), 2019 (**d**), and 2020 (**e**). See the dotted ellipses where n-AER successfully detects the effect of ground compaction around solar panels. Note the range of n-AER is now adjusted between 0.0 to 1.0.

Table 4. Distribution of low EVI areas and their n-AER values.

EVI Threshold (<EVI)	2007 LANDSAT 5 EVI			2016 LANDSAT 8 EVI		
	Ratio * in 2007	Mean n-AER Value in 2007	n-AER Stdev in 2007	Ratio in 2016	Mean n-AER Value in 2016	n-AER Stdev in 2016
0.2	99.34	0.355	0.122	81.96	0.349	0.148
0.15	92.26	0.363	0.122	50.7	0.396	0.161
0.1	76.88	0.377	0.126	6.56	0.6	0.095
EVI Threshold (<EVI)	**2007 MODIS EVI**			**2016 MODIS EVI**		
	Ratio in 2007	**Mean n-AER Value in 2007**	**n-AER Stdev in 2007**	**Ratio in 2016**	**Mean n-AER Value in 2016**	**n-AER Stdev in 2016**
0.2	98.48	0.356	0.122	95.2	0.337	0.144
0.15	87.29	0.365	0.123	80	0.348	0.15
0.1	43.8	0.428	0.135	30.4	0.467	0.162

* Percentage of the area where EVI value less than EVI threshold.

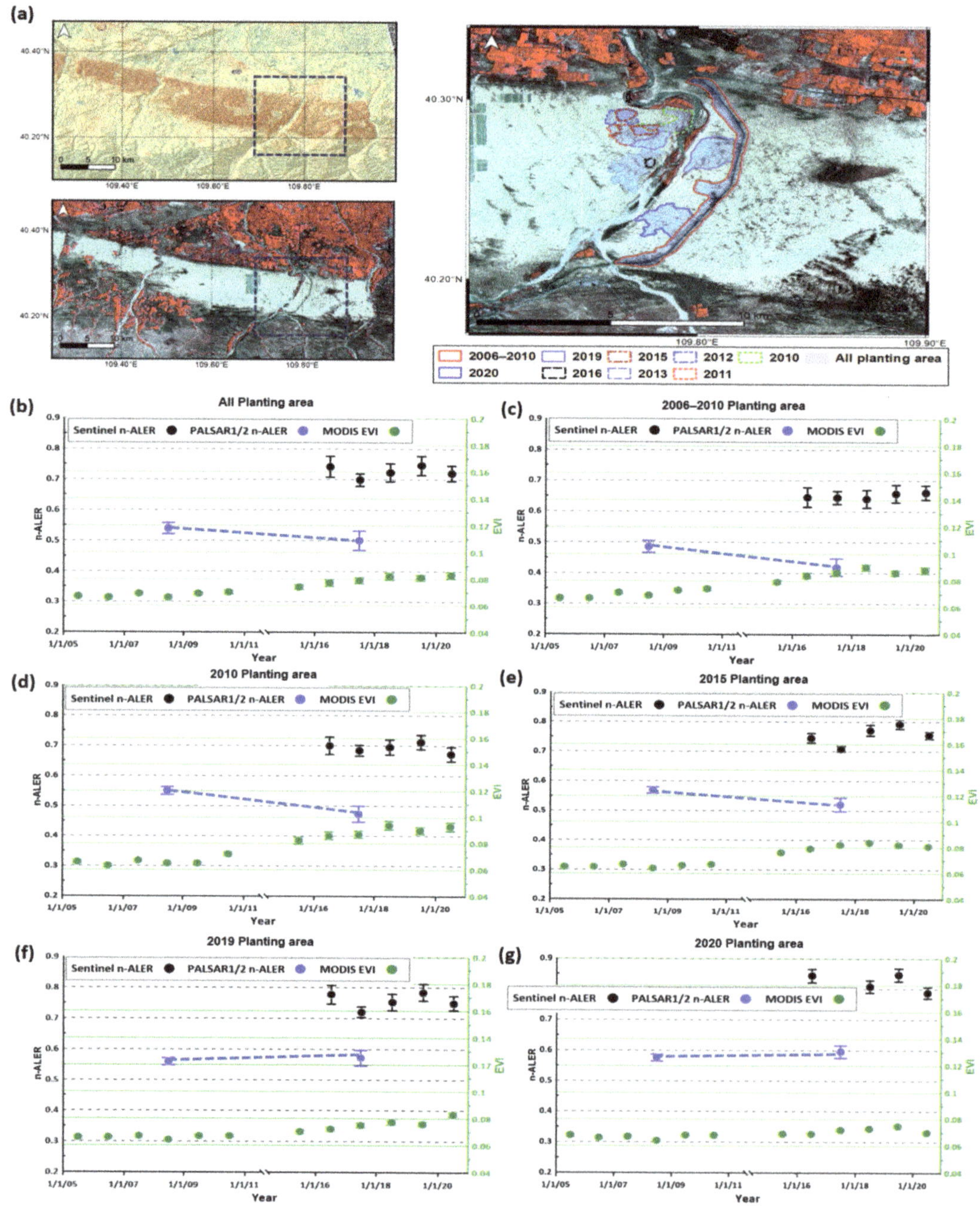

Figure 13. n-AER values using Sentinel-1 and PALSAR-1/2 over combat desertification areas (outlined by black dash line in (**a**)). The performance over all planting areas are shown in (**b**). Note the effects of combat desertification activities are highly obvious in the observations of very early planting such as 2006–2010 in (**c**), 2010 in (**d**) and possibly 2015 (**e**) employing PALSAR-1 and 2. However, the n-AER changes after 2016 measured by Sentinel-1 observations, such as 2019 (**f**) and 2020 (**g**), are not clear.

All these results support that the phase coherence time-series analysis developed in this study is effective in tracing erosion. It has also shown great potential for identifying sources of dust and ongoing erosion by replacing existing approaches such as AOD analysis. Compared to 25 km resolution dust source tracing using AOD analysis, the InSAR phase coherence method provided an improved opportunity to directly observe erosion with a few decameters spatial resolution and better temporal resolution regardless of climate condition. However, it should be noted that the limitations of the phase coherence time-series analysis for dust source mapping are also clear. As mentioned above, the values observed by phase coherence time-series analysis are only relative intensities. Therefore, a calibration process between erosion maps in each epoch using land control points is necessary. Another significant problem of phase coherence analysis is the reliability of the decomposition of the phase coherence contribution by other factors such as vegetation, soil moisture, snow cover, and other land changes. In this respect, we propose a more detailed interpretation of phase coherence time-series that combines high-resolution remote sensing data products such as drone and space-borne ultra-high-resolution products. Cross-comparison studies using climatic factors such as soil moisture, wind speed, and precipitation will provide insight into the improvement of the phase coherence AER.

6. Conclusions and Future Work

This study proved that precise long-term monitoring of aeolian erosion and progress of involved surface condition is feasible using time series InSAR and medium resolution vegetation index products. A scheme employing InSAR phase coherence time-series successfully traced a temporal consequence in which activity to combat desertification caused the increase of aerodynamic roughness length and the decrease of sand dune mobility. In spite of the favorable natural environment for intensive aeolian land degradation in N.W Chinese deserts, the efforts to combat desertification and other anthropogenic factors led to the overall reduction of erosion in Kubuqi during the period 2000–2020. Our scheme successfully reconstructed such a long-term and steady-going combating desertification consequence on the establishment of a blocking wall leading to the effective mitigation of aeolian erosion. Thus, the effectiveness of the phase coherence approach for erosion monitoring was well verified in both aspects of precision and spatio-temporal resolution.

However, it has appeared that a few issues remained to be solved in the proposed scheme before it is to be employed as a standard approach for AER measurement. First, the driven InSAR AER only extracted the relative strength of aeolian erosion; thus, it should be calibrated compared to the metrics of quantitative erosion. On the other hand, the InSAR signal still needs to be further inspected to decompose other governing factors of erosion, such as climatic factors and the footprints of the urban area, rock surface, and concrete resistant land cover that exists against erosion. It means the procedure needs to be more updated considering how surface conditions for erosion are very complicated.

For future work, in order to carry out the quantitative analysis which is more generally applicable for erosion monitoring in major sand dust source areas, it is essential to perform additional measurement over the target area, such as UAV stereo, high-resolution satellite analysis, and close-range photogrammetry. Improved time-series data analysis together with high-resolution products will be of great help for the planning of more effective activity to combat desertification and better prediction and observation of erosion in the arid desert. Future space-borne sensors such as *NASA-ISRO Synthetic Aperture Radar* (NiSAR) [73], Geostationary Ocean Color Imager-2 (GOCI-2) [74], and Panchromatic Remote-sensing Instrument for Stereo Mapping-2 (PRISM-2) will provide highly valuable data sets to update or precisely validate proposed algorithms.

Author Contributions: J.-R.K. developed algorithms and wrote major part of draft. C.-W.L. conducted InSAR processing and time series analysis of other data sets. S.-Y.L. was charge of normalization and corresponding writing together with summarizing of draft. All authors have read and agreed to the published version of the manuscript.

Funding: The authors acknowledge support by the Ministry of Science and Technology (MOST) in Taiwan. This research received partial funding for S.-Y.L. under grant number MOST 110-2119-M-004-001 to support APC.

Data Availability Statement: The data in this study are available on request from the corresponding author.

Acknowledgments: This study was conducted by the support of Korean Forest service. The research activities in the target area were kindly supported by Future Forest (http://www.futureforest.org/ (accessed on 1 June 2021)), United Nations Convention to Combat Desertification (UNCCD) and China Youth League. ALOS PALSAR data is kindly provided by 6th Research Announcement of Japan Aerospace Exploration Agency (JAXA).

Conflicts of Interest: The authors declare no conflict of interest. The funders had no role in the design of the study; in the collection, analyses, or interpretation of data; in the writing of the manuscript, or in the decision to publish the results.

References

1. Tegen, I.; Fung, I. Contribution to the atmospheric mineral aerosol load from land surface modification. *J. Geophys. Res. Atmos.* **1995**, *100*, 18707–18726. [CrossRef]
2. Xuan, J.; Sokolik, I.N.; Hao, J.; Guo, F.; Mao, H.; Yang, G. Identification and characterization of sources of atmospheric mineral dust in East Asia. *Atmos. Environ.* **2004**, *38*, 6239–6252. [CrossRef]
3. Wang, X.; Huang, J.; Ji, M.; Higuchi, K. Variability of East Asia dust events and their long-term trend. *Atmos. Environ.* **2008**, *42*, 3156–3165. [CrossRef]
4. Kurosaki, Y.; Mikami, M. Regional difference in the characteristic of dust event in East Asia: Relationship among dust outbreak, surface wind, and land surface condition. *J. Meteorol. Soc. Japan. Ser. II* **2005**, *83*, 1–18. [CrossRef]
5. Zhang, B.; Tsunekawa, A.; Tsubo, M. Contributions of sandy lands and stony deserts to long-distance dust emission in China and Mongolia during 2000–2006. *Glob. Planet. Chang.* **2008**, *60*, 487–504. [CrossRef]
6. Kim, J. Transport routes and source regions of Asian dust observed in Korea during the past 40 years (1965–2004). *Atmos. Environ.* **2008**, *42*, 4778–4789. [CrossRef]
7. Bryant, R.G.; Bigg, G.R.; Mahowald, N.M.; Eckardt, F.D.; Ross, S.G. Dust emission response to climate in southern Africa. *J. Geophys. Res. Atmos.* **2007**, *112*. [CrossRef]
8. Kaskaoutis, D.; Kosmopoulos, P.; Nastos, P.; Kambezidis, H.; Sharma, M.; Mehdi, W. Transport pathways of Sahara dust over Athens, Greece as detected by MODIS and TOMS. *Geomat. Nat. Hazards Risk* **2012**, *3*, 35–54. [CrossRef]
9. Prasad, A.K.; Singh, R.P. Changes in aerosol parameters during major dust storm events (2001–2005) over the Indo-Gangetic Plains using AERONET and MODIS data. *J. Geophys. Res. Atmos.* **2007**, *112*. [CrossRef]
10. Ginoux, P.; Prospero, J.M.; Gill, T.E.; Hsu, N.C.; Zhao, M. Global-scale attribution of anthropogenic and natural dust sources and their emission rates based on MODIS Deep Blue aerosol products. *Rev. Geophys.* **2012**, *50*. [CrossRef]
11. United Nations Convention to Combart Desertification Addressing Sand and Dust Strom in SDG Implementation. Available online: https://knowledge.unccd.int/publications/addressing-sand-and-dust-storms-sdg-implementation (accessed on 1 June 2021).
12. Du, H.; Wang, T.; Xue, X.; Li, S. Modelling of sand/dust emission in Northern China from 2001 to 2014. *Geoderma* **2012**, *330*, 162–176. [CrossRef]
13. Du, H.; Xue, X.; Wang, T. Estimation of saltation emission in the Kubuqi Desert, North China. *Sci. Total Environ.* **2014**, *479*, 77–92. [CrossRef]
14. Sun, J.; Zhang, M.; Liu, T. Spatial and temporal characteristics of dust storms in China and its surrounding regions, 1960–1999: Relations to source area and climate. *J. Geophys. Res. Atmos.* **2001**, *106*, 10325–10333. [CrossRef]
15. Zhang, B.; Tsunekawa, A.; Tsubo, M. Identification of dust hot spots from multi-resolution remotely sensed data in eastern China and Mongolia. *Water Air Soil Pollut.* **2015**, *226*, 117. [CrossRef]
16. Wang, X.; Eerdun, H.; Zhou, Z.; Liu, X. Significance of variations in the wind energy environment over the past 50 years with respect to dune activity and desertification in arid and semiarid northern China. *Geomorphology* **2007**, *86*, 252–266. [CrossRef]
17. Wang, X.; Yang, Y.; Dong, Z.; Zhang, C. Responses of dune activity and desertification in China to global warming in the twenty-first century. *Glob. Planet. Chang.* **2009**, *67*, 167–185. [CrossRef]
18. Yu, X.; Zhuo, Y.; Liu, H.; Wang, Q.; Wen, L.; Li, Z.; Liang, C.; Wang, L. Degree of desertification based on normalized landscape index of sandy lands in inner Mongolia, China. *Glob. Ecol. Conserv.* **2020**, *23*, e01132. [CrossRef]
19. Liu, L.; Skidmore, E.; Hasi, E.; Wagner, L.; Tatarko, J. Dune sand transport as influenced by wind directions, speed and frequencies in the Ordos Plateau, China. *Geomorphology* **2005**, *67*, 283–297. [CrossRef]
20. Yun, J.; Kim, J.; Choi, Y.; Yun, H. Monitoring of desert dune topography by multi angle sensors. *AGUFM* **2011**, *2011*, EP31A-0794.
21. Yang, X.; Forman, S.; Hu, F.; Zhang, D.; Liu, Z.; Li, H. Initial insights into the age and origin of the Kubuqi sand sea of northern China. *Geomorphology* **2016**, *259*, 30–39. [CrossRef]

22. Hermas, E.; Leprince, S.; Abou El-Magd, I. Retrieving sand dune movements using sub-pixel correlation of multi-temporal optical remote sensing imagery, northwest Sinai Peninsula, Egypt. *Remote Sens. Environ.* **2012**, *121*, 51–60. [CrossRef]
23. Yao, Z.; Wang, T.; Han, Z.; Zhang, W.; Zhao, A. Migration of sand dunes on the northern Alxa Plateau, Inner Mongolia, China. *J. Arid Environ.* **2007**, *70*, 80–93. [CrossRef]
24. Mahmoud, A.M.A.; Novellino, A.; Hussain, E.; Marsh, S.; Psimoulis, P.; Smith, M. The use of SAR offset tracking for detecting sand dune movement in Sudan. *Remote Sens.* **2020**, *12*, 3410. [CrossRef]
25. Solazzo, D.; Sankey, J.B.; Sankey, T.T.; Munson, S.M. Mapping and measuring aeolian sand dunes with photogrammetry and LiDAR from unmanned aerial vehicles (UAV) and multispectral satellite imagery on the Paria Plateau, AZ, USA. *Geomorphology* **2018**, *319*, 174–185. [CrossRef]
26. Laporte-Fauret, Q.; Marieu, V.; Castelle, B.; Michalet, R.; Bujan, S.; Rosebery, D. Low-Cost UAV for high-resolution and large-scale coastal dune change monitoring using photogrammetry. *J. Mar. Sci. Eng.* **2019**, *7*, 63. [CrossRef]
27. Grohmann, C.H.; Garcia, G.P.; Affonso, A.A.; Albuquerque, R.W. Dune migration and volume change from airborne LiDAR, terrestrial LiDAR and Structure from Motion-Multi View Stereo. *Comput. Geosci.* **2020**, *143*, 104569. [CrossRef]
28. Liu, J.G.; Black, A.; Lee, H.; Hanaizumi, H.; Moore, J.M. Land surface change detection in a desert area in Algeria using multi-temporal ERS SAR coherence images. *Int. J. Remote Sens.* **2001**, *22*, 2463–2477. [CrossRef]
29. Wegmuller, U.; Strozzi, T.; Farr, T.; Werner, C.L. Arid land surface characterization with repeat-pass SAR interferometry. *IEEE Trans. Geosci. Remote Sens.* **2000**, *38*, 776–781. [CrossRef]
30. Gómez, D.; Salvador, P.; Sanz, J.; Casanova, C.; Casanova, J.L. Detecting Areas Vulnerable to Sand Encroachment Using Remote Sensing and GIS Techniques in Nouakchott, Mauritania. *Remote Sens.* **2018**, *10*, 1541. [CrossRef]
31. Havivi, S.; Amir, D.; Schvartzman, I.; August, Y.; Maman, S.; Rotman, S.R.; Blumberg, D.G. Mapping dune dynamics by InSAR coherence. *Earth Surf. Process. Landf.* **2018**, *43*, 1229–1240. [CrossRef]
32. Ullmann, T.; Sauerbrey, J.; Hoffmeister, D.; May, S.M.; Baumhauer, R.; Bubenzer, O. Assessing Spatiotemporal Variations of Sentinel-1 InSAR Coherence at Different Time Scales over the Atacama Desert (Chile) between 2015 and 2018. *Remote Sens.* **2019**, *11*, 2960. [CrossRef]
33. Santoro, M.; Askne, J.I.; Wegmuller, U.; Werner, C.L. Observations, modeling, and applications of ERS-ENVISAT coherence over land surfaces. *IEEE Trans. Geosci. Remote Sens.* **2007**, *45*, 2600–2611. [CrossRef]
34. Askne, J.I.; Dammert, P.B.; Ulander, L.M.; Smith, G. C-band repeat-pass interferometric SAR observations of the forest. *IEEE Trans. Geosci. Remote Sens.* **1997**, *35*, 25–35. [CrossRef]
35. Askne, J.; Santoro, M. Multitemporal repeat pass SAR interferometry of boreal forests. *IEEE Trans. Geosci. Remote Sens.* **2005**, *43*, 1219–1228. [CrossRef]
36. Wang, T.; Liao, M.; Perissin, D. InSAR coherence-decomposition analysis. *IEEE Geosci. Remote Sens. Lett.* **2009**, *7*, 156–160. [CrossRef]
37. Zalite, K.; Antropov, O.; Praks, J.; Voormansik, K.; Noorma, M. Monitoring of agricultural grasslands with time series of X-band repeat-pass interferometric SAR. *IEEE J. Sel. Top. Appl. Earth Obs. Remote Sens.* **2015**, *9*, 3687–3697. [CrossRef]
38. Ahmed, R.; Siqueira, P.; Hensley, S.; Chapman, B.; Bergen, K. A survey of temporal decorrelation from spaceborne L-Band repeat-pass InSAR. *Remote Sens. Environ.* **2011**, *115*, 2887–2896. [CrossRef]
39. Zebker, H.A.; Villasenor, J. Decorrelation in interferometric radar echoes. *IEEE Trans. Geosci. Remote Sens.* **1992**, *30*, 950–959. [CrossRef]
40. Just, D.; Bamler, R. Phase statistics of interferograms with applications to synthetic aperture radar. *Appl. Opt.* **1994**, *33*, 4361–4368. [CrossRef]
41. Tamm, T.; Zalite, K.; Voormansik, K.; Talgre, L. Relating Sentinel-1 interferometric coherence to mowing events on grasslands. *Remote Sens.* **2016**, *8*, 802. [CrossRef]
42. Wei, M.; Sandwell, D.T. Decorrelation of L-band and C-band interferometry over vegetated areas in California. *IEEE Trans. Geosci. Remote Sens.* **2010**, *48*, 2942–2952.
43. Rosenqvist, A.; Shimada, M.; Ito, N.; Watanabe, M. ALOS PALSAR: A pathfinder mission for global-scale monitoring of the environment. *IEEE Trans. Geosci. Remote Sens.* **2007**, *45*, 3307–3316. [CrossRef]
44. Rosenqvist, A.; Shimada, M.; Suzuki, S.; Ohgushi, F.; Tadono, T.; Watanabe, M.; Tsuzuku, K.; Watanabe, T.; Kamijo, S.; Aoki, E. Operational performance of the ALOS global systematic acquisition strategy and observation plans for ALOS-2 PALSAR-2. *Remote Sens. Environ.* **2014**, *155*, 3–12. [CrossRef]
45. Geudtner, D.; Torres, R.; Snoeij, P.; Davidson, M.; Rommen, B. Sentinel-1 system capabilities and applications. In Proceedings of the 2014 IEEE Geoscience and Remote Sensing Symposium, Quebec City, QC, Canada, 13–18 July 2014; pp. 1457–1460.
46. Torres, R.; Snoeij, P.; Geudtner, D.; Bibby, D.; Davidson, M.; Attema, E.; Potin, P.; Rommen, B.; Floury, N.; Brown, M. GMES Sentinel-1 mission. *Remote Sens. Environ.* **2012**, *120*, 9–24. [CrossRef]
47. Lee, C.-W.; Lu, Z.; Jung, H.-S. Simulation of time-series surface deformation to validate a multi-interferogram InSAR processing technique. *Int. J. Remote Sens.* **2012**, *33*, 7075–7087. [CrossRef]
48. Yun, H.-W.; Kim, J.-R.; Choi, Y.-S.; Lin, S.-Y. Analyses of Time Series InSAR Signatures for Land Cover Classification: Case Studies over Dense Forestry Areas with L-Band SAR Images. *Sensors* **2019**, *19*, 2830. [CrossRef]
49. Gallagher, N.B.; O'Sullivan, D.; Palacios, M. The Effect of Data Centering on PCA Models. 2020. Available online: https://eigenvector.com/ (accessed on 1 June 2021).

50. Kim, J.; Dorjsuren, M.; Choi, Y.; Purevjav, G. Reconstructed Aeolian Surface Erosion in Southern Mongolia by Multi-Temporal InSAR Phase Coherence Analyses. *Front. Earth Sci* **2020**, *8*, 531104. [CrossRef]
51. Santoro, M.; Wegmuller, U.; Askne, J.I. Signatures of ERS–Envisat interferometric SAR coherence and phase of short vegetation: An analysis in the case of maize fields. *IEEE Trans. Geosci. Remote Sens.* **2009**, *48*, 1702–1713. [CrossRef]
52. Zhengxing, W.; Chuang, L.; Alfredo, H. From AVHRR-NDVI to MODIS-EVI: Advances in vegetation index research. *Acta Ecol. Sin.* **2003**, *23*, 979–987.
53. Huete, A.R.; Liu, H.; van Leeuwen, W.J. The use of vegetation indices in forested regions: Issues of linearity and saturation. In Proceedings of the IGARSS'97, 1997 IEEE International Geoscience and Remote Sensing Symposium Proceedings. Remote Sensing-A Scientific Vision for Sustainable Development, Singapore, 3–8 August 1997; pp. 1966–1968.
54. Tian, X.; Li, Z.; Van der Tol, C.; Su, Z.; Li, X.; He, Q.; Bao, Y.; Chen, E.; Li, L. Estimating zero-plane displacement height and aerodynamic roughness length using synthesis of LiDAR and SPOT-5 data. *Remote Sens. Environ.* **2011**, *115*, 2330–2341. [CrossRef]
55. Faivre, R.; Colin, J.; Menenti, M. Evaluation of methods for aerodynamic roughness length retrieval from very high-resolution imaging lidar observations over the Heihe Basin in China. *Remote Sens.* **2017**, *9*, 63. [CrossRef]
56. Colin, J.; Faivre, R. Aerodynamic roughness length estimation from very high-resolution imaging LIDAR observations over the Heihe basin in China. *Hydrol. Earth Syst. Sci. Discuss.* **2010**, *7*, 2661–2669. [CrossRef]
57. Bastiaanssen, W.G.; Menenti, M.; Feddes, R.; Holtslag, A. A remote sensing surface energy balance algorithm for land (SEBAL). 1. Formulation. *J. Hydrol.* **1998**, *212*, 198–212. [CrossRef]
58. Choudhury, B.J.; Monteith, J. A four-layer model for the heat budget of homogeneous land surfaces. *Q. J. R. Meteorol. Soc.* **1988**, *114*, 373–398. [CrossRef]
59. Hautecœur, O.; Leroy, M.M. Surface bidirectional reflectance distribution function observed at global scale by POLDER/ADEOS. *Geophys. Res. Lett.* **1998**, *25*, 4197–4200. [CrossRef]
60. Wanner, W.; Li, X.; Strahler, A. On the derivation of kernels for kernel-driven models of bidirectional reflectance. *J. Geophys. Res. Atmos.* **1995**, *100*, 21077–21089. [CrossRef]
61. Lucht, W.; Roujean, J.L. Considerations in the parametric modeling of BRDF and albedo from multiangular satellite sensor observations. *Remote Sens. Rev.* **2000**, *18*, 343–379. [CrossRef]
62. Marticorena, B.; Chazette, P.; Bergametti, G.; Dulac, F.; Legrand, M. Mapping the aerodynamic roughness length of desert surfaces from the POLDER/ADEOS bi-directional reflectance product. *Int. J. Remote Sens.* **2004**, *25*, 603–626. [CrossRef]
63. Chappell, A.; Webb, N.P.; Guerschman, J.P.; Thomas, D.T.; Mata, G.; Handcock, R.N.; Leys, J.F.; Butler, H.J. Improving ground cover monitoring for wind erosion assessment using MODIS BRDF parameters. *Remote Sens. Environ.* **2018**, *204*, 756–768. [CrossRef]
64. Waggoner, D.G.; Sokolik, I.N. Seasonal dynamics and regional features of MODIS-derived land surface characteristics in dust source regions of East Asia. *Remote Sens. Environ.* **2010**, *114*, 2126–2136. [CrossRef]
65. Refice, A.; Bovenga, F.; Nutricato, R. MST-based stepwise connection strategies for multipass radar data, with application to coregistration and equalization. *IEEE Trans. Geosci. Remote Sens.* **2006**, *44*, 2029–2040. [CrossRef]
66. Strahler, A.H.; Muller, J.; Lucht, W.; Schaaf, C.; Tsang, T.; Gao, F.; Li, X.; Lewis, P.; Barnsley, M.J. MODIS BRDF/albedo product: Algorithm theoretical basis document version 5.0. *MODIS Doc.* **1999**, *23*, 42–47.
67. Lee, S.J.; Kim, J.R.; Choi, Y.S. The extraction of forest CO_2 storage capacity using high-resolution airborne LiDAR data. *GIScience Remote Sens.* **2013**, *50*, 154–171. [CrossRef]
68. Li, M.; Im, J.; Quackenbush, L.J.; Liu, T. Forest biomass and carbon stock quantification using airborne LiDAR data: A case study over Huntington Wildlife Forest in the Adirondack Park. *IEEE J. Sel. Top. Appl. Earth Obs. Remote Sens.* **2014**, *7*, 3143–3156. [CrossRef]
69. Narine, L.L.; Popescu, S.; Neuenschwander, A.; Zhou, T.; Srinivasan, S.; Harbeck, K. Estimating aboveground biomass and forest canopy cover with simulated ICESat-2 data. *Remote Sens. Environ.* **2019**, *224*, 1–11. [CrossRef]
70. Duncanson, L.; Neuenschwander, A.; Hancock, S.; Thomas, N.; Fatoyinbo, T.; Simard, M.; Silva, C.A.; Armston, J.; Luthcke, S.B.; Hofton, M. Biomass estimation from simulated GEDI, ICESat-2 and NISAR across environmental gradients in Sonoma County, California. *Remote Sens. Environ.* **2020**, *242*, 111779. [CrossRef]
71. Sadeghi, V.; Ahmadi, F.F.; Ebadi, H. A new automatic regression-based approach for relative radiometric normalization of multitemporal satellite imagery. *Comput. Appl. Math.* **2017**, *36*, 825–842. [CrossRef]
72. Hall, F.G.; Strebel, D.E.; Nickeson, J.E.; Goetz, S.J. Radiometric rectification: Toward a common radiometric response among multidate, multisensor images. *Remote Sens. Environ.* **1991**, *35*, 11–27. [CrossRef]
73. Rosen, P.; Hensley, S.; Shaffer, S.; Edelstein, W.; Kim, Y.; Kumar, R.; Misra, T.; Bhan, R.; Sagi, R. The NASA-ISRO SAR (NISAR) mission dual-band radar instrument preliminary design. In Proceedings of the 2017 IEEE international geoscience and remote sensing symposium (IGARSS), Fort Worth, TX, USA, 23–28 July 2017; pp. 3832–3835.
74. Kim, J.; Jeong, U.; Ahn, M.-H.; Kim, J.H.; Park, R.J.; Lee, H.; Song, C.H.; Choi, Y.-S.; Lee, K.-H.; Yoo, J.-M. New era of air quality monitoring from space: Geostationary Environment Monitoring Spectrometer (GEMS). *Bull. Am. Meteorol. Soc.* **2020**, *101*, E1–E22. [CrossRef]

remote sensing

MDPI

Article

Assessment of the Effectiveness of Sand-Control and Desertification in the Mu Us Desert, China

Jie Li, Weicheng Wu *, Xiao Fu, Jingheng Jiang, Yixuan Liu, Ming Zhang, Xiaoting Zhou, Xinxin Ke, Yecheng He, Wenjing Li, Cuimin Zhou, Yuan Li, Yifei Song, Hongli Yang and Qihong Tu

Key Laboratory of Digital Lands, Resources and Faculty of Earth Sciences, East China University of Technology, Nanchang 330013, China; 201910705016@ecut.edu.cn (J.L.); 201910705015@ecut.edu.cn (X.F.); 201910818023@ecut.edu.cn (J.J.); 201910853001@ecut.edu.cn (Y.L.); 201910818003@ecut.edu.cn (M.Z.); 201900818004@ecut.edu.cn (X.Z.); 2020120009@ecut.edu.cn (X.K.); 2020120001@ecut.edu.cn (Y.H.); 2020110085@ecut.edu.cn (W.L.); 2020110012@ecut.edu.cn (C.Z.); 2020110084@ecut.edu.cn (Y.L.); 2020120019@ecut.edu.cn (Y.S.); 2019210151@ecut.edu.cn (H.Y.); qhtu@ecut.edu.cn (Q.T.)

* Correspondence: wuwch@ecut.edu.cn or wuwc030903@sina.com; Tel.: +86-18970847278

Abstract: The first successful sand-control was achieved in the Mu Us Desert by local people in the 1950–1960s, and their experience and approach have been extended to the whole Ordos and Northern China since then. The objective of this paper is to assess comprehensively the effectiveness of sand-control in 15 counties in and around Mu Us using multitemporal satellite images and socioeconomic data. After atmospheric correction, Landsat TM and OLI images were harnessed for land cover classification based on the ground-truth data and for derivation of the GDVI (generalized difference vegetation index) to extract the biophysical changes of the managed desert and desertification. Climatic, socioeconomic, environmental and spatial factors were selected for coupling analysis by multiple linear and logistic regression models to reveal the driving forces of desertification and their spatial determinants. The results show that from 1991 to 2020, 8712 km^2 or 63% of the desert has been converted into pastures and shrublands with a greenness increase of 0.3509 in GDVI; the effectiveness of sand-control is favored by the rational agropastoral activities and policies; though desertification occurs locally, it is associated with both climatic and socioeconomic factors, such as wind speed, precipitation, water availability, distance to roads and animal husbandry.

Keywords: post-classification differencing; generalized difference vegetation index (GDVI); multiple linear regression; logistic regression

Citation: Li, J.; Wu, W.; Fu, X.; Jiang, J.; Liu, Y.; Zhang, M.; Zhou, X.; Ke, X.; He, Y.; Li, W.; et al. Assessment of the Effectiveness of Sand-Control and Desertification in the Mu Us Desert, China. *Remote Sens.* **2022**, *14*, 837. https://doi.org/10.3390/rs14040837

Academic Editor: Ioannis Gitas

Received: 28 December 2021
Accepted: 8 February 2022
Published: 10 February 2022

1. Introduction

Desertification is a phenomenon of land degradation resulting from by human activities and climate variations in arid, semi-arid and partially sub-humid areas [1,2]. With more than about four million km^2 of dryland, China has been suffering from desertification since centuries ago. Though sand-control through afforestation and plantation of shrubs and herbaceous vegetation has been widely effectuated in the dryland areas since the 1950s, there is still large area of desert in China, reaching 739,200 km^2 [3]. This poses a serious threat to the sustainable socioeconomic development in the arid region.

Remote sensing (RS) technology has become an important data source and technical means for desertification monitoring due to its wide detection range and observation periodicity since the 1970s [2,4–10]. It has been successfully applied to distinguish desert encroachment or climate events of the south border movement of the Saharan Desert [5], desertification in Sahel [4], in western Rajasthan in India in 1990s [11], and in Mu Us [2,8]. Remote sensing approaches include, but are not limited to, differencing-and-thresholding technique and supervised classification [7,8,12–14], in which the quantitative evaluation of desertification based on remote sensing indicators is made possible [15,16]. Commonly-used indicators are the normalized difference vegetation index (NDVI), fractional vegetation coverage (FVC), vegetation net primary productivity (NPP), rainfall-use efficiency (RUE), etc. [17–20]. Both FVC and NDVI are able to identify land degradation and distinguish the degree of governance of sand land [8,21]. However, it is difficult to discern subtle changes in vegetation when applied to desert areas with sparse vegetation. Recently, Wu (2014) proposed the generalized difference vegetation index (GDVI) for arid areas [22], and Xie et al. (2020) developed a set of monitoring and restoration assessment indicators (MRAIs) for assessing the mining impacts and restoration effectiveness [23]. Since these two indicators were developed for sparse vegetation areas, both of them are promising for land degradation and desertification study. In addition, supervised classification is also widely used in land cover mapping and dynamic monitoring of desertification [5,8,12,24]. With improvement of classification algorithm and technique, land cover mapping including desertification can be achieved with high accuracy and reliability [24]. Therefore, combination of the supervised classification with differencing-and-thresholding technique will allow the advantages of each to complement the other, and hence achieve a comprehensive quantitative evaluation of the effectiveness of sand-control [8].

Actually, for desertification monitoring and assessment, we have to mention the European megaproject, the Mediterranean desertification and land use (MEDALUS) projects I, II and III from 1989 to 1999 [25], which proposed a comprehensive assessment approach using four main quality indices, e.g., climate quality index (CQI), soil quality index (SQI), vegetation quality index (VQI), and the management (human influence) quality index (MQI). These indices were employed to constitute an environmental sensitive area index (ESAI) to identify the vulnerable area of desertification. The DesertWatch projects funded by the European Space Agency (ESA) (2004–2008) followed this and applied Earth Observation (EO)-based information to derive environmental elements to generate a land degradation index and develop the desertification scenarios [26]. Lee et al. (2019), Bouhata and Bensekhria (2021), and Abuzaid and Abdelatif (2022) respectively applied the MEDALUS approach for desertification vulnerability assessment in Mongolia, Algeria and Egypt [27–29].

Analysis on the driving forces of desertification is an indispensable part of desert monitoring and assessment [9]. Most of the driving factors are related to climate and human activities [30], e.g., precipitation, temperature, wind speed, and sunshine hours, and socioeconomic factors such as gross domestic product (GDP), population, and livestock numbers [18,31,32]. At present, the research on the driving force mainly focuses on two aspects: (1) Using correlation analysis, principal component analysis, stepwise linear regression, and other methods to explore the influence of human activity and climate on land cover change and desertification [2,6], and these methods have been proved effective. However, desertification is a complicated process, and simple linear models cannot fully reveal its driving mechanism [33,34]. (2) In recent years, several studies on the driving forces of the spatial variability of desertification have been carried out employing the geographically weighted regression, geographic detectors, logistic regression models, etc., and all of these approaches may lead to promising results [35–38]. With the development in socioeconomy, construction of roads, expansion of settlements, protection of farmland and other human activities have also become important factors for desertification reversal [36,38]. However, currently, the existing research mainly focuses on a fixed time point or single phase, and few studies have incorporated human activity into desertification modeling, and this may fail to rationally uncover the real mechanism and causes of desertification in different periods. Even the MEDALUS and DesertWatch projects focused on prediction of the vulnerable areas of desertification, but not on its driving forces.

Based on the above understanding, the objective of the study is to conduct a comprehensive monitoring and assessment of the desertification progress and the effectiveness of sand-control in space and time, taking the Mu Us Desert in China as an example, so as to provide an extension of or a complement to the MEDALUS and DesertWatch projects and put forward feasible suggestions for decision-makers to manage the remaining active deserts. One specific objective of this study is to build spatially explicit models of desertification and sand-control activity. Multitemporal Landsat images and socioeconomic data will be employed for achieving these purposes.

2. Materials and Methods

2.1. The Study Area

The study area is located at the junction of three provinces, namely Inner Mongolia, Shaanxi and Ningxia, in northern China. The main part is the Mu Us Desert (or Sandy Land) with a geographical extent from 105°20′17″ E to 109°32′41″ E in longitude and 36°49′12″ N to 39°49′31″ N in latitude and covering an area of 73,737 km^2 (Figure 1). The terrain is relatively flat, with an average elevation of 1380 m. Apart from a few mountains (Mts) in the northwest (e.g., Helan Mts and Zhuozi Mts), most of the study area is sandy land, desert in Mu Us, and farmland in the Yinchuan Plain. It borders the Loess Plateau in the south and southeast.

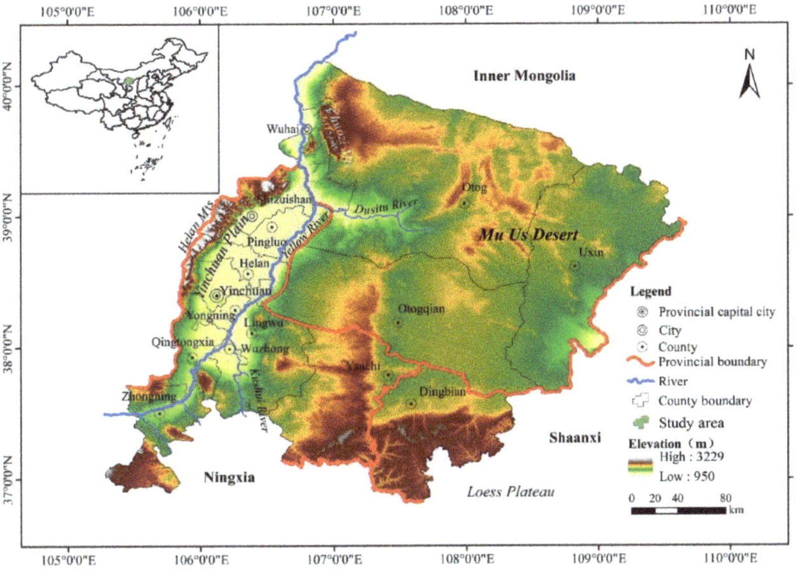

Figure 1. Location of the study area.

The average annual rainfall is about 317 mm, mainly concentrated in July–September [2,8], while annual evaporation reaches 2500 mm, much stronger than precipitation. Winds from the northwest blow on average 230 days per year, and those exceeding the speed of 17 m/s (Beaufort Scale-8) may occur during more than 40 days [8]. The mean annual temperature is about 6.0–8.5 °C, but there is a significant difference between summer and winter and between day and night.

The vegetation type is mainly shrubs and herbs in the sandy land, and rice and maize are the main cultivation in the Yinchuan Plain with which the greenness reaches its maximum in July and August. The special terrain and climate conditions, as well as the overgrazing in the early periods, have made the study area largely desertified or susceptible to desertification.

There are rich mineral resources including coal, oil and gas in the study area, and that is why the latter was designated as one of the energy bases of China by the central government in 1999 [8].

Since 1959, local people started to conduct sand-control activities and many large-scale combating desertification campaigns have been successfully undertaken by individual/household companies and national teams since then.

2.2. Data and Processing

To achieve our purposes, a combined method of post-classification differencing and GDVI differencing was proposed to quantitatively evaluate the spatial distribution and degree of sand-control. A set of socioeconomic, climatic and environmental factors were selected to quantitatively analyze the driving forces of the spatial heterogeneity of desert governance in different periods using multiple linear regression and logistic regression models.

2.2.1. Data

Satellite Data

Landsat TM, ETM+ and OLI images from 1987 to 2020 with path/row numbers 128/33-34 and 129/33-34 and 16 tiles of digital elevation model (DEM) data (ASTERV003, 30 m resolution) of the study area were obtained from the USGS data server (glovis.usgs.gov, accessed on 16 June 2020) and NASA (earthdata.nasa.gov, accessed on 16 June 2020). Landsat images were mostly acquired in summer from July to September with cloud cover less than 1% (Table 1) in terms of the phenological features of crops and vegetation in the continental climate zone. Due to frequent cloudy weather in summer, it is difficult to acquire cloud-free images for the adjacent scenes of different paths in the same year. Hence, selection of images depended on the availability of cloud-free images and there was difference in acquisition dates of the adjacent scenes of images. The obtained Landsat images were utilized for the following supervised classification and derivation of different vegetation index, GDVI, and land surface temperature (LST).

Table 1. Landsat images used in the study.

Captors	Scene Path/Row No	Acquisition Dates	Spatial Resolution	Source
Landsat 5 TM Landsat 5 TM Landsat 5 TM Landsat 8 OLI	128/33	23 August 1991 28 July 1999 27 August 2010 5 July 2020	30 m	USGS https://glovis.usgs.gov (last accessed on 20 May 2021)
Landsat 5 TM Landsat 5 TM Landsat 5 TM Landsat 8 OLI	128/34	23 August 1991 26 June 1999 10 July 2010 7 September 2020		
Landsat 5 TM Landsat 5 TM Landsat 5 TM Landsat 8 OLI	129/33	30 August 1991 20 August 1999 28 June 2009 26 July 2019		
Landsat 5 TM Landsat 5 TM Landsat 5 TM Landsat 8 OLI	129/34	30 August 1991 22 August 2000 17 July 2010 26 July 2019		

High resolution images available on Google Earth were harnessed to derive road vector data for 1991,1999, 2010 and 2020.

Socioeconomic Data

It is difficult to monitor and quantify human activity, and the best way is to use annual socioeconomic data, the finally quantified expression of human activity during a one-year period, for research [2]. For the study area, county-level socioeconomic data are available from the Statistic Yearbooks published by the government. The data include the total sown area, meat product, total number of livestock at the end of year, sheep stock, per capita net income of farmers and herdsmen, gross output of agriculture, forestry, animal husbandry and fishery, gross domestic product (GDP) and total population. For this reason, the Statistic Yearbooks of Inner Mongolia, Ningxia and Shaanxi from 1990 to 2020 were collected. Due to the change of administrative divisions, we combined the data of Xixia,

Xingqing and Jinfeng districts into Yinchuan City; Wuda, Hainan and Haibowan districts were merged into Wuhai City; and Dawukou and Huinong districts were incorporated into Shizuishan City. Finally, socioeconomic data of 15 counties and cities were prepared for analysis.

Meteorological Data

The monthly average temperature, average precipitation, average wind speed, maximum wind speed and sunshine duration of four stations, e.g., Yulin, Etog, Otogqian and Uxin, were obtained from the website of National Meteorological Science Data Center of China (http://data.cma.cn, accessed on 18 June 2021) [39]. Those of Yinchuan, Shizuishan, Pingluo, Huinong, Yongning, Helan, Wuzhon, Lingwu, Yanchi, Qingtongxia and Zhongning were extracted from the Ningxia Statistic Yearbooks.

Field Data

Field survey was conducted in August 2000, June 2002, May 2005, October 2020 and July 2021 to (1) understand land use/cover and its change; (2) coal mining in five coal fields, namely Qipanjing, Zhuozi Mts, Shanghaimiao, Shitanjing and Ruqigou coal fields; (3) oil and gas exploitation in the banners Otog, Otogqian and Uxin; and (4) sand-control achievement by local and national companies in the whole study area. In total, more than 2000 observation points were collected with GPS location. One surprising thing observed is that sandy land has greatly decreased from 2000 to 2021 and converted mostly into shrublands and forests in dunes and grasslands in interdunes thanks to the sand-control activity.

2.2.2. Processing Procedures

Satellite Data Processing

Atmospheric correction of Landsat images was conducted using the COST model developed by Chavez (1996) [40], where the band minimum was employed to remove the haze effect and at-satellite radiance of all bands was converted into surface reflectance.

Land surface temperature (LST) was derived from the thermal band of Landsat TM and OLI data [41,42].

The dryland-tailored vegetation index, GDVI, is shown in Equation (1). When $n = 1$, GDVI = NDVI; when $n = 2$, it is suitable for characterizing dryland biomes including shrubland, woodland and forest; when $n = 3$, it can be used for monitoring the degradation and desertification in sparsely vegetated area [22,23]. Compared with NDVI and other vegetation indices, GDVI has higher sensitivity and dynamic range, and is able to identify more effectively subtle differences in vegetation greenness in low vegetated areas [22]. Therefore, GDVI of $n = 3$ was produced for the successive analysis.

$$\mathrm{GDVI} = (\rho_{\mathrm{NIR}}^{n} - \rho_{\mathrm{R}}^{n}) / (\rho_{\mathrm{NIR}}^{n} + \rho_{\mathrm{R}}^{n}) \quad (1)$$

where, ρ_{NIR} is the spectral reflectance of the near-infrared band, and ρ_{R} is that of the red band.

Slope and aspect were produced from the DEM data and used for land cover classification [24] and desertification modeling as a part of the environmental factors.

Land Cover Classification

A dataset of eleven bands, including elevation, slope, aspect, LST, GDVI and six spectral bands (blue, green, red, near infrared, shortwave infrared 1 (SWIR 1) and SWIR 2) through layer-stacking function was composed for each observation year [24].

Based on field survey and knowledge, we defined 30–32 initial classes of ground-truth samples (regions of interest, ROIs) for different scenes and observation years, and half of the samples (about 5–6% of the total scene) were used for training, i.e., training samples (TS), and the remaining half for validation, or rather, validation samples (VS). Sand-control and desertified land (or deserts) were the important classes to be classified out. After classification, all initial classes were merged into 12 classes in terms of the similarity, namely sandy land, shrublands, grasslands, waters, artificial (built up) areas, bare soil, loess land, woodlands, coal mines, saline land, croplands and fallows. Forests were sporadic and less than several pixels, and merged into woodlands. Finally, the sandy land was extracted as a mask, and all the remaining classes were merged into non-sand land for further analysis.

Supervised classification using maximum likelihood approach was conducted, and the results were validated with VS.

Processing of the Driving Factors

The distance from the random points in the managed areas and desertified land to roads, water bodies, residential areas and farmland was obtained by proximity analysis. The change of water bodies and mining area were obtained by the statistics of classified images. Population density was calculated by Equation (2).

$$Y = X/A \tag{2}$$

where, Y is the population density, X is total population, and A is the area.

2.3. *Effectiveness Assessment of Sand-Control*

2.3.1. Post-Classification Differencing

Post-classification differencing is a method for identification of land cover change proposed by Wu [12]. The same type of land cover of the different observation time, deserts and controlled sandy land in this case, was extracted from the classified images and a differencing was applied to obtain the change in space. Equation (3) was used to calculate the difference. Taking desertification as an example, when $\Delta S = 0$, it means no change in sandy land during this period; when $\Delta S = 1$, the desert or sandy land has expanded; when $\Delta S = -1$, the sandy land is reduced and controlled [8,12]. This method can be used to monitor the overall change of sandy land and identify accurately the area and spatial distribution of sandy land, and at the same time, quantify which land cover types have been converted into sandy land or vice versa. This will allow to produce the gain/loss change matrix of land cover of the study area in different observation periods.

$$\Delta S = S_{T2} - S_{T1} \tag{3}$$

2.3.2. ΔGDVI

Based on the calculation of GDVI of different time images, a differencing was effectuated under the masks of deserts or sand-control extracted from the classified images:

$$\Delta GDVI = GDVI_{T2} - GDVI_{T1} \tag{4}$$

When $\Delta GDVI < 0$, desertification has occurred as vegetation vigor has decreased; if $\Delta GDVI > 0$, indicating a vegetation increase, or rather, sandy land has been controlled and converted into vegetated area such as grasslands and shrublands.

2.3.3. Multiple Linear Regression Analysis

Multiple linear stepwise regression is an approach to analyze the relationship between a dependent variable and several independent variables, and to find which independent variable(s) contribute(s) most to the dependent or which factor(s) has(have) played the most important role in the event occurrence (Equation (5)) [2,6,43].

$$Y_i = \beta_0 + \beta_1 X_{1i} + \cdots + \beta_k X_{ki} + U_i \tag{5}$$

where, Y_i is the dependent variable, $X_{1i} \ldots X_{ki}$ are the independent variables, and $\beta_0, \beta_1 \ldots, \beta_k$ are the regression coefficients of their corresponding variables, and U_i is the random error. The dependent variable in this case is the desertified land or sand-control area at county-level, and the independent variables are county-level socioeconomic and climatic factors.

2.3.4. Logistic Regression Analysis

Logistic regression model (LRM) is capable of dealing with both categorical and continuous variables and can effectively achieve spatially explicit analysis to reveal the probability of certain change in space (i.e., with probability of 1 indicating event occurred, and 0 indicating it has not occurred) associated with multiple independent variables, i.e., spatial determinants. It was first used in the field of disease diagnosis and has been extended to land change research to reveal the spatial determinants of land cover change in environmental geography and geological disaster prediction in recent decades [2,44–46]. However, it has been rarely used in the quantitative analysis of desertification driving force because of the particularity of its dependent variable. The formula (6) was used to establish the LRM.

$$P = \frac{1}{1 + e^{-(\beta_0 + \beta_1 X_1 + \cdots + \beta_i X_i)}} \quad (6)$$

where, P is the probability, $X_1 \ldots, X_i$ are the independent variables, and β_0, $\beta_1 \ldots, \beta_i$ are the regression coefficients of their corresponding variables.

Before logistic regression analysis, it is necessary to undertake the collinearity diagnosis to check whether there is collinearity among the independent variables using either tolerance (TOL) or variable inflation factor (VIF) as an indicator. If VIF < 10 or TOL > 0.1, there is no evident collinearity and the result of the logistic regression analysis shall be reliable.

In order to examine the importance of factors influencing sand-control and desertification, the following preprocessing is required. First, rasterization of the independent variables, e.g., GDP, per capita rural net income, meet product and so on are converted into county-level raster; then, creation of the random points in the controlled and uncontrolled areas of sandy lands, the desertified areas and the unchanged areas in line with the ratio of 1:1. The points in the desertified areas and sand-control areas were assigned a value of 1, and in non-desertification and non-control areas were assigned a value of 0, which were used as dependent variables. At last, the values of independent and dependent variables were both extracted and input into SPSS, a package for statistic analysis, for logistic regression modeling.

The model was tested in two aspects: (1) the Hosmer–Lemeshow (H-L) test was used to evaluate the degree of fit of the model; if H-L > 0.05, the model was considered to have a good fit; and (2) using significance (Sig), e.g., Sig < 0.05, to test whether the independent variable in the model has a significant effect on the dependent variable. Finally, the absolute values of the regression coefficients or odds ratios (OR) were used for judging the influence degree of the independent variables on the dependent variable.

3. Results

3.1. Dynamic Situation of the Desert

As shown in Table 2, the overall accuracy (OA) of the land cover classification vs. the validation samples is about 90.3–92.8% with a kappa coefficient of 0.87–0.90. The desert or sandy land is mainly distributed in the banners Otog, Otogqian and Uxin, among which the last has the largest area (Table 3 and Figure 2). In the past 30 years, sandy land has decreased by 8712.23 km^2, accounting for 63.05% of the total sandy land though there were slight fluctuation in some counties in different periods, e.g., from 1991 to 1999, sandy land increased by 291.6 km^2 in Yanchi, and from 1999 to 2010, sandy land in Yinchuan, Otogqian and Wuzhong had an increase of 32.22 km^2, 65.13 km^2 and 3.59 km^2, respectively, while a decrease appeared in other areas in the whole period.

Table 2. Accuracy of land cover classification.

Year	Overall Accuracy	Kappa Coefficient
1991	92.35%	0.9009
1999	92.87%	0.9086
2010	90.33%	0.8767
2020	92.598	0.9061

Table 3. Area of sandy land.

City/County or Banner		Sandy Land			
		1991	1999	2010	2020
WuhaiCity	Area (km^2)	232.60	114.73	43.36	5.93
	Proportion (%)	1.68	1.10	0.60	0.12
Otogqian	Area (km^2)	3164.50	1925.91	1991.04	1611.14
	Proportion (%)	22.90	18.52	27.75	31.55
Otog	Area (km^2)	3262.41	2362.18	2118.26	1271.96
	Proportion (%)	23.61	22.71	29.53	24.91
Uxin	Area (km^2)	4693.44	3861.74	2159.62	1848.40
	Proportion (%)	33.97	37.13	30.10	36.20
Dingbian	Area (km^2)	224.60	99.40	56.16	55.20
	Proportion (%)	1.63	0.96	0.78	1.08
Yinchuan	Area (km^2)	172.50	33.16	65.38	4.45
	Proportion (%)	1.25	0.32	0.91	0.09
Yongning	Area (km^2)	10.31	1.10	0.56	0.88
	Proportion (%)	0.07	0.01	0.01	0.02
Helan	Area (km^2)	10.15	4.79	2.29	0.07
	Proportion (%)	0.07	0.05	0.03	0.00
Lingwu	Area (km^2)	703.12	614.50	323.03	143.80
	Proportion (%)	5.09	5.91	4.50	2.82
Shizuishan	Area (km^2)	2.95	0.87	0.41	0.04
	Proportion (%)	0.02	0.01	0.01	0.00
Pingluo	Area (km^2)	215.35	148.06	134.74	44.64
	Proportion (%)	1.56	1.42	1.88	0.87
Wuzhong	Area (km^2)	20.09	5.85	9.44	0.77
	Proportion (%)	0.15	0.06	0.13	0.01
Yanchi	Area (km^2)	740.98	1032.56	222.99	103.33
	Proportion (%)	5.36	9.93	3.11	2.02
Qingtongxia	Area (km^2)	235.02	116.77	7.96	3.91
	Proportion (%)	1.70	1.12	0.11	0.08
Zhongning	Area(km^2)	130.41	79.41	38.63	11.71
	Proportion (%)	0.94	0.76	0.54	0.23
Total	Area (km^2)	13,818.45	10,401.03	7173.87	5106.22

3.2. *Effectiveness of the Combating Desertification*

Between 1991–2020, a total of 9140.44 km^2 of sandy land has been managed and converted into pastures, shrublands, and even forests (Table 4). The main sand-control areas are located in the banners (counties) Uxin, Otog and Otogqian, with an area of 3022.76 km^2, 2090.48 km^2 and 1637.36 km^2 respectively. As shown in Figures 3 and 4, the area of sand-control was much larger than the desertified land.

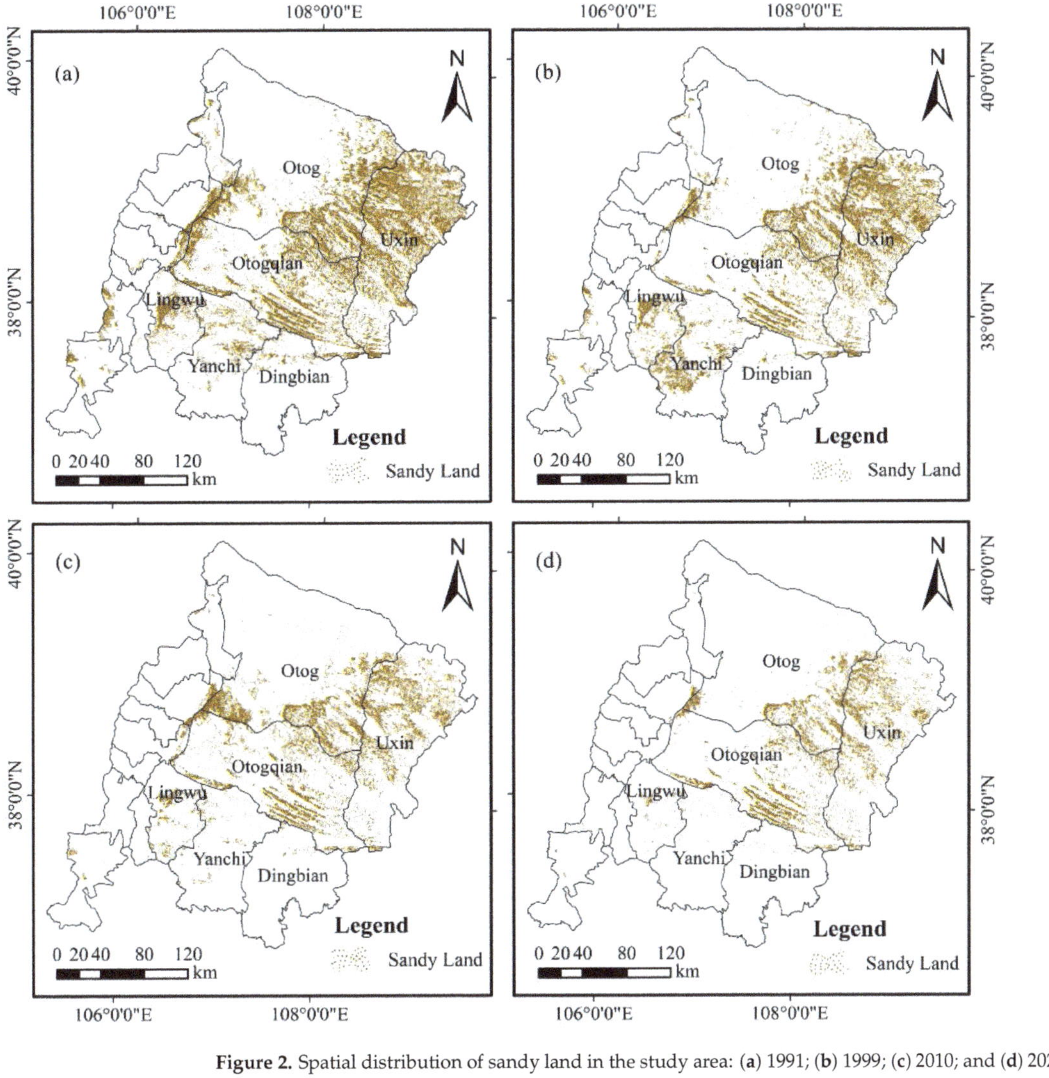

Figure 2. Spatial distribution of sandy land in the study area: (**a**) 1991; (**b**) 1999; (**c**) 2010; and (**d**) 2020.

Table 4. Areas of sand-control and desertification.

Period	Sand-Control				Desertification			
	1991–1999	1999–2010	2010–2020	1991–2020	1991–1999	1999–2010	2010–2020	1991–2020
Area (km^2)	5537.61	4969.28	3420.24	9140.44	2120.36	1742.40	1352.53	428.13
Proportion (%)	40.07	47.78	47.68	66.15	3.36	2.62	1.94	0.68
ΔGDVI	0.1575	0.1806	0.1456	0.3518	−0.0921	−0.1237	−0.0997	−0.0691

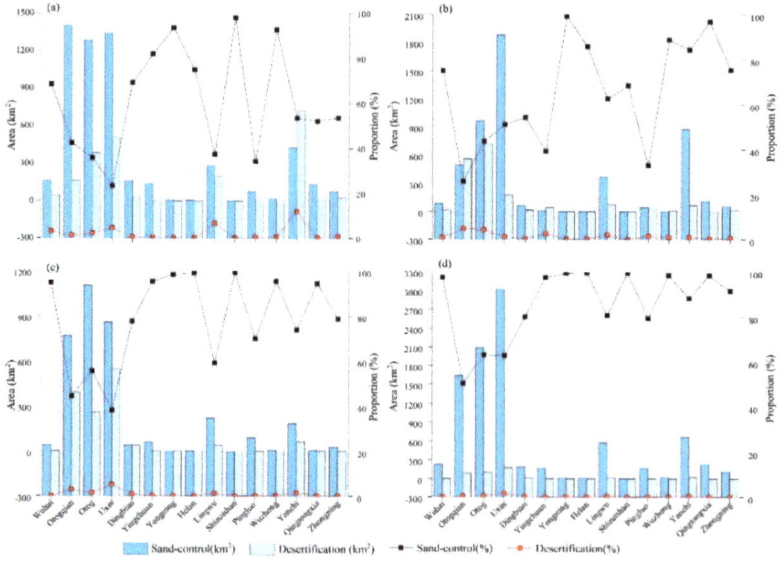

Figure 3. Area and proportion of the sand-control and desertification at county- and/or city-level: (**a**) 1991–1999; (**b**) 1999–2010; (**c**) 2010–2020; (**d**) 1991–2020.

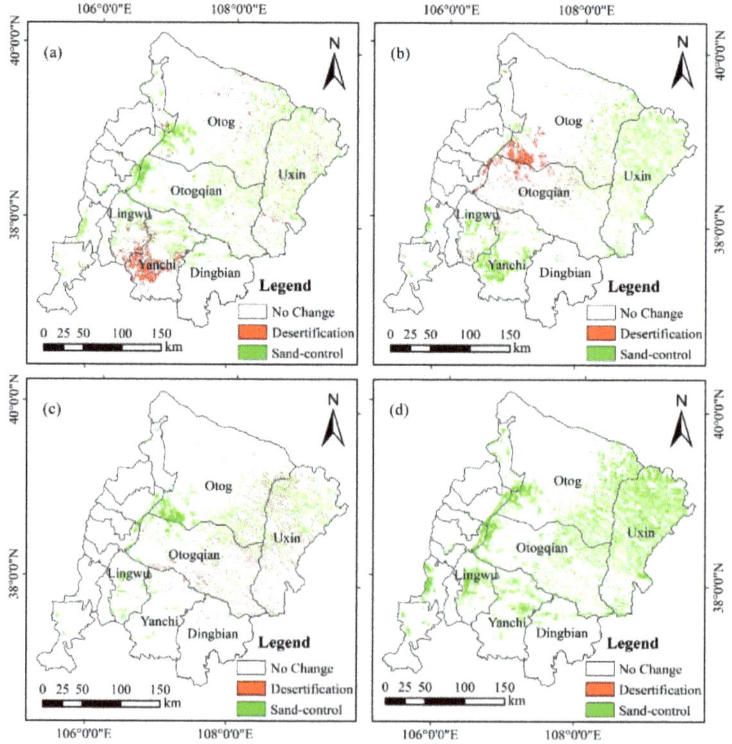

Figure 4. Spatial distribution of sand-control and desertification: (**a**) 1991–1999; (**b**) 1999–2010; (**c**) 2010–2020; (**d**) 1991–2020.

Most of the sand-control was effectuated in the western margin of the desert, i.e., the west part of the banners Otog and Otogqian, followed by the southern area in Dingbian, Lingwu and Yanchi (Figures 3 and 4). The desertification occurred mainly in Yanchi and Lingwu in 1991–1999 (Figure 4a), but it was concentrated in the west of Otogqian and Otog in 1999–2010 (Figure 4b), and it appeared as small dots in the east of the study area in 2010–2020, and a large piece of desertification was not observed in the last period (Figure 4c).

As seen in Table 4, the vegetation greenness represented by ΔGDVI in the controlled areas in the past 30 years has gained an increase of 0.3518 with a minimum of 0.1456 in 2010–2020, while the greenness had decreased by 0.0691 in the desertified area, and the biggest decrease (−0.1237) appeared in 1999–2010.

Presented in Tables 5–8, sandy land was mostly transferred into shrublands and grasslands with an area of 6359.79 km^2, accounting for 69.58%, from 1991 to 2020. This conversion took up 4943.51 km^2, 4088.33 km^2 and 2606.92 km^2, accounting for 89.27%, 82.27% and 76.22%, respectively, in the three observed periods 1991–2000, 2000–2010 and 2010–2020. It is obvious that sand-control by planting shrubs such as *Salix cheilophila, Tamarix ramosissima, Hedysarum mongolicum, Caragana korshinskii Kom, Hippophae rhamnoides L.* and *Lycium chinense Miller* and herbaceous vegetation including *Medicago sativa L., Artemisia desertorum, Artemisia sieversiana, Astragalus adsurgens Pall.* has achieved a remarkable output.

Figures 3–5 show that the extent of sand-control was much larger than that of desertification, whereas it was the opposite in Yanchi during 1991–1999. From 1991 to 2020, sand-control seems to have had the best effectiveness in Uxin with an increase of 0.0897 in GDVI, and the lowest effectiveness in Shizuishan, with an increase of 0.0002. Meanwhile, ΔGDVI of the desertified area was close to 0 (Figure 5).

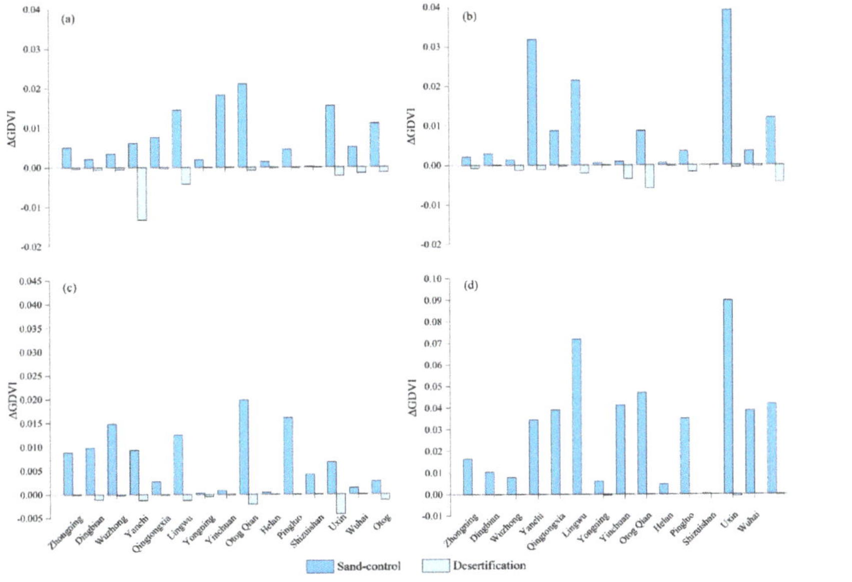

Figure 5. The mean values of ΔGDVI of the sand-control and desertification area in each county: (**a**) 1991–1999; (**b**) 1999–2010; (**c**) 2010–2020; (**d**) 1991–2020.

3.3. Determinants of Sand-Control and Desertification

As shown in Table 9, 15 factors from three aspects were used for stepwise linear regression modeling with the area of sand-control and desertification as the dependent variable. Taking $R^2 > 0.60$ as the model testing standard, seven models were finally obtained (Table 10).

Table 5. Land use transfer matrix of the study area from 1991 to 1999 (km^2).

	1991	Sandy Land	Shrubs	Saline	Cropland	Water	Grassland	Building	Woodland	Loess	Bare Soil	Coal Mine	Gain
1999	Sandy Land	8275.67	499.35	59.94	136.86	19.26	1274.39	1.26	20.37	30.85	78.03	0.05	2120.36
	Shrubs	2253.4	8250.17	258.6	564.14	65.52	4906.12	19.94	40.48	91.86	514.37	4.3	8718.73
	Saline	229.32	221.6	744.4	97.18	25.94	261.19	3.24	27.36	18.26	318.12	0.33	1202.54
	Cropland	213.24	326.93	200.48	5937.56	563.39	1582.78	70.77	493.51	412.97	350.06	0.99	4215.12
	Water	38.06	115.75	102.51	323.55	896.71	231.81	75	28.55	28.9	143.73	13.47	1101.33
	Grassland	2476.09	5752.67	214.73	1000.56	72.61	9852.77	26.3	83.28	135.12	472.95	0.97	10,235.28
	Building	23.39	93.29	16.64	238.93	79.12	40.55	92.21	3.05	8.35	113.68	15.33	632.33
	Woodland	0.78	3.18	16.52	228.48	1.08	49.49	0.32	394.15	150.39	36.43	0.46	487.13
	Loess	23.2	7.72	90.41	394.35	4.92	39.88	0.23	996.12	1461	214.76	0	1771.59
	Bare soil	278.71	745.14	174.94	252.38	41.28	381.84	9.12	88.26	133.16	4370.87	38.52	2143.35
	Coal mine	1.42	27.5	1.5	0.95	3.73	1.77	4.34	0.99	0	40.76	38.53	82.96
	Loss	5537.61	7793.13	1136.27	3237.38	876.85	8769.82	210.52	1781.97	2320.47	2282.89	74.42	
	Net Change	−3417.25	925.6	66.27	977.74	224.48	1465.46	421.81	−1294.84	−548.88	−139.54	8.54	

Table 6. Land use transfer matrix of the study area from 1999 to 2010 (km^2).

	1999	Sandy Land	Shrubs	Saline	Cropland	Water	Grassland	Building	Woodland	Loess	Bare Soil	Coal Mine	Gain
2010	Sandy Land	5430.03	590.29	40.22	23.73	5.23	1046.43	1.58	0.46	1.13	33.33	0	1742.4
	Shrubs	2202.47	10,266.32	460.17	826.95	185.41	9653.9	63.1	14.69	79.68	735.44	4.62	14,226.43
	Saline	162.9	360.85	476.33	53.4	42.26	510.44	5.63	1.42	29.46	318.46	0.6	1485.42
	Cropland	102.32	464.16	120.67	5878.66	459.82	985.43	136.96	311.18	408.84	118.27	1.13	3108.78
	Water	26.49	87.23	89.75	162.16	660.52	85.4	27.66	0.88	4.45	28.28	1.97	514.27
	Grassland	1885.86	3955.27	167.59	1079.92	80.18	6746.54	15.51	29.08	28.05	80.34	0.05	7321.85
	Building	101.66	361.7	162.77	756.31	418.78	384.72	412.51	6.25	20.16	381.41	15.28	2609.04
	Woodland	1.82	10.73	2.34	191.17	6.82	23.98	0.74	359.32	786.72	14.16	0.15	1038.63
	Loess	157.1	168.95	81.81	877.57	26.91	328.69	10.12	111.33	1779.78	170.55	0	1933.03
	Bare soil	325.93	685.61	434.22	385.57	117.55	340.43	68.57	117.28	387.01	4599.71	29.93	2892.1
	Coal mine	2.73	64.05	16.01	5.2	14.21	7.05	27.66	0.68	0.02	142.6	68.06	280.21
	Loss	4969.28	6748.84	1575.55	4361.98	1357.17	13,366.47	357.53	593.25	1745.52	2022.84	53.73	
	Net Change	−5006.44	7477.59	−90.13	−1253.2	−842.9	−6044.62	2251.51	445.38	187.51	869.26	226.48	

Table 7. Land use transfer matrix of the study area from 2010 to 2020 (km^2).

	2010	Sandy Land	Shrubs	Saline	Cropland	Water	Grassland	Building	Woodland	Loess	Bare Soil	Coal Mine	Gain
2020	Sandy Land	3753.44	643.64	36.24	27.78	6.35	598.53	14.17	0.14	4.21	21.46	0.01	1352.53
	Shrubs	2028.74	15,292.54	487.92	342.77	54	6020.56	137	33.33	272.44	503.24	0.87	9880.87
	Saline	134.77	380.96	350.77	55.83	107.8	159.92	20.94	1.62	0.99	34.46	0.73	898.02
	Cropland	277.54	1200.4	196	6208.44	179.03	1799.9	884.76	267.71	669.18	475.64	4.73	5954.89
	Water	17.68	109.23	43.59	133.26	604.69	71.2	76.54	1.53	3.79	62.93	6.01	525.76
	Grassland	578.18	4145.24	106.36	354.77	62.85	4146.06	81.64	37	95.62	101.29	0.25	5563.2
	Building	141.14	692.7	260.23	921.41	114.57	357.92	1523.51	15.04	66.65	706.38	66.89	3342.93
	Woodland	51.17	218.46	1.76	435.71	7.08	491.66	9.55	172.03	64.64	80.09	0.4	1360.52
	Loess	21.29	419.05	45.02	428.81	10.13	222.12	22.34	844.28	2351.67	663.06	0.02	2676.12
	Bare soil	166.82	1339.93	422.8	75.33	23.62	199.46	202.91	24.69	183.61	4610.54	41.64	2680.81
	Coal mine	2.91	47.58	11.06	3.32	4.69	1.06	48.19	0.56	0	232.71	226.73	352.08
	Loss	3420.24	9197.19	1610.98	2778.99	570.12	9922.33	1498.04	1225.76	1361.13	2881.26	121.55	
	Net Change	−2067.71	683.68	−712.96	3175.9	−44.36	−4359.13	1844.89	134.76	1314.99	−200.45	230.53	

Table 8. Land use transfer matrix of the study area from 1991 to 2020(km^2).

	1991	Sandy Land	Shrubs	Saline	Cropland	Water	Grassland	Building	Woodland	Loess	Bare Soil	Coal Mine	Gain
2020	Sandy Land	4677.05	111.58	13.69	49.05	6.45	240.85	0.41	1.02	1.08	3.99	0.01	428.13
	Shrubs	5109.79	9696.53	340.29	788.02	57.74	8469.22	4.8	49.28	102.58	461.43	0.58	15,383.73
	Saline	216.93	179.9	336.83	38.72	41.87	308.03	4.4	42.52	3.24	75.5	0.44	911.55
	Cropland	892.83	781.66	366.31	5118.01	654.07	2606.05	105.92	555.63	381.29	515.73	4.72	6864.21
	Water	60.23	58.97	83.91	134.14	555.17	144.84	11.74	7.32	3.28	56.96	4.39	565.78
	Grassland	1250	2905.03	140.72	506.66	33.29	4660.11	4.45	42.94	40.69	116.96	0.32	5041.06
	Building	508.39	802.95	211.96	1245.39	326.56	735.01	155.42	34.38	36.18	678.49	28.22	4607.53
	Woodland	322.7	197.8	5.2	383.24	42.46	259.67	4	243.93	29.92	31.75	0.17	1276.91
	Loess	82.59	85.36	153.79	682.65	12.82	484.15	0.64	1052.73	1756.75	420.72	0.02	2975.47
	Bare soil	671.34	1051.78	214.48	223.71	37.67	686.21	5.92	144.33	115.86	4010.18	27.21	3178.51
	Coal mine	25.64	164.21	13.49	5.36	5.48	28.46	5.04	2.09	0.01	282.05	46.88	531.83
	Loss	9140.44	6339.24	1543.84	4056.94	1218.41	13,962.49	147.32	1932.24	714.13	2643.58	66.08	
	Net Change	−8712.31	9044.49	−632.29	2807.27	−652.63	−8921.43	4460.21	−655.33	562.78	331.89	465.75	

Table 9. Socio-environmental factors for multiple linear regression modeling.

Type	Factors	Symbol	Data Preprocessing
Social and Economic Factors	Total Sown Area	X_1	Z-Score
	Meat Product	X_2	
	Sheep Number	X_3	
	Total Number of Livestock at the Year-end	X_4	
	Per capita Net Income of Farmers and Herdsmen	X_5	
	Gross Output of Farming, Forestry, Animal Husbandry and Fishery	X_6	
	Gross Domestic Product (GDP)	X_7	
	Population Density	X_8	
Meteorological Factors	Annual Precipitation	X_9	
	Mean Temperature	X_{10}	
	Maximum Wind Speed	X_{11}	
	Average Wind Speed	X_{12}	
	Sunshine Duration	X_{13}	
Environmental Factors	Area of Water	X_{14}	
	Area of Coal Mines	X_{15}	

Table 10. Multiple linear regression models.

Model	Period	Expression	R^2	
Sand-Control	1991–1999	$Y_1 = -0.370 - 0.679{*}\Delta X_{10} - 0.420{*}\Delta X_{15}$	0.898	(7)
	1999–2010	$Y_2 = -8.009 \times 10^{-17} + 0.855{*}\Delta X_2 - 0.392{*}\Delta X_{13}$	0.764	(8)
	2010–2020	$Y_3 = 1.410 \times 10^{-16} + 0.570{*}\Delta X_5 + 0.520{*}\Delta X_3$	0.835	(9)
	1991–2020	$Y_4 = -1.03 \times 10^{-15} + 0.801{*}\Delta X_5 + 0.465{*}\Delta X_9$	0.764	(10)
Desertification	1991–1999	$Y_5 = 0.206 - 1.058{*}\Delta X_5$	0.600	(11)
	2010–2020	$Y_6 = 1.646 \times 10^{-16} + 0.573{*}\Delta X_5 - 0.404{*}\Delta X_{14}$	0.638	(12)
	1991–2020	$Y_7 = -1.091 \times 10^{-15} + 0.637{*}\Delta X_5 + 0.411{*}\Delta X_{12}$	0.628	(13)

Note: Here Y_1, Y_2, Y_3 and Y_4 are the areas of sand-control in each county in 1991–1999, 1999–2010, 2010–2020 and 1991–2020, respectively, while Y_5, Y_6, and Y_7 are the areas of desertification in each county in 1991–1999, 2010–2020 and 1991–2020, respectively. Of the corresponding period, ΔX_{10} is the difference in annual mean temperature, ΔX_{15} the difference in coal mine area, ΔX_2 the difference in meat product, ΔX_{13} the difference in sunshine duration, ΔX_5 the difference in per capita net income of farmers and herdsmen, ΔX_3 the difference in sheep number, ΔX_9 the difference in annual precipitation, ΔX_{12} the difference in average wind speed, and ΔX_{14} the difference in area of water body.

In terms of the multiple linear regression models in Table 10, it is seen that sand-control area is closely associated with the difference in temperature, mining area, meat product, sunshine hours, per capita net income of farmers and herdsmen, sheep number, and precipitation between the two observation years. More concretely, the increases in mining area, temperature and sunshine hours are negatively correlated with the sand-controlled area, while the increase in meat (pork, beef and mutton) product, per capita net income of farmers and herdsmen, sheep number and precipitation are positively correlated with the sand-controlled area.

The desertified area is related to the average wind speed and water. More exactly, the increase in average wind speed is the main driving force of desertification, while the increase in water areas are the favorable factors to reverse desertification, that is, lead to a decrease in desertified area.

3.4. Spatial Variability of Sand-Control

Different from those used for multiple linear regression analysis, here 19 socio-environmental factors shown in Table 11 were used for logistic regression modeling to understand the different spatial distribution of sand-control and desertification in different periods, and the obtained models that have passed the Hosmer–Lemeshow (HL) test are presented in Table 12. It is worth mentioning that the spatial factors are the different distances at the start of the observed period, and the socioeconomic and climate factors are the increments of the same period which have been rastorized to county-level as depicted before. The optimal discretization method was used to discretize the meteorological and spatial factors, and Z-Score was applied to normalize the socioeconomic factors, which were all taken as independent variables.

Table 11. Socio-environmental factors for logistic regression (LR) modeling.

Type	Factors		Data Preprocessing
Social and Economic Factors	Total Sown Area	X_1	Z-Score
	Meat Product	X_2	
	Sheep Number	X_3	
	Total Number of Livestock at the End of Year	X_4	
	Per capita Net Income of Farmers and Herdsmen	X_5	
	Gross Output of Farming, Forestry, Animal Husbandry and Fishery	X_6	
	Gross Domestic Product (GDP)	X_7	
Climatic Factors	Precipitation	X_8	The optimal discretization
	Temperature	X_9	
	Maximum Wind Speed	X_{10}	
	Average Wind Speed	X_{11}	
	Sunshine Duration	X_{12}	
Spatial Factors	Distance from Road	X_{13}	
	Distance from City	X_{14}	
	Distance from Water	X_{15}	
	Distance from Cropland	X_{16}	
Terrain Factors	Elevation	X_{17}	
	Aspect	X_{18}	
	Slope	X_{19}	

Temperature, meat product, per capita net income of farmers and herdsmen and sheep number, elevation, distance from road and distance from cropland are the important factors driving spatial differentiation of sand-control. In general, low elevation, increases in temperature, and increase in per capita net income of farmers and herdsmen and proximity to roads created a favorable condition for sand-control activity. It is worth mentioning that prior to 1999, increases in meat product and sheep number constrained sand-control, or rather, led to desertification, but after 1999 they have become favorable factors as controlled deserts have served for cropping and forage production. Prior to 2010, the closer to cropland, the higher the possibility of sandy land to be controlled; while it was the opposite after 2010, as combating desertification activity gradually moved into the heart areas of deserts.

Based on the regression coefficients and odds ratio (OR), it is possible to distinguish the importance of the socio-environmental factors (independent variables) in the sand-control or desertification event. As seen in Tables 10 and 12–14 the important factors causing the spatial variability of desertification include temperature, average wind speed, precipitation, per capita net income of farmers and herdsmen, GDP, total sown area and year-end number of big livestock. The areas far from cities and roads with low elevation and slope seem more susceptible to desertification than those close to cities and roads because the latter is easier to be managed and controlled. The increase in GDP, net income of farmers and herdsmen, the year-end number of big livestock and the total sown area for grain production are the driving forces of desertification. Additionally, the decrease in temperatures and reduction in water availability and increase in wind speeds increase the probability of desertification.

Table 12. Logistic regression models.

Model	Period	Expression	HL Test	
Sand-Control	1991–1999	$P_1 = \frac{exp(42.666 - 0.362\Delta X_2 - 0.107\Delta X_3 + \cdots - 0.047X_{18})}{1 + exp(42.666 - 0.362\Delta X_2 - 0.107\Delta X_3 + \cdots - 0.047X_{18})}$	0.184	(14)
	1999–2010	$P_2 = \frac{exp(3.464 + 0.34\Delta X_2 - 0.693\Delta X_5 + \cdots - 0.647X_{16})}{1 + exp(3.464 + 0.34\Delta X_2 - 0.693\Delta X_5 + \cdots - 0.647X_{16})}$	0.567	(15)
	2010–2020	$P_3 = \frac{exp(3.274 + 0.268\Delta X_3 + 0.522\Delta X_4 + \cdots - 0.326X_{17})}{1 + exp(3.274 + 0.268\Delta X_3 + 0.522\Delta X_4 + \cdots - 0.326X_{17})}$	0.723	(16)
	1991–2020	$P_4 = \frac{exp(5.005 + 0.425\Delta X_1 - 1.067\Delta X_3 + \cdots - 0.029X_{19})}{1 + exp(5.005 + 0.425\Delta X_1 - 1.067\Delta X_3 + \cdots - 0.029X_{19})}$	0.079	(17)
Desertification	1991–1999	$P_5 = \frac{exp(-12.257 + 0.458\Delta X_3 - 0.493\Delta X_4 + \cdots - 0.069X_{19})}{1 + exp(-12.257 + 0.458\Delta X_3 - 0.493\Delta X_4 + \cdots - 0.069X_{19})}$	0.277	(18)
	1999–2010	$P_6 = \frac{exp(1.627 - 0.261\Delta X_3 + 0.31\Delta X_5 + \cdots - 2.744X_{17})}{1 + exp(1.627 - 0.261\Delta X_3 + 0.31\Delta X_5 + \cdots - 2.744X_{17})}$	0.155	(19)
	2010–2020	$P_7 = \frac{exp(11.825 - 0.238\Delta X_1 - 0.805\Delta X_2 + \cdots - 0.059X_{19})}{1 + exp(11.825 - 0.238\Delta X_1 - 0.805\Delta X_2 + \cdots - 0.059X_{19})}$	0.476	(20)
	1991–2020	$P_8 = \frac{exp(-7.632 + 1.755\Delta X_8 + 1.741\Delta X_{11} + 0.84X_{13} - 0.081X_{19})}{1 + exp(-7.632 + 1.755\Delta X_8 + 1.741\Delta X_{11} + 0.84X_{13} - 0.081X_{19})}$	0.276	(21)

Note: P_1, P_2, P_3 and P_4 are the probability of sand-control occurrence respectively in 1991–1999, 1999–2010, 2010–2020 and 1991–2020, while P_5, P_6, P_7, and P_8 are the probability of desertification appearance respectively in 1991–1999, 1999–2010, 2010–2020, and 1991–2020. Of the corresponding period, ΔX_1 is the difference in total sown area, ΔX_2 the difference in meat product, ΔX_3 the difference in sheep number, ΔX_4 the difference of year-end big livestock, ΔX_5 the difference of per capita net income of farmers and herdsmen, ΔX_6 the difference of gross output of farming, forestry, animal husbandry and fishery, ΔX_7 the difference of gross domestic product (GDP), ΔX_8 the difference of precipitation, ΔX_9 the difference of temperature, ΔX_{10} the difference of maximum wind speed, ΔX_{11} the difference of average wind speed, ΔX_{12} the difference of sunshine duration, X_{13} the distance from road, X_{14} the distance from city, X_{15} the distance from water resource, X_{16} the distance from cropland, X_{17} the elevation, X_{18} the aspect, and X_{19} the slope.

Table 13. Logistic regression models for sand-control.

Factor	Sand-Control																			
	1991–1999					1999–2010					2010–2020					1991–2020				
	β	SE	Wals	Sig	OR	β	SE	Wals	Sig	OR	β	SE	Wals	Sig	OR	β	SE	Wals	Sig	OR
X_1																0.425	0.141	9.113	0.003	1.530
X_2	−0.362	0.055	42.557	00.000	0.696	0.340	0.060	32.301	0.000	1.405										
X_3	−0.107	0.056	3.654	0.056	0.898						0.268	0.077	12.091	0.001	1.307	−1.067	0.173	38.104	0.000	0.344
X_4											0.522	0.061	73.807	0.000	1.686	0.750	0.105	50.774	0.000	2.117
X_5						−0.693	0.089	61.035	0.000	0.500										
X_6						−0.286	0.065	19.267	0.000	0.752						−0.974	0.132	54.407	0.000	0.378
X_7						0.151	0.074	4.182	0.041	1.163	0.210	0.071	8.675	0.003	1.234	0.268	0.094	8.095	0.004	1.308
X_8						0.801	0.209	14.761	0.000	2.228	−1.014	0.169	36.107	0.000	0.363					
X_9	−0.711	0.136	27.178	0.000	0.491	0.400	0.145	7.577	0.006	1.492	0.532	0.138	14.970	0.000	1.703	0.270	0.115	5.507	0.019	1.310
X_{10}						−0.658	0.194	11.563	0.001	0.518						0.235	0.135	3.033	0.082	1.265
X_{11}	0.421	0.076	31.035	0.000	1.524											−0.206	0.110	3.530	0.060	0.814
X_{12}	−19.515	7842.123	0.000	0.998	0.000						−0.259	0.064	16.424	0.000	0.772					
X_{13}	−0.294	0.096	9.447	0.002	0.745	−0.776	0.093	69.097	0.000	0.460	−0.687	0.128	28.944	0.000	0.503	−0.527	0.067	61.855	0.000	0.590
X_{14}	−0.385	0.140	7.524	0.006	0.681						−0.246	0.110	4.942	0.026	0.782					
X_{15}						−0.637	0.157	16.395	0.000	0.529						−0.333	0.077	18.564	0.00	0.717
X_{16}	−0.376	0.077	23.601	0.000	0.686	−0.647	0.119	29.633	0.000	0.523	0.356	0.093	14.697	0.000	1.428	−0.597	0.061	94.860	0.000	0.551
X_{17}	−0.566	0.105	29.117	0.000	0.568						−0.326	0.075	18.896	0.000	0.722	−0.723	0.078	85.185	0.000	0.485
X_{18}	−0.047	0.017	7.510	0.006	0.954															
X_{19}																−0.029	0.013	5.349	0.021	0.971
Const	42.666	15684.246	0.000	0.998	3.386E18	3.464	0.705	24.143	0.000	31.955	3.274	0.536	37.246	0.000	26.419	5.005	0.585	73.242	0.00	149.184

Note: (1) β denotes the regression coefficient and SE standard error; Wals means Wald test, and Sig significance; OR is odds ratio. (2) Factors or variables whose variable inflation factor (VIF) was largely less than 2.5 were used for logistic regression modeling, and those with VIF > 10 were removed prior to modeling.

Table 14. Logistic regression models for desertification.

Factor	Desertification																			
	1991–1999					1999–2010					2010–2020					1991–2020				
	β	SE	Wals	Sig	OR	β	SE	Wals	Sig	OR	β	SE	Wals	Sig	OR	β	SE	Wals	Sig	OR
X_1											−0.238	0.127	3.473	0.062	0.789					
X_2											−0.805	0.222	13.105	0.000	0.447					
X_3	0.458	0.106	18.831	0.000	1.581	−0.261	0.124	4.397	0.036	0.770	0.707	0.203	12.152	0.000	2.027					
X_4	−0.0493	0.108	20.845	0.000	0.611															
X_5	−0.688	0.126	29.675	0.000	0.503	0.310	0.119	6.806	0.009	1.363										
X_6	0.368	0.122	9.077	0.003	1.444															
X_7						0.260	0.124	4.424	0.035	1.297	0.346	0.125	7.689	0.006	1.414					
X_8	0.449	0.121	13.635	0.000	1.566	−1.052	0.197	28.409	0.000	0.349	2.157	0.378	32.625	0.000	8.648	1.755	0.450	15.233	0.000	5.784
X_9	3.410	0.603	31.954	0.000	30.255	−0.491	0.203	5.876	0.015	0.612	−18.066	6138.015	0.000	0.998	0.000					
X_{10}	0.391	0.127	9.530	0.002	1.479															
X_{11}	−0.741	0.189	15.350	0.000	0.477						0.746	0.207	12.923	0.000	2.108	1.741	0.228	58.118	0.000	5.703
X_{12}						−0.524	0.153	11.641	0.001	0.592										
X_{13}						0.798	0.142	31.579	0.000	2.222	0.619	0.179	11.890	0.001	1.857	0.840	0.222	14.376	0.000	2.316
X_{14}	0.930	0.164	32.229	0.000	2.535	0.387	0.172	5.064	0.024	1.472	0.705	0.214	10.875	0.001	2.025					
X_{15}	1.327	0.415	10.228	0.001	3.769															
X_{16}						1.153	0.156	54.779	0.000	3.169										
X_{17}	0.232	0.138	2.832	0.092	1.261	−2.744	0.624	19.349	0.000	0.064	−0.903	0.323	7.809	0.005	0.405					
X_{18}																				
X_{19}	−0.069	0.020	12.197	0.000	0.934						−0.059	0.026	5.160	0.023	0.943	−0.081	0.034	5.720	0.017	0.922
Const	−12.257	1.587	59.674	0.000	0.000	1.627	1.066	2.331	0.127	5.090	11.825	6138.015	0.000	0.998	136669.368	−7.632	0.959	63.292	0.000	0.000

Note: (1) β denotes the regression coefficient and SE standard error; Wals means Wald test, and Sig significance; OR is odds ratio. (2) Factors or variables whose VIF was largely less than 2.5 were used for logistic regression modeling, and those with VIF > 10 were removed prior to modeling.

4. Discussion

4.1. Spatiotemporal Variability of the Sand-Control Effectiveness

As demonstrated above, the sand-control exhibits obvious spatiotemporal variability. This is probably because of the difference in regional implementation of the ecological restoration projects or environmental protection policies [8,33,47,48]. In 1999, the Grain for Green Program (GGP), or the Conversion of Cropland to Forest and Grassland Program, was promulgated by the Central Government and first implemented in Shaanxi, Ningxia and Inner Mongolia while the third phase of the Three-North Shelter Forest Program (TNSFP) was reaching its end. In addition, the local governments, e.g., the banners Otogqian and Otog, issued the "Grazing ban and rotation policy with a subsidy system" in 2001 [8]. At the same time, a large number of families or household companies started to devote themselves to the sand-control activities. The fourth and fifth phase of the TNSFP, which were focused on desertification prevention and control, was achieved respectively in 2010 and 2020. Moreover, the Central Government invested another 13.6 billion yuan in 2011 to eight main pastoral areas in Northern China for implementation of a protection policy with subsidy and incentive mechanisms, including subsidies for the grazing ban, incentives for balancing forage productivity and livestock, and for good performance of herdsmen (http://www.moa.gov.cn, accessed on 8 June 2021) [49]. These programs, policies and spontaneous sand-control activities have played a positive role in desertification reversal since the 1990s, but with a spatial variability in space as there was a difference in implementation progress of these initiatives. For this reason, the proportion of sand-control has increased significantly, despite of its spatial variability. Actually, a large piece of the degraded grassland in the banners Uxin, Otog and Otogqian has been restored.

4.2. Driving Forces for Desertification and Sand-Control

Some authors consider that climate is the main driving force of desertification [50], while others believe that human activities exert more impact on the latter [18]. Our research indicates that both human activities and climate condition including animal husbandry (i.e., increase in big livestock and, in particular, sheep number), GDP growth, variation in wind speed and temperature (Tables 10, 12 and 14) have played an important role in this process. More concretely, linear regression analysis reveals that desertification is positively correlated with the maximum and average wind speed and negatively correlated with the area of water (Table 10).

These analyses demonstrate that variation in strong wind speed, coal mining, grazing and reduction in water availability may have driven the occurrence of desertification.

As demonstrated in Table 14, logistic regression modeling uncovers that in different periods, the probability of desertification in space is different and the roles of the socio-environmental factors are also different. However, in the past 30 years, the contribution of ΔX_8, local reduction in precipitation, ΔX_{11}, increase in wind speed, and X_{13}, distance to roads, have played a positive role, while X_{19}, slope, takes a negative part in the desertification process, meaning that steep slope does not favor desertification. Therefore, desertification is associated with both climate condition and human activity, and this finding confirms the correctness of the definition of "desertification" by UNCCD in 1994.

Increase in precipitation is conducive to vegetation growth, thus promoting sand-control and vegetation recovery. It is worth mentioning that previous studies have shown that over-reclamation and over-stock will inevitably exacerbate desertification [2,8,30], whereas our study reveals that since 1999, the increase in meat product including pork, beef and mutton is a favorable sand-control indicator, and since 2010, the increase of per capita income of farmers and herdsmen has been also conducive for combating desertification. This is because in the frame of the TNSFP, GGP (1999), the Ecological Grassland Protection Project (2011) and other national ecological projects, sand-control has not only promoted vegetation recovery to reverse desertification, but also produced economic value as forages such as *Medicago sativa* and *Astragalus adsurgens pall* for dry-lot feeding [8], and plantations of *Elaeagnus angustifolia Linn.*, *Lycium barbarum* L., *Ephedra sinica Stapf* [2], etc., have brought a significant improvement in net rural income. Hence, sand-control favors both agricultural and pastoral activities.

4.3. Spatially Explicit Probability of Sand-Control and Desertification

Logistic regression modeling discovers that the area with flat terrain and low elevation is convenient for transportation and irrigation of planted trees, shrubs and herbage, and thus shows high probability of sand-control events. This is consistent with our field survey (Figure 6). Owing to the developed transportation, high accessibility and convenience for management, the areas close to cities, roads and farmland have a high probability of being managed. Meanwhile, those far from cities, roads and farmland are prone to desertification due to less human management and costly transport; this is similar to the results of Feng et al. (2021) [33]. Water source is essential for vegetation growth, and the closer to water, the lower the cost for irrigation and the higher the possibility of being controlled. The areas with irrational agricultural activity and animal husbandry had a high probability of desertification before 1999 because profit-seeking did not go with environmental protection. As Wu et al. (2013) revealed [8], reclamation for cropping and abandonment after 2–3 years of cultivation, land was left with erosion by both wind and water, and herdsmen preferred to breed the maximum possible number of sheep and big livestock to compete for the public grassland resource to maximize their economic profit, which had been discussed by Hardin in his article "The Tragedy of the Commons" [51]. However, since the large-scale implementation of policies, such as rotational grazing and the grazing-ban policy with a subsidy mechanism in 2001, the probability of successful sand-control in these areas has been significantly increased, in particular after 2010. This is in agreement with our field survey in July 2021 (Figure 7), which shows the difference between the rotational grazing areas and the grazing-ban area.

Figure 6. Landscape of interdune (grassland and pasture) and sand dune. (Photo was taken on 15 July 2021).

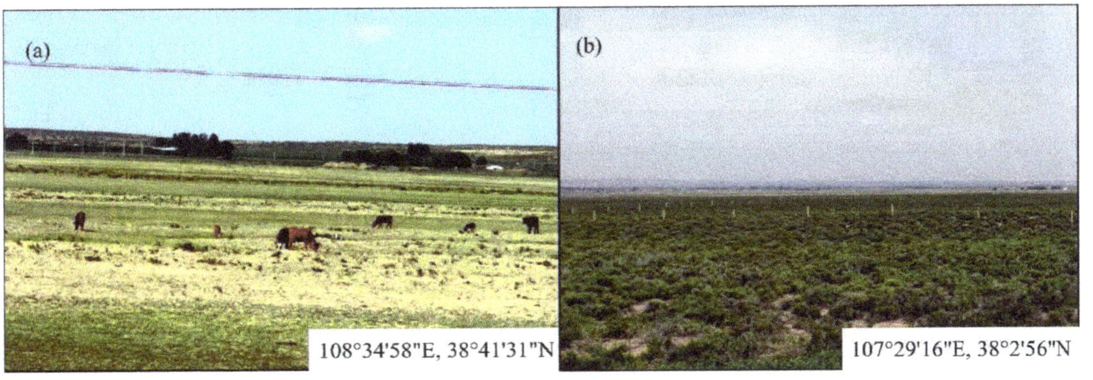

Figure 7. Vegetation in areas with measure of rotational grazing (**a**) and implementation of the policy grazing-ban (**b**) (Photos were taken on 16 July 2021 and 11 July 2021, respectively).

5. Conclusions

This paper presents an integrated remote sensing-based study on the effectiveness of sand-control and desertification in the Mu Us Desert, and spatially explicit models aiding to explain spatial determinants of sand-control and desertification were developed. We found that a total of 8712.23 km^2 of sandy land (63.05% of the study area) has been controlled and converted into shrublands and grasslands with different greenness, where GDVI has increased by 0.3509 on average from 1991 to 2020. This activity has brought not only recovery of vegetation cover, but also growth in rural income and meat products, allowing an improvement of the livelihood of herdsmen, and thus created a win–win situation. This is a great success of combating desertification by local people and the Chinese governments, and this experience deserves to be extended to other desert control in Northern China and even to other dryland countries across the world. However, despite the decrease in desert area, desertification is still taking place locally and more than 5000 km^2 of deserts remain unmanaged in the study area.

Multiple linear regression models illustrate that the rural per capita income, wind speed and water source play a role in desertification process, and logistic regression models reveal that the local reduction in precipitation, increase in wind speed and the distance to roads are the key factors influencing the desertification in the past three decades. Anyway, desertification is associated with both climate and human activity, such as wind speed, precipitation, water availability, distance to roads, rural per capita income and animal husbandry. The climatic part, such as variation in wind speed and precipitation. can be hardly managed, but it is possible to control animal husbandry (e.g., rational sheep number and livestock), and to make relatively remote areas accessible by road construction and available water. In this way, existing desertification may be reversed and future desertification could be avoided. The findings of this study may provide advice for local decision-makers in taking measures to conduct sand-control of the next step in Mu Us or elsewhere.

Different from the existing MEDALUS and DesertWatch projects, which were aimed at desertification vulnerability assessment and scenario modeling, this study attempted to reveal the desertification mechanism by two kinds of regression modeling, and achieved a quantitative desertification assessment with a deepened understanding on its driving forces. We can say that this study is an extension or a complement of the two mentioned projects in the continental climate dryland environment. Probably, a combination of these two approaches may lead to a more comprehensive research on desertification. This will be the topic of our next research in the Ordos region.

Author Contributions: Conceptualization, J.L. and W.W.; methodology, W.W.; software, W.W.; validation, J.L., X.F., M.Z., Y.L. (Yixuan Liu), J.J., W.L., C.Z., Y.H., X.K., Y.L. (Yuan Li), Y.S., H.Y. and X.Z.; formal analysis, J.L.; investigation, J.L, X.F., W.W., Q.T., J.J., Y.L. (Yixuan Liu), Y.L. (Yuan Li), C.Z. and W.L.; resources, W.W.; data curation, J.L.; writing—original draft preparation, J.L.; writing—review and editing, W.W.; visualization, J.L., X.F., X.K., Y.S. and Y.H.; supervision, W.W.; project administration, W.W. and Q.T.; funding acquisition, W.W. All authors have read and agreed to the published version of the manuscript.

Funding: This research was supported by the Start-up Fund for Scientific Research of the East China University of Technology, grant number DHTP2018001, and the fund of the Jiangxi Talent Program, grant number 900/2120800004, both granted to Weicheng Wu.

Data Availability Statement: Data used for this research will be available upon reasonable request to the corresponding author.

Acknowledgments: Landsat images were freely obtained from the USGS data server (https://glovis.usgs.gov, last accessed on 20 May 2021) and DEM data (ASTERV003 30m) from NASA (www.earthdata.nasa.gov, last accessed on 16 June 2020); Google is acknowledged for making the very high-resolution images available on Google Earth.

Conflicts of Interest: The authors declare no conflict of interest.

References

1. UNCCD, UNCCD Terminology. 1994. Available online: https://www.unccd.int (accessed on 27 December 2021).
2. Wu, W. Application de la Geomatique au Suivi de la Dynamique Environnementale en Zones Arides. Doctoral Dissertation, Université Panthéon-Sorbonne-Paris I, Paris, France, 2003.
3. The State Forestry Administration of the People's Republic of China. *Atlas of Desertified and Sandified Land in China*, 1st ed.; Science Press: Beijing, China, 2009; pp. 36–37.
4. Hellden, U. Desertification monitoring: Is the desert encroaching? *Desertif. Control. Bull* **1988**, *17*, 8–12.
5. Tucker, C.J.; Dregne, H.E.; Newcomb, W.W. Expansion and contraction of Sahara Desert from 1980 to 1990. *Science* **1991**, *253*, 299–300. [CrossRef]
6. Wu, W.; Zhang, W. Present land use and cover patterns and their development potential in North Ningxia. *J. Geogr. Sci.* **2003**, *13*, 54–62. [CrossRef]
7. Symeonakis, E.; Drake, N. Monitoring desertification and land degradation over sub-Saharan Africa. *Int. J. Remote Sens.* **2004**, *25*, 573–592. [CrossRef]
8. Wu, W.; De Pauw, E.; Zucca, C. Using remote sensing to assess impacts of land management policies in the Ordos rangelands in China. *Int. J. Digit. Earth* **2013**, *6* (Suppl. 2), 81–102. [CrossRef]
9. Wang, Z.; Xu, D.; Peng, D.; Zhang, Y. Quantifying the influences of natural and human factors on the water footprint of afforestation in desert regions of northern China. *Sci. Total Environ.* **2021**, *780*, 146577. [CrossRef]
10. Kaushalya, R. Monitoring the impact of desertification in western Rajasthan using remote sensing. *J. Arid. Environ.* **1992**, *22*, 293–304. [CrossRef]
11. Kumar, M.; Goossens, E.; Goossens, R. Assessment of sand dune change detection in Rajasthan (Thar) Desert, India. *Int. J. Remote Sens.* **1993**, *14*, 1689–1703. [CrossRef]
12. Wu, W. Monitoring Land Degradation in Drylands by Remote Sensing. In *Desertification and Risk Analysis Using High and Medium Resolution Satellite Data*; Marini, A., Talbi, M., Eds.; Springer: Berlin/Heidelberg, Germany, 2009; pp. 157–170. [CrossRef]
13. Kim, J.W.; Song, C.H.; Park, E.B.; Lee, J.Y.; Choi, S.E.; Lee, W.K. A study for Desertification Monitoring and Assessment based on satellite imagery in Tunisia. *J. Korean Assoc. Geogr. Inf. Stud.* **2018**, *21*, 91–107. [CrossRef]
14. Duan, H.; Wang, T.; Xue, X.; Yan, C. Dynamic monitoring of aeolian desertification based on multiple indicators in Horqin Sandy Land, China. *Sci. Total Environ.* **2019**, *650*, 2374–2388. [CrossRef]
15. Cui, G.; Lee, W.K.; Kwak, D.A.; Choi, S.; Park, T.; Lee, J. Desertification monitoring by LANDSAT TM satellite imagery. *For. Sci. Technol.* **2011**, *7*, 110–116. [CrossRef]
16. Sun, D.; Richard, D.; Li, H.; Li, B. Modeling desertification change in Minqin County, China. *Environ. Monit. Assess.* **2005**, *108*, 169–188. [CrossRef] [PubMed]
17. Kang, W.; Liu, S. A Review of Remote Sensing Monitoring and Quantitative Assessment of Aeolian Desertification (in Chinese with English abstract). *J. Desert Res.* **2014**, *34*, 1222–1229. [CrossRef]
18. Zhang, C.; Wang, X.; Li, J.; Hua, T. Identifying the effect of climate change on desertification in northern China via trend analysis of potential evapotranspiration and precipitation. *Ecol. Indic.* **2020**, *112*. [CrossRef]
19. Han, X.; Sun, Y.; Chen, Y.; Feng, Z.; Gong, Y.; Cheng, Y. Monitoring in Dynamic Change of Desertification in the Region of Central Asia Based on NOAA/AVHRR Image. *Key Eng. Mater.* **2012**, *500*, 592–597. [CrossRef]

20. Lamchin, M.; Lee, W.; Jeon, S.W.; Lee, J.; Song, C.; Piao, D.; Lim, C.H.; Khaulenbek, A.; Navaandorj, I. Correlation between Desertification and Environmental Variables Using Remote Sensing Techniques in Hogno Khaan, Mongolia. *Sustainability* **2017**, *9*, 581. [CrossRef]
21. Eckert, S.; Hüsler, F.; Liniger, H.; Hodel, E. Trend analysis of MODIS NDVI time series for detecting land degradation and regeneration in Mongolia. *J. Arid. Environ.* **2015**, *113*, 16–28. [CrossRef]
22. Wu, W. The Generalized Difference Vegetation Index (GDVI) for Dryland Characterization. *Remote Sens.* **2014**, *6*, 1211–1233. [CrossRef]
23. Xie, L.; Wu, W.; Huang, X.; Ou, P.; Lin, Z.; Wang, Z.; Song, Y.; Lang, T.; Huangfu, W.; Zhang, Y.; et al. Mining and Restoration Monitoring of Rare Earth Element (REE) Exploitation by New Remote Sensing Indicators in Southern Jiangxi, China. *Remote Sens.* **2020**, *12*, 3558. [CrossRef]
24. Wu, W.; Zucca, C.; Karam, F.; Liu, G. Enhancing the performance of regional land cover mapping. *Int. J. Appl. Earth Obs. Geoinf.* **2016**, *52*, 422–432. [CrossRef]
25. European Commission. Directorate-General for Research and Innovation. In *The Medalus Project: Mediterranean Desertification and Land Use: Manual on Key Indicators of Desertification and Mapping Environmentally Sensitive Areas to Desertification*; Publications Office: Brussel, Belgium, 1999; pp. 1–85.
26. European Space Agency (ESA). DesertWatch Project Final Report. 2009. Available online: http://due.esrin.esa.int/files/131-176-149-30_2009430103852.pdf (accessed on 27 December 2021).
27. Lee, E.; Piao, D.; Song, C.; Kim, J.; Lim, C.; Kim, E.; Moon, J.; Kafatos, M.; Lamchin, M.; Jeon, S.; et al. Assessing environmentally sensitive land to desertification using MEDALUS method in Mongolia. *For. Sci. Technol.* **2019**, *15*, 210–220. [CrossRef]
28. Bouhata, R.; Bensekhria, A. Adaptation of MEDALUS method for the analysis depicting desertification in Oued Labiod valley (Eastern Algeria). *Arab. J. Geosci.* **2021**, *14*. [CrossRef]
29. Abuzaid, A.S.; Abdelatif, A.D. Assessment of desertification using modified MEDALUS model in the north Nile Delta, Egypt. *Geoderma* **2022**, *405*, 115400. [CrossRef]
30. Xu, D.; Song, A.; Tong, H.; Ren, H.; Hu, Y.; Shao, Q. A spatial system dynamic model for regional desertification simulation–A case study of Ordos, China. *Environ. Model. Softw.* **2016**, *83*, 179–192. [CrossRef]
31. Feng, K.; Yan, C.; Xie, J.; Qian, D. Spatial-temporal Evolution of Aeolian Desertification Process in Ordos City during 1975–2015 (in Chinese with English abstract). *J. Desert Res.* **2018**, *38*, 233–242. [CrossRef]
32. Bai, Z.; Cui, J.; Ding, X. Desertification and its driving factors in the Ordos Plateau from 1986 to 2015 (in Chiness with English abstract). *Arid. Zone Res.* **2020**, *37*, 749–756. [CrossRef]
33. Feng, K.; Wang, T.; Liu, S.; Yan, C.; Kang, W.; Chen, X.; Guo, Z. Path analysis model to identify and analyse the causes of aeolian desertification in Mu Us Sandy Land, China. *Ecol. Indic.* **2021**, *124*, 107386. [CrossRef]
34. Liu, Q.; Zhang, Q.; Yan, Y.; Zhang, X.; Niu, J.; Svenning, J.C. Ecological restoration is the dominant driver of the recent reversal of desertification in the Mu Us Desert (China). *J. Clean. Prod.* **2020**, *268*, 122241. [CrossRef]
35. Hu, Y.; Han, Y.; Zhang, Y. Land desertification and its influencing factors in Kazakhstan. *J. Arid. Environ.* **2020**, *180*, 104203. [CrossRef]
36. Han, J.; Wang, J.; Chen, L.; Xiang, J.; Ling, Z.; Li, Q.; Wang, E. Driving factors of desertification in Qaidam Basin, China: An 18-year analysis using the geographic detector model. *Ecol. Indic.* **2021**, *124*, 107404. [CrossRef]
37. Hua, D.; Hao, X. Spatiotemporal change and drivers analysis of desertification in the arid region of northwest China based on Geographic Detector. *Environ. Chall.* **2021**, *4*, 100082. [CrossRef]
38. Tong, L.; Ning, X.; Zhang, J.; Zhang, X. Spatial-temporal variation and driving mechanism of desertification in Hunshandake (Otindag) Sandy Land in recent 30 years (in Chinese with English abstract). *Arid. Land Geogr.* **2021**, *44*, 992–1002. [CrossRef]
39. National Meteorological Science Data Center of China. Available online: http://data.cma.cn/data/cdcindex/cid/0b9164954813c573.html (accessed on 18 June 2021).
40. Chavez, P.S., Jr. Image-Based Atmospheric Correction—Revisited and Improved. *Photogramm. Eng. Remote Sens.* **1996**, *62*, 1025–1036.
41. USGS. Landsat 8 (L8) Data Users Handbook (V1.0). 2015. Available online: http://landsat.usgs.gov/documents/Landsat8DataUsersHandbook.pdf (accessed on 30 June 2015).
42. Chander, G.; Markham, B.L.; Helder, D.L. Summary of current radiometric calibration coefficients for Landsat MSS, TM, ETM+, and EO-1 ALI sensors. *Remote Sens. Environ.* **2009**, *113*, 893–903. [CrossRef]
43. Snee, R. Validation of Regression Models: Methods and Examples. *Technometrics* **1977**, *19*, 415–428. [CrossRef]
44. Serneels, S.; Lambin, E.F. Proximate cause of land-use change in Narok District, Kenya: A spatial statistical model. *Agric. Ecosyst. Environ.* **2001**, *85*, 65–81. [CrossRef]
45. Ou, P.; Wu, W.; Qin, Y.; Zhou, X.; Huangfu, W.; Zhang, Y.; Xie, L.; Huang, X.; Fu, X.; Li, J.; et al. Assessment of Landslide Hazard in Jiangxi Using Geo-information Technology. *Front. Earth Sci.* **2021**, *9*. [CrossRef]
46. Huangfu, W.; Wu, W.; Zhou, X.; Lin, Z.; Zhang, G.; Chen, R.; Song, Y.; Lang, T.; Qin, Y.; Ou, P.; et al. Landslide Geo-Hazard Risk Mapping Using Logistic Regression Modeling in Guixi, Jiangxi, China. *Sustainability* **2021**, *13*, 4830. [CrossRef]
47. Su, B.; Wang, Y.; Shuangguan, Z. Analysis on the Scale of a New Period of Returning Farmland to Forestland and Grassland in Northwest China (in Chinese with English abstract). *Res. Soil Water Conserv.* **2017**, *24*, 59–65. [CrossRef]

48. Yao, W.; Yu, J.; Kang, H. Problems and Prospects of the Construction Effectiveness of the Three North Shelterbelt Project in Shaanxi Province (in Chinese with English abstract). *Protection For. Sci. Technol.* **2020**, *38*, 70–71. [CrossRef]
49. Ministry of Agriculture and Rural Affairs of the People's Republic of China. Available online: http://www.moa.gov.cn/ztzl/lszczc/201103/t20110325_1955106.htm (accessed on 8 December 2021).
50. Zhao, Z.; Wu, Q. Study on desertification monitoring from 2000 to 2014 and its driving factors through remote sensing in Ningxia, China. *Int. Arch. Photogramm. Remote Sens. Spat. Inf. Sci.* **2018**, *XLII-3*, 2439–2477. [CrossRef]
51. Hardin, G. The Tragedy of the Commons. *Science* **1968**, *162*, 1243–1248. [CrossRef] [PubMed]

remote sensing

MDPI

Article

High-Resolution Mining-Induced Geo-Hazard Mapping Using Random Forest: A Case Study of Liaojiaping Orefield, Central China

Yaozu Qin [1,2,*], Li Cao [3], Ali Darvishi Boloorani [1,4] and Weicheng Wu [1,2]

1 Key Laboratory of Digital Land and Resources, East China University of Technology, Nanchang 330013, China; ali@ecut.edu.cn or ali.darvishi@ut.ac.ir (A.D.B.); wuwch@ecut.edu.cn (W.W.)
2 Faculty of Earth Sciences, East China University of Technology, Nanchang 330013, China
3 Key Laboratory of Natural Resources Monitoring and Supervision in Southern Hilly Region, Ministry of Natural Resources, Changsha 430103, China; cl@img.net
4 Department of Remote Sensing and GIS, Faculty of Geography, University of Tehran, Tehran 1417853933, Iran
* Correspondence: qyz60010@ecut.edu.cn; Tel.: +86-186-7311-0286

Abstract: Mining-induced geo-hazard mapping (MGM) is a critical step for reducing and avoiding tremendous losses of human life, mine production, and property that are caused by ore mining. Due to the restriction of the survey techniques and data sources, high-resolution MGM remains a big challenge. To overcome this problem, in this research, such an MGM was conducted using detailed geological exploration and topographic survey data as well as Gaofen-1 satellite imagery as multi-source geoscience datasets and machine learning technique taking Liaojiaping Orefield, Central China as an example. First, using Gaofen-1 panchromatic and multispectral (PMS) sensor data and Random Forest (RF) non-parametric ensemble classifier, a seven-class land cover map was generated for the study area with an overall accuracy (OA) and Kappa coefficient (KC) of 99.69% and 98.37%, respectively. Next, several environmental drivers including land cover, topography (aspect and slope), lithology, distance from fault, elevation difference between surface and underground excavation, and the difference of spectral information from PMS multispectral data of different years were integrated as predictors to construct an RF-based MGM model. The constructed model showed an excellent prediction performance, with an OA of 98.53%, KC of 97.06%, and AUC of 0.998, and the 85.60% of the observed geo-disaster that have occurred in the predicted high susceptibility class (encompassing 2.82% of the study area). The results suggested that the changes in environmental factors in the high susceptibility areas can be used as indicators for monitoring and early-warning of the geo-disaster occurrence.

Keywords: geo-hazard mapping; Gaofen-1 satellite; land cover; environmental factors; susceptibility

Citation: Qin, Y.; Cao, L.; Darvishi Boloorani, A.; Wu, W. High-Resolution Mining-Induced Geo-Hazard Mapping Using Random Forest: A Case Study of Liaojiaping Orefield, Central China. *Remote Sens.* **2021**, *13*, 3638. https://doi.org/10.3390/rs13183638

Academic Editor: Andrea Ciampalini

Received: 26 July 2021
Accepted: 7 September 2021
Published: 11 September 2021

1. Introduction

Mining-induced geo-disasters (MG) are a type of disaster related to geological processes induced by natural and/or man-made factors [1,2]. These disasters, which include debris flow, landslide, collapse, ground fissure, and subsidence, are usually caused by intensive mining activities with tremendous damage to the natural and man-made environment, such as water bodies, farmlands, roads, and pipelines. More importantly, mining-induced disasters lead to mining accidents and losses of human life and property and even reduce the sustainability and stability of development among human beings, resources, and the environment. Hence, some useful prevention measures and technology of MG must be proposed [3–5]. Mining-induced geo-hazard mapping (MGM) based on determining the relative probability of geo-disaster occurrence is essential for real-time monitoring and prediction of the spatial patterns of geological disasters and subsequently protection of the ecological resources and human health in the mining areas [6,7].

Qualitative or semi-quantitative estimation of the occurrence possibility is considered a common procedure for evaluating geo-disaster susceptibility, especially for individual disasters. This can be implemented by studying the mechanism of geo-disaster occurrences, identifying triggering factors, and then using these factors to simulate the deformation progress of the related geological bodies, especially for the single landslide triggered by rainfall or earthquake [8–12]. Various geo-disasters may occur concurrently by the same type of environmental factors, such as rainfall, geological structures, and excavation activity. Moreover, trigger factors caused by geo-disasters should be used for predicting and evaluating geo-disaster susceptibility. The characteristics of geological structures are one of the important factors in the field of MGM. In this regard, Wang et al. [13] developed a disaster-area prediction model that is based on analyzing the correlation of geo-disaster with mining-induced activity, lithology, and geological structure. In another study, Segoni et al. [14] performed a landslide susceptibility mapping approach using various geological data including structural, lithology, chronologic, genetic units, and paleogeography. These triggering and triggered factors, as well as the geological and geographical conditions and environmental factors, can be obtained from field-based disaster investigation, geological survey, and remote sensing (RS), taking advantage of the earth observation satellite data, geographic information system (GIS) technique, and machine learning modeling [15].

In order to quantitatively conduct MGM, it is necessary to first consider all causes of previous events and accordingly analyze the association between disasters with different environmental drivers using data-driven methods in the GIS platform [16–19]. In the literature, various multi-source geospatial data, i.e., topographic features, geological information, rainfall conditions, and vegetation indexes (VIs) from field survey and satellite imagery were used as environmental predictive factors for MGM using powerful data-driven methods, such as support vector machine (SVM) [20], logistic regression (LR) [21], artificial neural networks (ANN) [22], random forest (RF) [23], decision tree (DT) [24], weights of evidence (WofE) [20], frequency ratio (FR) [25], analytic hierarchy process (AHP), and linear combination (LC) [26,27]. Overall, a wide variety of approaches have been used for MGM, among which supervised machine learning algorithms have shown high efficiency and reliability. In recent years, these methods have been successfully applied in the field of geoscience, especially for mineral prospectivity mapping (MPM) and MGM [15,28–38]. MG occurs suddenly within/around mining areas with the characteristic of small scale, high density, and frequency. Due to the vital need for more detailed mining activity and geological exploration data, the implementation of MGM is associated with some restrictions [8,39]. Despite numerous studies in this field, due to the restriction of the survey techniques and data sources, MGM with high-resolution remains a major challenge.

Preparation of land cover map is a preliminary to analyzing physiognomy characteristics and evaluating land resources, and it also definitely facilitates the prediction and evaluation of MG. Under normal circumstances, different land cover types indicate the different levels of human activities as the triggering factors of MG. Utilizing multispectral and multi-temporal RS datasets is a momentous approach to mining geospatial information. For example, a great number of researchers obtain the land use/cover maps based on RS image classification techniques by taking advantage of the capabilities of supervised machine learning methods (e, g., RF, SVM, ANN, and LR) [40–42].

Nowadays, thanks to the development of high-spatial- and spectral-resolution RS technology, it has become feasible to extract more precise and comprehensive geospatial information. In the same context, Youssef [43] generated predictive geo-disaster drivers by integrating 15 m resolution satellite imagery and 10 m contour maps to obtain the landslide susceptibility indices. Pachuau [44] identified the areas susceptible to landslide occurrence with a variety of high spatial resolution satellite datasets, i.e., Quick Bird, IRS, and Cartosat-I imagery. Arabameri et al. [45] used RS datasets with different spatial resolutions to assess landslide susceptibility based on combined FR and RF approaches. In their study, the sample data were collected from various resources, such as extensive field surveys, historical records, aerial photo interpretation, and high-spatial-resolution Google Earth images.

The Liaojiaping Orefield, which is located in Hunan province, Central China, is an important part of the gold (Au) and antimony-tungsten (Sb-W) polymetallic metallogenic belt in the southern branch of the middle Xuefeng Arcuate Tectonic Belt (XATB). The main deposits hosted in this orefield have been indiscriminately mined for decades. Coupled with the complex geological and structural setting of the mining areas, this has led to the frequent occurrence of different MGs such as landslip, collapse, land subsidence, and fissure. It should be noted that these MGs directly restrict mine exploitation and pose serious threats to human life and property. In the absence of systemic research on susceptibility, these disasters are difficult to prevent. Accordingly, the main purpose of this study is to perform a high-resolution MGM in Liaojiaping Orefield based on multi-source high spatial resolution geo-environmental data using data-driven methods, taking the main environmental factors that are associated with MG into account.

2. Study Area and Materials

2.1. Geological Setting

Liaojiaping Orefield, covering an area of 41.25 km^2 and located in the central Hunan province, China, is situated in the southern margin of the middle XATB, which is developed between the Dongting Basin and the Gui-Xiang subsidence belt in the Yangtze Block and consists of Northeastern Hunan fault-uprising belt and the Xuefeng thrust belt (Figure 1). The approximately EW- and NE-striking faults and the secondary anticlines with the NE direction axis in this tectonic setting form the basic structural framework of the orefield (Figure 2). These multi-phase geological structures intricately crisscross and lead to the dip and steep landform.

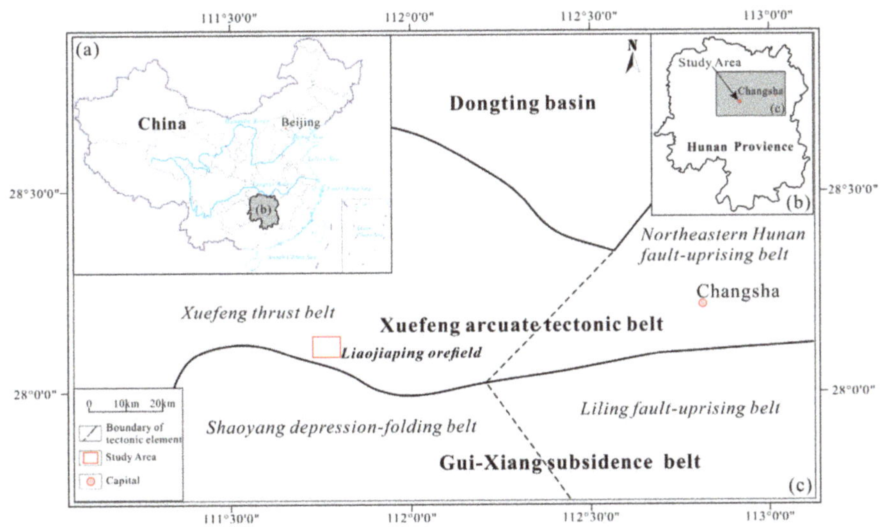

Figure 1. Location of the Hunan province in China (**a**), the study area in Hunan province (**b**) and the geological settings of the Liaojiaping Orefield (**c**).

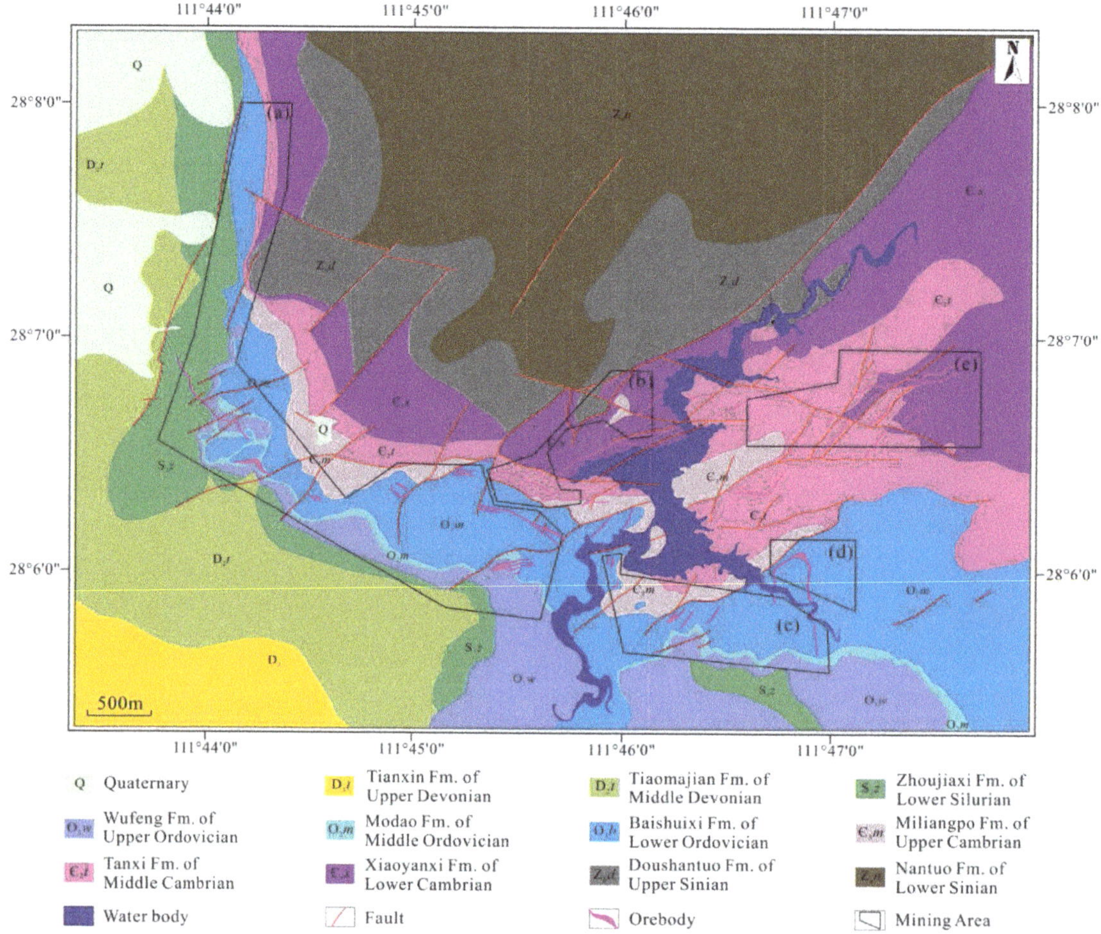

Figure 2. Geological map of the Liaojiaping Orefield, showing the main stratigraphic units, faults, water body, and the main mining areas, including (**a**) Taiping–Tanchelun, (**b**) Babaoshan, (**c**) Xiaojiawan, (**d**) Niejialing, and (**e**) Tianshenghe.

The fine clastic rocks intercalated with carbonate rocks that were deposited in the epicontinental rift basin environment from Lower Proterozoic to Upper Paleozoic Era and the carbonate rocks intercalated with clastic rocks in the epicontinental basin environment (later Paleozoic) form the stratigraphic assemblage of this region. The strata from Sinian to Devonian are well exposed in this orefield, and the Quaternary sediments are mainly deposited in the northwest corner (Figure 2 and Table 1). The outcrops of different strata have been experiencing various degrees of weathering and splintering; for example, the fine sandstone in the Upper Zhoujiaxi Formation of Lower Silurian presents a bead shape as a result of an intense spheroidal weathering process.

Table 1. Detailed stratigraphy of the Liaojiaping Orefield.

Epoch	Lithological Unit	Code	Thickness and Lithological Composition
Quaternary	No	Q	1~3 m. Eluvium and alluvium: thin clay and clayey soil.
Upper Devonian	Tianxin Formation	D_3t	180~400 m. Thin-bedded siltstone, silty shale.
Middle Devonian	Tiaomajian Formation	D_2t	More than 660 m. Silty shale with siltstone interblended, thick fine-grained quartzose sandstone intercalated with siltstone and celadon shale.
Lower Silurian	Zhoujiaxi Formation	S_1z	64~375 m. Medium-thick fine sandstone intercalated with thin layered silty shale, carbonaceous fine sandstone with interlayers of the siltstone.
Upper Ordovician	Wufeng Formation	O_3w	5~28 m. Medium-bedded silty carbonaceous platy shale with intercalated siliceous bands.
Middle Ordovician	Modao Formation	O_2m	48~80 m. Carbon-bearing silicate with thin silty shale interblended.
Lower Ordovician	Baishuixi Formation	O_1b	150~520 m. Gray plate shale locally intercalated with carbon-bearing mudstone and siliceous bands.
Upper Cambrian	Miliangpo Formation	ϵ_3m	140~320 m. Crystal powder limestone intercalated with siliceous bands.
Middle Cambrian	Tanxi Formation	ϵ_2t	110-280 m. Gray banded marlstone and globular crystal powder limestone.
Lower Cambrian	Xiaoyanxi Formation	ϵ_1x	158~368 m. Carbonaceous mudstone intercalated with poor coal seam and siliceous bands.
Upper Sinian	Doushantuo Formation	Z_bd	70~121 m. Thin-bedded carbon-bearing mudstone, biomicrite, and silicate layered clearly.
Lower Sinian	Nantuo Formation	Z_an	100~680 m. Moraine conglomerate, conglomerate, and carbonate with the character of glaciomarine deposit.

2.2. Geological Disasters

This orefield has been mined on and off for more than half a century. Early unauthorized and later wasteful mining activities led to a series of environmental problems in these mining areas, such as ground deformation, water, and soil pollution. The MG often occurring next to each other cause serious damage to human life and property, although the mining has been conducted in a more scientific and cautious way in the last decade. For example, the landslide that occurred in July 2018 caused one death and two injuries in one family in the Tianchelun mine of this orefield. This highlights the need for MGM using multi-source environmental factors that are related to the geological setting and mining activities.

It took several months to investigate the MG that occurred in Liaojiaping Orefield, and the survey results showed that the landslide, collapse, land subsidence, and fissure erratically took place in this orefield, especially in case of heavy rainfall. The main characteristic of MG is that they usually occur at a different scale around mining and excavated areas. The difference in lithology and physical environment leads to different degrees of outcrop weathering, and in this circumstance, various MGs are triggered in these outcrop areas by various types and scales of human activities. The MGs that occurred (e.g., Figure 3) are mainly medium–small in size in the Liaojiaping Orefield. In this regard, the detached mass of landslides in Figure 3a,b is less than 1000 m^3, the biggest collapsed area (Figure 3c) is no more than 500 m^2, and other common collapsed areas (e.g., Figure 3d) are about 10 to 100 m^2. Most of the collapsed blocks (e.g., Figure 3e) are only several m^3, and the ground fissures are normally tens of centimeters in width and several meters in length (e.g., Figure 3f). Some of these MG are interconnected in terms of occurrence; e.g., the ground fissures above the mining areas always occur before the subsidence, and the places often affected by the collapse may concurrently produce landslides.

Figure 3. Mining-induced geo-disasters (MG) in the Liaojiaping Orefield, showing the different types and scales, (**a**,**b**): landslide, (**c**,**d**): subsidence, (**e**): collapse, and (**f**): ground fissure.

2.3. *Multi-Source Geo-Environmental Data*

The occurrences of MG in the Liaojiaping Orefield are often related to different factors including underground mining, geological structures, topographic features, near-surface excavation, and rock weathering. In addition to these factors, the land cover information and surface spectral characteristics can be also used for MGM. MG investigation, geological survey, mineral exploration, and RS are vital and common techniques that can provide all mentioned multi-source geo-environmental data necessary for MGM.

The dataset composed of the above factors is actually an integration of different variable layers that are rasterized into the same grid size, and the sample set, an important part of this grid dataset containing the target variable, is used for training the prediction model and its validation. The determination of features (for the whole dataset) and the target variable (for the sample set) plays an important role in the construction of the prediction model, and these variables, which are used as predictive factors [46,47], need to be explored by different methods, and their spatial autocorrelation must be reduced [48].

Three Au and two Sb-W deposits have been mined for more than 20 years in the study area. The data supporting this study can be sourced accordingly: (1) the detailed geological data acquired by continuous geological survey and exploration, i.e., the main stratigraphic units and faults presented in the geological map of the study area (Figure 2);

(2) the exploration data from mining activities, such as tunnels and stopes implemented in the mining areas; (3) the topographic features such as aspect and slope values extracted from the high-precision topographic map on the scale of 1:5000; (4) the minor structures and the surface spectral characteristics (e.g., VIs) obtained and interpreted using high-resolution RS imagery, in this case, Gao Fen-1 (GF-1) satellite, which was launched on April 26, 2013 by CNSA (China National Space Administration) [49]. Two panchromatic and multispectral sensors (PMS) and four wide field-of-view (WFV) sensors are aboard the GF-1 satellite [50]. The present study took advantage of the PMS sensor data. The specifications of GF-1/PMS are presented in Table 2.

Table 2. Imagery parameters of the GF-1/PMS [50].

Sensor	Spectral Band		Wavelength Range (μm)	Spatial Resolution (m)
PMS	Panchromatic	B–1 (PAN)	0.45–0.90	2
	Multispectral	B–2 (Blue)	0.45–0.52	8
		B–3 (Green)	0.52–0.59	
		B–4 (Red)	0.63–0.69	
		B–5 (NIR)	0.77–0.89	

3. Methodology

The results of different statistic-based prediction models for MGM are quite different [51–53]. For a certain algorithm, it may achieve good prediction accuracy/performance in one case but perform poorly in another. The intrinsic structure of samples must be the decisive factor that causes this situation. In the same context, Kalantar et al. [54] and Qin et al. [37] have also pointed out that the determination of the sample dataset has a direct effect on the model prediction accuracy. Accordingly, to increase the generalizability of the predictive model, it is adequate to combine the classical and popular mathematical methods to construct a robust prediction model as long as the relevant dataset is well prepared.

3.1. GF-1 Image Processing

Band ratio operation, multispectral transformation, and image filtering are important techniques for image enhancement and extraction of spectral information of the ground objects after preprocessing, including ortho-rectification, radiometric calibration, and atmospheric correction [55]. For the GF-1 imagery, the spatial resolution of the multispectral bands can be improved to 2 m by fusing them with the panchromatic band so that it can meet the requirements of this study despite its low spectral resolution.

3.1.1. Band Ratio Operation

All kinds of VIs that can detect spatiotemporal patterns of vegetation can be used as an important factor for land cover classification [32]. Kaufman and Tanré [56] proposed a VI named soil-adjusted atmospherically resistant vegetation index (*SARVI*) based on the soil-adjusted vegetation index (SAVI) [57], which can be written as Equation (1),

$$SARVI = (1+L)\frac{B_{NIR} - (2 \times B_{RED} - B_{BLUE})}{B_{NIR} + (2 \times B_{RED} - B_{BLUE}) + L} \tag{1}$$

where L is a constant that is used to reduce the soil effect as much as possible, and it is suggested to be set as 1; B_{NIR}, B_{RED}, and B_{BLUE} are, respectively, the reflectance of the near-infrared (NIR), red, and blue bands. *SARVI* is suitable for the strongly vegetated areas from various satellite sensors, and it also can be employed for vegetation analysis based on Gaofen-1/PMS data.

3.1.2. Image Transformations

With the help of color-space conversions and principal component analysis (PCA), the spectral information can be enforced while the noise is reduced to a certain extent. Munsell HSV transformation, which converts a three-layer color space of red (R), green (G), and blue (B), known as RGB, into another three-layer color space, including hue (H), saturation (S), and value (V), known as HSV, facilitates the description and distinction of the color features of soil and rock [58]. The theoretical model of the Munsell HSV transformation is presented as follows:

$$H = \begin{cases} 0 & R = G = B \\ 60 \times \left(\frac{G-B}{\max(R,G,B)-\min(R,G,B)} + 1 \right) & \max(R,G,B) = R \\ 60 \times \left(\frac{B-R}{\max(R,G,B)-\min(R,G,B)} + 3 \right) & \max(R,G,B) = G \\ 60 \times \left(\frac{R-G}{\max(R,G,B)-\min(R,G,B)} + 5 \right) & \max(R,G,B) = B \end{cases} \quad (2)$$

$$S = \begin{cases} \frac{\max(R,G,B)-\min(R,G,B)}{\max(R,G,B)} & \max(R,G,B) \neq 0 \\ 0 & \max(R,G,B) = 0 \end{cases} \quad (3)$$

$$V = \max(R,G,B) \quad (4)$$

where R, G, and B are the reflectance values of the corresponding RGB combined band, H is a range from 0 to 360, and S and V range from 0 to 1.

PCA, which is also known as the Karhunen–Loeve (K–L) transform [59], is used to generate a new spectral space F from the original space X that consists of n samples with p dimensions. The dimensions p of the space X can be reduced to m using a linear transformation matrix A, which contains m multi-feature vectors. The first few principal components of the new space F usually contain the vast majority of the spectral information. This process can be described as Equation (5):

$$X = \begin{Bmatrix} x_{11} & x_{12} & \cdots & x_{1p} \\ x_{21} & x_{22} & \cdots & x_{2p} \\ \vdots & \vdots & \cdots & \vdots \\ x_{n1} & x_{n2} & \cdots & x_{np} \end{Bmatrix} \xrightarrow{F=AX} F = \begin{Bmatrix} F_{11} & F_{12} & \cdots & F_{1m} \\ F_{21} & F_{22} & \cdots & F_{2m} \\ \vdots & \vdots & \cdots & \vdots \\ F_{n1} & F_{n2} & \cdots & F_{nm} \end{Bmatrix} \quad (5)$$

3.1.3. Filtering

The purpose of image filtering is to highlight useful spatial information and depress the noise of a single image using various filters [60]. Convolutions and morphology are two common filtering methods. The convolution filtering intensity depends on the parameter-setting transform kernels, and the morphology filtering is generally used for effectively eliminating the noise in single bands.

3.2. *RF-Based Classification Scheme and Prediction Model*

3.2.1. RF Background

Developed by Breiman [61], RF is a type of ensemble learning algorithm and is constructed by multiple decision trees. A decision tree is a typical supervised learning approach that can be used to categorize or regress something based on the data we have [62]. Classification and regression trees (CART), which is an important dichotomy algorithm, are used to generate binary decision trees [63]. Determining the optimal feature for splitting and providing a condition to stop splitting are two critical processes of tree generation. For the classification tree, the Gini coefficient (Gini) is used to measure the impurity of the node splitting, and the feature with the minimum Gini can be used for splitting in the generation of decision trees (Equations (6) and (7)). For the regression tree, the minimum squared error (MSE) is used for splitting in decision tree generation [63]. The Gini criterion for node splitting is defined as:

$$Gini\ (t) = 1 - \sum [p(c_k|t)]^2 \tag{6}$$

where $p(c_k|t)$ is the probability of the class c_k in the node t for a decision tree. There are two assemblies (D_L and D_R) corresponding to the left and right child nodes around the parent node, and the Gini after splitting can be defined as Equation (7):

$$Gini(D, A) = \frac{|D_L|}{|D|} Gini(D_L) + \frac{|D_R|}{|D|} Gini(D_R) \tag{7}$$

In general, two random processes, namely bootstrap aggregating (bagging) [64] and stochastic subspace [65], are employed to construct RF. These two processes can help to ensure the accuracy of every tree and effectively avoid its overfitting. More details on the generation procedure of the RF are given in Qin et al. [37].

3.2.2. RF-Based Classifier

Each sample has only one single attribute class, both for the case of binary- and multi-class classification, i.e., all the attribute classes of the sample set are separately and exclusively present in one sample. For a sample set with n $(1, 2 \ldots, n)$ attribute classes, it can be classified by n binary classifiers; every classifier has two classes, e.g., class 1 with classes $(2, 3 \ldots,$ and $n)$ or class 2 with classes $(1, 3 \ldots,$ and $n)$. In this way, one classifier can be learned for binary classification, while n classifiers will be learned for n-class problems from every training set.

The training and validation datasets are randomly determined using the bagging method from the sample dataset, and the ratio of these two sets is about 7 to 3 (i.e., 70% for training and 30% for validation). The RF-based classifier that was constructed based on multiple training sets will return a classification result based on the ratio of the votes provided by all the tree classifiers. In other words, the final attribute class is decided by the maximum of all the returned values (namely prediction probability) for every class.

The out-of-bag error (OBB error), F1 score, overall accuracy (OA), kappa coefficient (KC), and area under the receiver operating characteristic (ROC) curve are obtained from the generated confusion matrix based on the classification result and the validation dataset. These statistics can be used to evaluate the performance of the constructed classification and prediction model, and higher values indicate the higher prediction accuracy of the corresponding model [37,66]. The RF classifier can provide the relative importance of different features in the sample dataset, and this kind of importance value indicates their contribution to the decision tree, and thus, the correlation of every feature with the attribute class could be analyzed using other statistical methods.

3.3. *Sample-Improved WofE Method*

Weight of evidence (WofE), a multivariate statistical approach and fusion method based on probabilistic uncertainty and Bayes theorem, was developed for spatial correlation analysis and posterior probability prediction in mineral prospectivity mapping [67–69]. In the WofE analysis, the samples D (e.g., the MG occurrence) are used as training points, the geological factors that are related to the samples are used as evidential factors, and these themes should be generated as the grid file with a given unit cell size. In the study area T, the number of the grid cell is marked as N, and the prior probability of the sample occurrence is defined by Equation (8).

$$P\{D\} = \frac{N(D)}{N(T)} \tag{8}$$

According to the theorem, the conditional probability of the sample occurrence with the appearance of evidential factor B_j ($j = 1, 2 \ldots, n$) can be written as Equation (9):

$$P\{D|B_j\} = \frac{P\{D \cap B_j\}}{P\{B_j\}} \tag{9}$$

The positive and negative weights of the sample occurrence are defined as Equation (10) and Equation (11):

$$W_j^{+} = \ln \frac{P\{B_j|D\}}{P\{B_j|\overline{D}\}} \tag{10}$$

$$W_j^{-} = \ln \frac{P\{\overline{B_j}|D\}}{P\{\overline{B_j}|\overline{D}\}} \tag{11}$$

where the positive W_j^{+} and negative W_j^{-} indicates that the occurrence of sample D is positively related to the evidence B_j; otherwise it has a negative correlation. In addition, this degree of correlation can be measured with the contrast (C), in which a larger positive C value means a greater positive correlation. For the evidence B_j, its Cj is calculated by Equation (12):

$$C_j = W_j^{+} - W_j^{-} \tag{12}$$

In conventional WofE analysis, all the samples are abstracted as the training points regardless of their spatial size. This process is able to reduce the number of sample occurrences and directly affects the correlation based on probability analysis between samples and evidential factors. Therefore, the areas of the sample occurrence are firstly identified and then grided into the same cell size as other factors of the study area. In this way, every sample area is converted into a certain number of training points for spatial correlation analysis (Figure 4). In addition, this approach is also suitable for improving samples to train machine-learning-based prediction models.

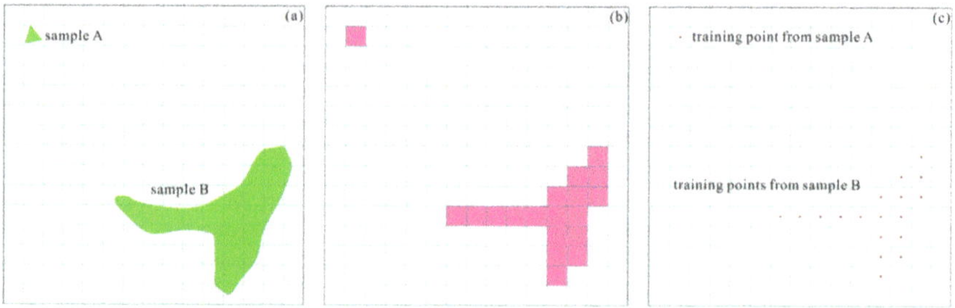

Figure 4. Improving the process of GM sample occurrence, (**a**) show the occurrence areas (vector), (**b**) are their grid form (raster), and (**c**) are the training points converted from samples.

4. Results

4.1. Land Cover Mapping

The study area encompasses 10,312,500 grid cells with a size of 2 × 2 m. In this study, RF classifier, as a non-parametric supervised machine learning algorithm, is employed for land cover mapping. The ground truth samples were determined based on GF-1/PMS (the year 2020) true-color image (TCI), composed of bands 3 (R), 2 (G), and 1 (B) (Table 2) based on the field disaster and land cover survey. The ground truth samples were randomly divided into two sets, i.e., training and validation sets, with a 7 to 3 ratio. Figure 5 shows the different ground truth land cover classes in the training and validation sets. As shown in Table 3, the training and validation sets occupy about 9.58% of the entire study area.

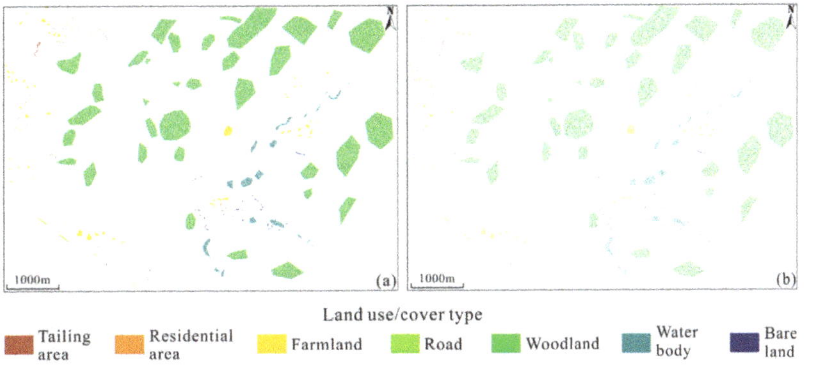

Figure 5. Ground truth datasets for land cover mapping: (**a**) training set and (**b**) validation set.

Table 3. Ground truth sample composition for land cover mapping.

Classification	No. of Samples for Training	Validation	Sample Proportion In Sample Set	In Study Area
Tailing area	1258	545	0.183	0.017
Residential area	3252	1412	0.472	0.045
Farmland	22,904	9618	3.293	0.315
Road	2718	1164	0.393	0.038
Woodland	621,245	266,578	89.902	8.609
Water body	35,801	15,346	5.179	0.496
Bare land	4108	1594	0.577	0.055

Aiming for precise land classification, four kinds of factors were considered for generating the classification dataset (Figure 6): (1) the SARVI calculated by Equation (1) and the vegetation and no-vegetation areas are distinguished in Figure 6a; (2) the first component (PC-1) of the PCA using bands 1, 2, 3, and 4 includes 87.42% of the eigenvalue (Figure 6b); (3) the HSV space image was generated by Munsell HSV transformation from the pseudo color image (PCI) composed of bands 3 (R), 2 (G), and 1 (B), and it can facilitate identifying soil and rocks as bare lands (Figure 6c); (4) the useless information of the TCI is depressed by convolution filtering, which helps distinguish between the land cover classes in the filtered image (Figure 6d).

The RF-based land cover classification model is constructed with the parameter of 168 trees and three randomly selected features within EnMap-Box [70]. The performance parameters and variable importance can be calculated by applying the constructed model to the validation set. Table 4 shows the obtained confusion matrix based on the classification result and validation set. The number of correctly classified grid cells in each class is displayed in bold on the diagonal matrix. The minimum F1 score calculated from this matrix is 92.28%, pointing to the remarkable performance of the classification model. The high OA of 99.69% and KC of 98.37% suggest that this RF-based model can be successfully used for classification.

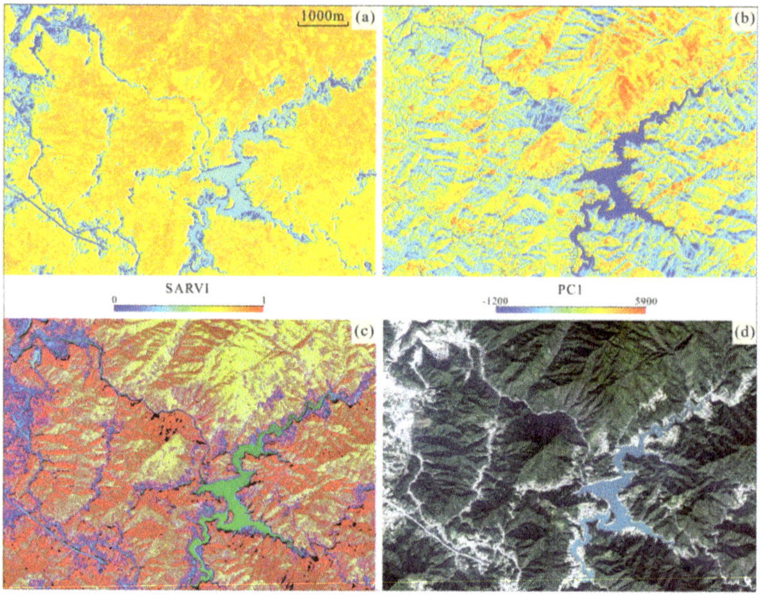

Figure 6. Determined classification factors for land cover mapping: (**a**) SARVI, (**b**) PC-1, (**c**) HSV_PCI, RGB color image from HSV space transformation (bands 4, 3, and 1), and (**d**) CF_TCI, RGB color image from convolution filtering (bands 3, 2, and 1).

Table 4. Accuracy assessment of RF-based land cover classification model.

Class	Confusion Matrix (No. of Grid Cells)								F1 Score (%)
	Class 1	Class 2	Class 3	Class 4	Class 5	Class 6	Class 7	Sum	
Tailing area	**536** *	1	0	2	0	0	6	545	98.80
Residential area	0	**1333**	1	14	0	4	60	1412	95.01
Farmland	0	7	**9356**	5	138	0	112	9618	96.33
Road	1	3	0	**1131**	0	0	29	1164	97.16
Woodland	0	22	439	0	**266,110**	1	6	266,578	99.89
Waters	0	7	0	0	0	**15,339**	0	15,346	99.96
Naked land	3	21	10	12	0	0	**1548**	1594	92.28
Sum	540	1394	9806	1164	266,248	15,344	1761	**296,257**	-

* Diagonal number highlighted in bold indicates the correctly classified cells.

The raw variable importance can indicate its contribution to the generation of every class. It can be seen that the filtering process on the TCI is most favorable for the identification of different land covers, while SARVI comes second, and HSV transformation also performs rather well (Figure 7).

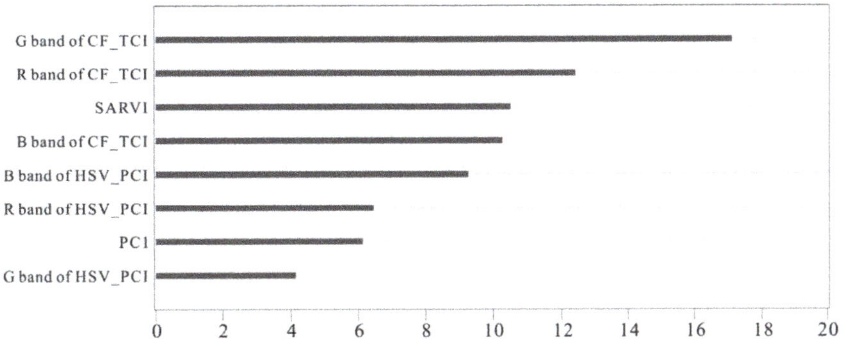

Figure 7. Contribution ranking of each classification factor to the RF-based classifier.

Finally, by applying the RF-based constructed model to the whole dataset, the classification result of the seven-class land cover map is presented in Figure 8. In this study area, the woodland has the highest proportion of up to 81.96%, farmland occupies 8.86%, and the other classes range from 1.19% to 3.42%, except for the tailing area (0.07%). This result is highly consistent with what has been observed in the recent field survey.

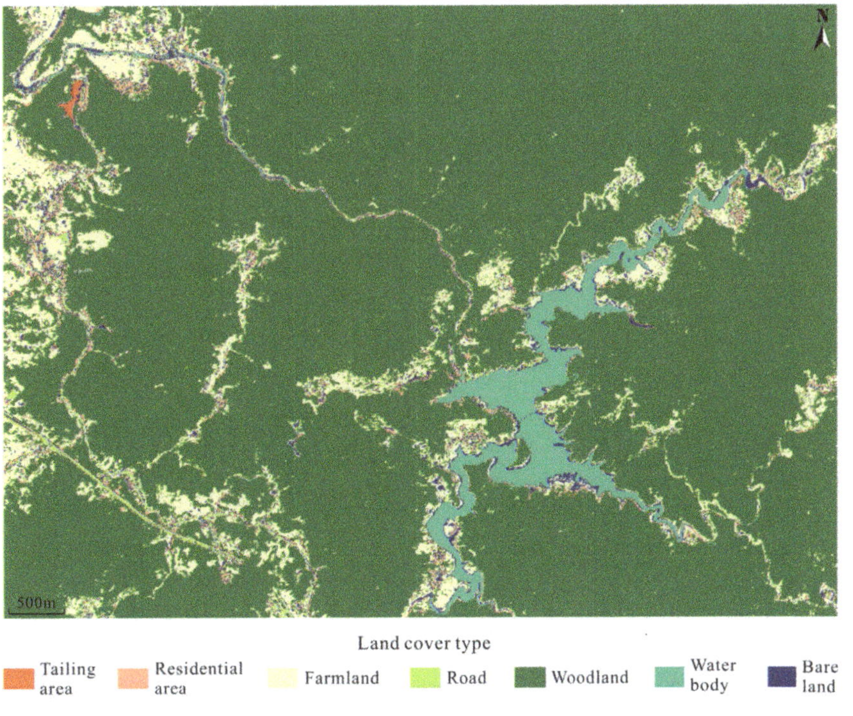

Figure 8. The obtained RF-based 2 m resolution land cover map of the study area using GF-1/PMS imagery.

4.2. Mining-Induced Geo-Hazard Mapping (MGM)

The actual distribution of the MG occurrences, which is obtained by a large amount of detailed fieldwork, is used as the positive samples, and the places with no MG occurrences are determined as the negative samples. It is important to note that the negative samples should be evenly selected in every land cover class and approximately equal to the positive samples. Here, 24,570 samples, containing 17,126 training samples and 7444 validation samples, are used for training and testing the RF-based prediction model (Figure 9).

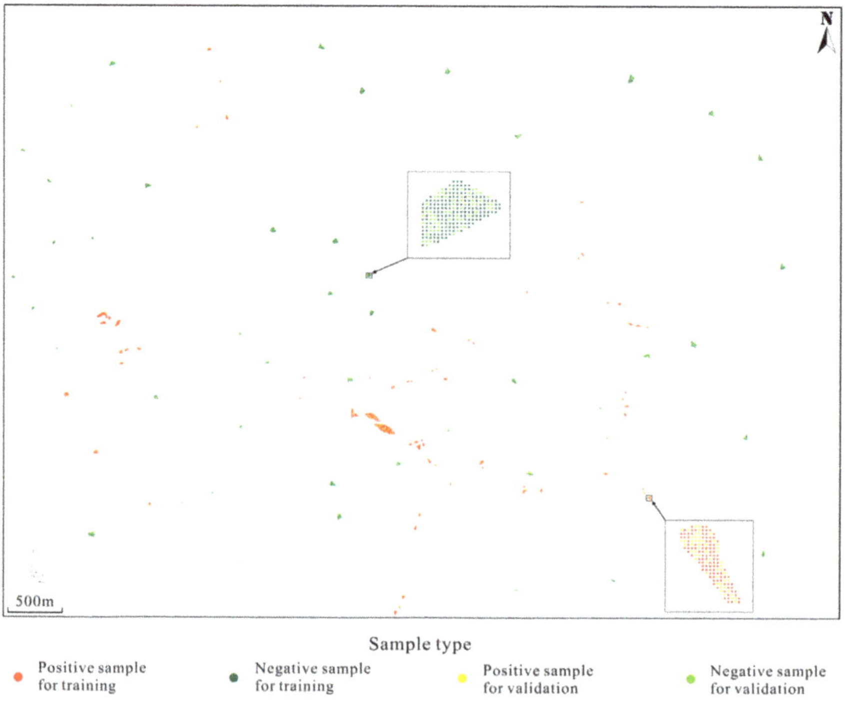

Figure 9. Spatial distribution of the acquired positive and negative samples in the training and validation sets for construction of RF-based prediction model for MGM of the study area.

Under the guidance of experts and former field investigations, stratigraphic lithology, geological structure, topographical features, road distribution, and rainfall rates are usually used as the predictive factors for MGM. There is no need for information on rainfall because the study area is only about 41 km^2, with no variation in rainfall. In addition, environmental factors related to mining activities should be considered as well as the different spectral information of the surface features. Accordingly, the eight predictive factor layers are determined as follows:

(1) **Lithology**: the lithology layer with twelve types of lithological information is generated from the geological map (Figure 2). Different lithology of the strata possesses different physical structures, resulting in different degrees of weathering and fragmentation.

(2) **Land cover map**: based on GF-1/PMS data and RF classifier, a land cover map was produced (Section 4.1), and this factor layer is shown in Figure 8.

(3) **Structure**: geological structure, especially the faults, has a strong relationship with MG. As Figure 2 shows, the identified structures are only distributed around mining areas, so the detailed structures of the whole study area need to be reinterpreted. Here, the three-dimension (3D) terrain surface is modeled using triangulated irregular network (TIN) and discrete smooth interpolation (DSI) within GOCAD platform based on a topographic map of 1:2000 on scale. Simultaneously, the noise of the spatial-resolution-improved multispectral bands (1, 2, 3, and 4) are depressed by morphological filtering, and the PC-1 of PCA that is carried out on the filtered result can be used to generate new PCI combining with the other two original bands. Finally, the TCI and two PCIs (enhanced in ENVI) are

displayed on a 3D terrain model within MICROMINE (Figure 10). In this way, the faults are easily extracted from these 3D displays through visual interpretation with the help of geologic recognition. The MG occurrences are associated with the distance to faults, and thus, the buffer zones of faults are constructed using three buffer radii of 10, 20, and 30 m, and the distance that is greater than 30 m is set to a value of 999 (Figure 11a) because this distance interval does not affect the MG occurrence under normal circumstances in the study area.

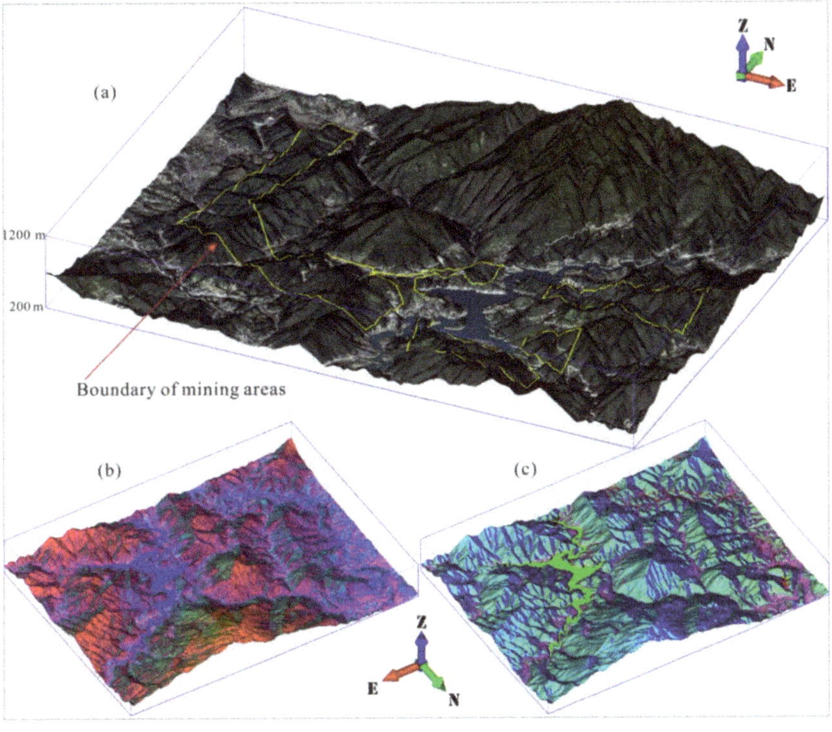

Figure 10. The 3D visualization of the terrain surface, showing the TCI (**a**) and two enhanced PCI (**b**,**c**) from GF-1/PMS imagery.

(4) **Elevation difference**: underground excavation, e.g., tunnels, stopes, and blasting area, will change the stability of strata in the mining areas, and this may lead to surface deformation. The minimum height difference between the surface and the underground mining sites is calculated from the field survey data (Figure 11b).

(5) **Aspect and slope**: these two property values from topographic features have been proven to be useful for the assessment of MG [27,46,71]. Constructed 3D terrain model can be transformed into a digital elevation model (DEM), and then the aspect and slope of every grid cell can be calculated from the DEM in ArcGIS (Figure 11c,d).

(6) **Difference netween the PC-1 and SARVI**: as mentioned before, most information of the multispectral bands can be presented in PC-1 using PCA. The SARVI is conducive to distinguish vegetation greenness between different land cover classes, and their difference from different years indicates the changes of the terrain surface. The GF-1/PMS data in the same acquisition phase of 2015 and 2020, in which spatial resolution is improved to 2m with the panchromatic band, are used to calculate SARVI and PC-1, and the difference between these two indexes is shown in Figure 11e,f.

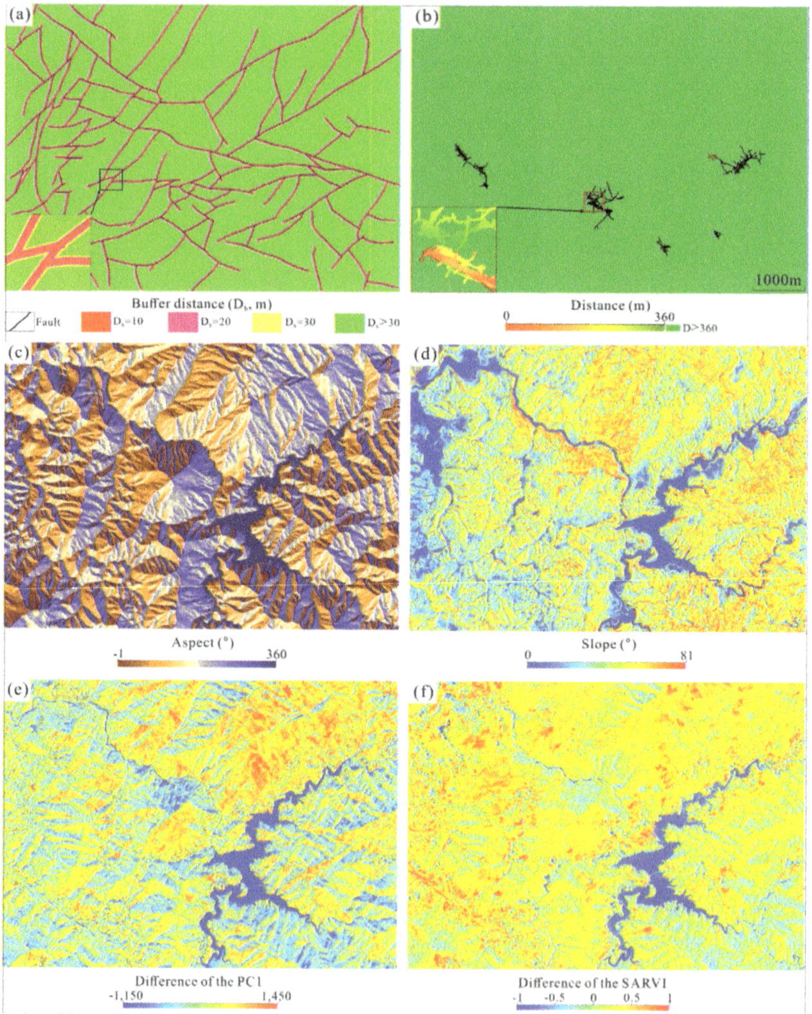

Figure 11. Predictive factor layers: (**a**) distance buffers of the structure, (**b**) difference in elevation between underground excavation and surface, (**c**,**d**) aspect and slope, and (**e**,**f**) difference between the PC-1 and SARVI.

The pixel-based values of every predictive factor layer with the samples layer were extracted as the data vector from their respective raster layers, and then these vectors were combined into a matrix, the dataset for training and prediction consisting of 10,312,500 rows and 9 columns in R. The RF-based prediction model was constructed using positive and negative sample sets (Figure 9) in the data matrix with optimal parameters, i.e., 108 trees and three randomly selected features. Meanwhile, its out-of-bag error (OOB Error) is 1.80%, which indicates a good classification performance. By applying the constructed model to the validation set, 3696 negative samples out of 3796 were correctly classified and 3705 positive samples out of 3722 were correctly predicted. Accordingly, the OA of 98.53% and KC of 97.06% were calculated. In addition, the acquired high AUC value of 0.998 suggests that this constructed model has high performance for MGM in this study.

The constructed RF-based prediction model was applied back to the whole data matrix, and every row returned a probability (P) value of classification, containing positive and negative classes. The returned positive class can be considered as the prediction result of MG occurrence probability or susceptibility. The whole dataset is ranked according to the probability values from high to low, and the cumulative percentage of the predictive cells and predicted sample cells can be calculated. Then, the prediction efficiency curve (PEC) and prediction probability curve (PPC) can be plotted (Figure 12). Three thresholds of 1, 2, and 3 were determined on the PEC (Figure 12a), and their corresponding probability values were 90.59%, 77.26%, and 50.20%, respectively (Figure 12b). According to these three thresholds, the whole study area, relative to the occurrence of MG, was divided into four susceptibility classes consisting of high, middle, low, and stable (Figure 13b and Table 5). For the high-susceptibility areas, 2.82% of the total grid cells hold 85.60% of the disaster samples. The stable areas occupy 79.79% of the study area, containing almost no disaster sample.

Table 5. Zonation of the MGM.

Susceptibility Class	Probability Interval (P, %)	Proportion of the Predictive Data (%)	Proportion of the Samples (%)	Occurrence Rate of the Samples (%)
High	$P \geq 90.59\%$	2.82	85.60	7.23
Middle	$90.59\% > P \geq 77.26\%$	5.28	8.08	0.36
Low	$77.26\% > P \geq 50.20\%$	12.11	6.07	0.12
Stable	$P < 50.20\%$	79.79	0.25	0

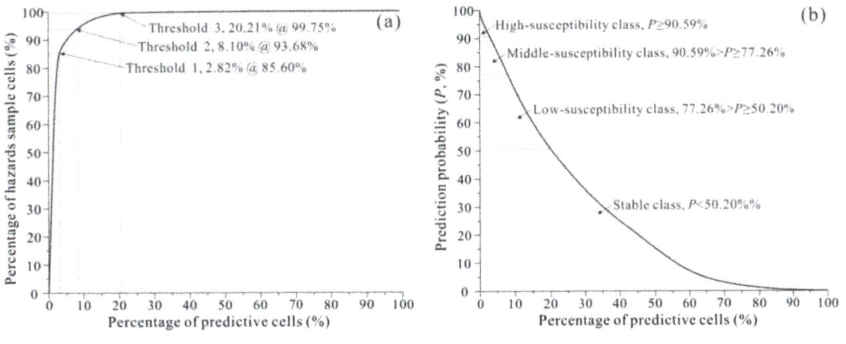

Figure 12. Analysis of capture-efficiency curve (**a**,**b**) prediction probability curve for zonation of the MGM.

By qualitatively comparing the terrain surface feature (Figure 10a), land cover map (Figure 8), and MGM (Figure 13), it can be seen that the probability distribution of the MG occurrence is closely related to the places of human activities, such as road excavation, residential area, and mining areas (Figure 13a). In particular, the high susceptibility areas to MG are distributed near the surface excavation and mining areas (Figure 13b).

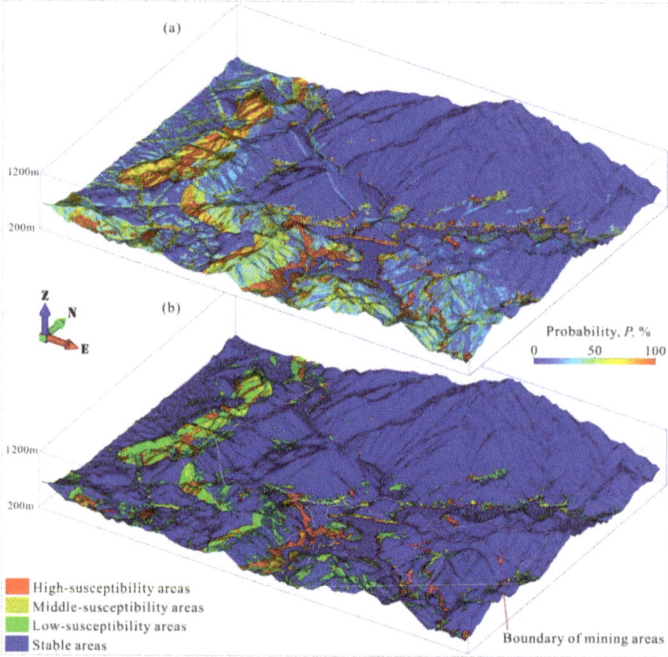

Figure 13. The 3D display of the MGM, showing (**a**) the probability distribution and (**b**) zonation of the MG susceptibility areas.

5. Discussion

5.1. Importance of the Feature Variable

Variable importance is regarded as the contribution to tree node splitting in the generation of the RF-based prediction model, i.e., the contribution of the predictive factors to sample occurrence. The mean decrease accuracy (MDA) and mean decrease Gini (MDG) are two common measures for estimating the variable importance of the RF model. The MDA rankings are more stable than those using MDG, although the higher value of these two indexes indicates the greater contribution of the factor to model construction [72]. Based on the performed importance ranking (Figure 14), we know that the lithology of the strata and the land cover map contributed to the occurrence of the MG more than the other six factors. The faults and underground excavation have been regarded as the critical ones for causing MG, but this result is contrary to our common sense. This highlights the need to quantitatively analyze the correlation of every factor with MG.

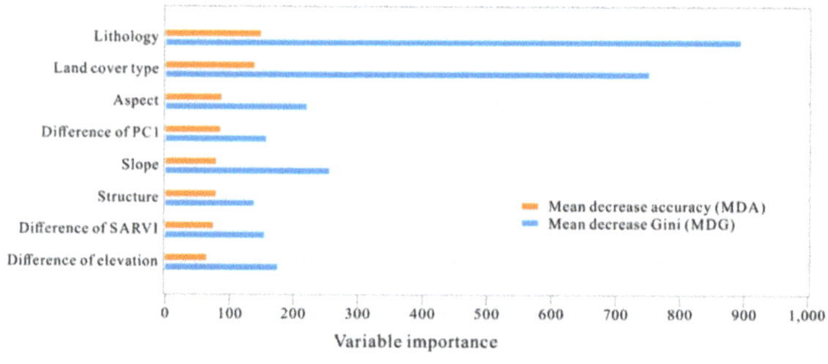

Figure 14. Ranking of predictive factors' contribution to RF-based prediction modeling.

5.2. Correlation of the Predictive Factors with MG Occurrence

Every predictive factor was divided into different intervals with its property categories (e.g., lithology of the strata, distance buffers of the faults, and the land cover map) or property value (e.g., elevation difference between the terrain surface and the underground excavation, SARVI difference, PC-1 difference, aspect, and slope). Then, the correlation indexes of every interval with MG occurrence, including positive and negative weights (W^+ & W^-) as well as the contrast (C), were calculated by the WofE method and presented in Figure 15.

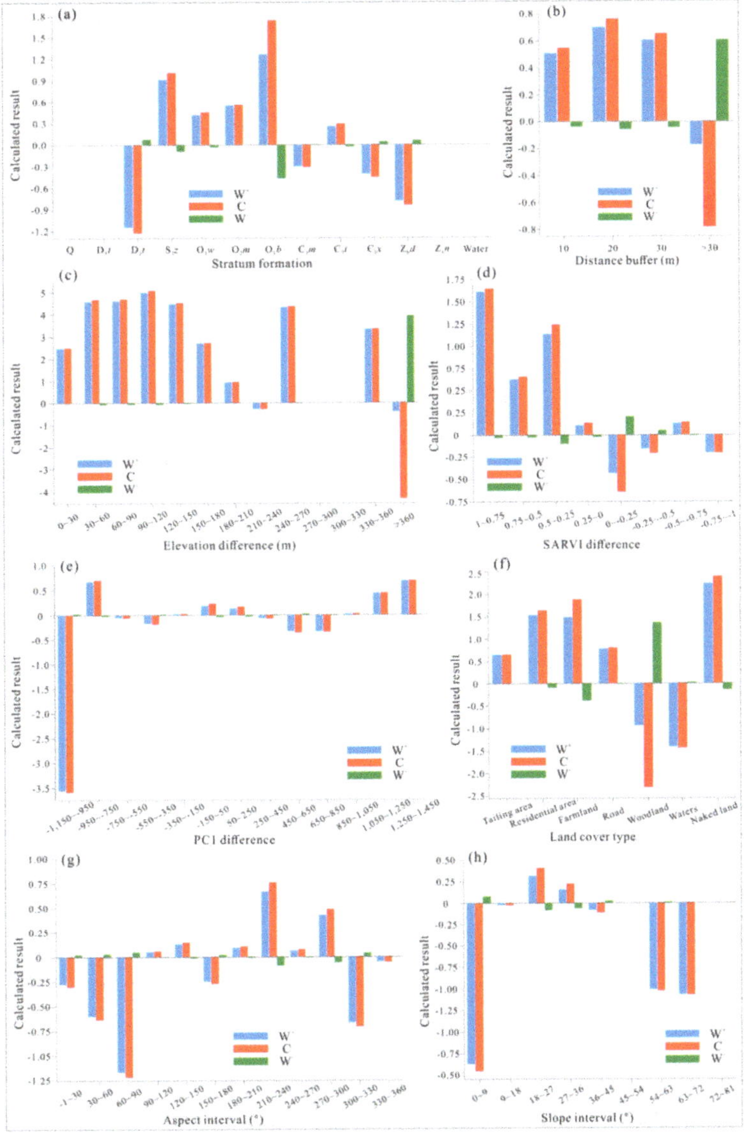

Figure 15. Calculated results of WofE, showing the different predictive factors: (**a**) lithology of the strata, (**b**) distance buffers of the faults, (**c**) elevation difference between the underground excavations and the surface, (**d**) SARVI difference, (**e**) PC-1 difference, (**f**) land cover, (**g**) aspect, and (**h**) slope.

The factors such as lithology and land cover map that are defined for MGM are closely related to MG occurrence (Figure 15). To be specific, in Figure 15a, the calculated W^+ and C values of the Baishuixi Formation are positive and the highest, followed by the Zhoujiaxi, Modao, and Wufeng Formation, illustrating that the stratum holding the shale and sandstone with intercalated carbon-bearing mudstones is the main geological disaster-bearing body. In Figure 15f, for bare land, farmland, residential area, road, and tailing area, their calculated W^+ and C values are all positive and greater than those for woodland and waters, showing a clear correlation between human activities and MG occurrence.

In Figure 15b,c, the generated buffers of the faults and the elevation difference between terrain surface and underground tunnels almost obtained greater W^+ and C than extremum areas. The buffer distance was more than 30 m, and the elevation difference was more than 360 m, suggesting that these two predictive factors are all in favor of MGM. The SARVI difference in the intervals between 0.25 and 1 showed a close correlation between the decreasing vegetation cover and MG occurrence (Figure 15d). In addition, a higher value of PC-1 difference indicates the increasing probability of the MG occurrence (Figure 15e). Figure 15g,h shows that the MG easily occurred in the surface areas with the features of aspect from 210° to 240° and 270° to 300° and slope from 18° to 36°. Comparative analysis of the results with the later field investigation showed that the aspect and slope of these areas are essentially in agreement with the spatial patterns of the strata outcrop. To sum up, the determination of the eight predictive factors above is reasonable and necessary for MGM.

5.3. MG Monitoring and Pre-Warning

The main purpose of MGM is to monitor and predict the occurrence of future MG, and this work should be continuously focused on the predicted high-susceptibility areas. In addition to monitoring the ground deformation and subsidence using professional GPS equipment and technology, the following precursor information should be captured by visual inspection for MG early-warning: (1) surface and underground excavation, (2) storage or flow changes of water, (3) suddenly bent trees and new fissures or bulges on the ground. Geologically, more attention should be paid to the spatial patterns of stratigraphic formation, especially for places that are highly consistent with the natural or side slope.

6. Conclusions

After more than half a century of mining activities in Liaojiaping Orefield, a series of mining-induced geo-disasters (MG) have been reported. One of the most effective strategies for managing and controlling MG in these mining areas is to identify and map their susceptibility. For this purpose, Gaofen-1 high-resolution satellite images, along with environmental factors identified through geological exploration and topographic survey, were used for mining-induced geo-hazard mapping (MGM) in Liaojiaping Orefield for the first time. RF classifier was used to model the relationship between environmental factors and actual MG events during the MGM, as well as to produce a land cover map. The main findings of this study are summarized as follows:

(1) Using Gaofen-1 high-resolution data, both RF-based binary and multi-class classifiers achieved good performance in land cover mapping and MGM. Some land cover types, e.g., tailing disposal sites, excavated sites, and MG, occupy a small land area. In such cases, a supervised learning algorithm can be used in tandem with high-resolution data to extract samples and detect ground targets.

(2) Based on variable importance analysis, the highest contribution to MGM is related to lithology and land cover among the observed environmental factors, which usually indicate the stability of geological bodies and should be employed to map the geo-disaster susceptibility. In addition, we are able to understand the contribution of variables to the risk modeling through importance analysis of variables; nevertheless, the quantitative analysis of the correlation between the geo-environmental factors and MG based on geostatistical

method will allow us to achieve a better understanding of their spatial correlation. In any cases, it is necessary and reasonable to involve all the predictive factors for MGM.

(3) Any changes in land cover, e.g., emerging excavation works and direct vegetation change or degradation, as well as rock bedding creeping in the high susceptibility areas need to be paid high attention to and shall be defined for monitoring and early-warning of MG.

Author Contributions: Y.Q. carried out the fieldwork, conducted the machine learning-based classification and prediction models, analyzed the results, and composed the draft of the paper; L.C. provided the preprocessed Gaofen-1 satellite imagery; A.D.B. revised this manuscript in terms of scientific and English writing; W.W. provided scientific recommendations for this research and revision of the paper. All authors have read and agreed to the published version of the manuscript.

Funding: This work was financially supported by the Start-up Fund for Scientific Research from the East China University of Technology (Grant No. DHBK2019040) and the Open-end Fund from the Key Laboratory of Digital Land and Resources of Jiangxi province (Grant No. DLLJ201901), both granted to Yaozu Qin.

Acknowledgments: The Anhua Xinfeng Mining Co., Ltd. of Hunan Province is acknowledged for providing financial and logistic support to the fieldwork and material collection of this research. Moreover, the constructive comments on this paper by two anonymous reviewers and timely editorial handling by Amiee Shi are highly appreciated.

Conflicts of Interest: We declare that we have no commercial or associative interest that represents a conflict of interest in connection with our work and this manuscript.

References

1. Marschalko, M.; Yilmaz, I.; Křístková, V.; Fuka, M.; Kubečka, K.; Bouchal, T. An indicative method for determination of the most hazardous changes in slopes of the subsidence basins in underground coal mining area in Ostrava (Czech Republic). *Environ. Monit. Assess.* **2013**, *185*, 509–522. [CrossRef]
2. Yang, Y.Y.; Xu, Y.S.; Shen, S.L.; Yuan, Y.; Yin, Z.Y. Mining-induced geo-hazards with environmental protection measures in Yunnan, China: An overview. *Bull. Eng. Geol. Environ.* **2015**, *74*, 141–150. [CrossRef]
3. Li, S.L. Study on the geological hazard in metal mines and its prevention countermeasures. *Chin. J. Geol. Hazard Control* **2002**, 46–50+54, (In Chinese with English Abstract)
4. Yi, H.P.; Jiang, Z. Causes and prevention measures of mine geological disasters. *Sci. Technol. Innov. Her.* **2010**, *4*, 126. (In Chinese)
5. Fan, L.M.; Li, C.; Chen, J.P.; Ning, J.M. Geological hazards and prevention technology in high-intensity mining area of mineral resources. *Sci. Press* **2016**, *28*, 8. (In Chinese)
6. Liu, L.; Chen, L.Q.; Tang, J.X. Present Situation and Future Prospects of Geologic Environment Issues in Mines in China. *Disaster Adv.* **2010**, *3*, 563–566.
7. Shao, L.F. Geological disaster prevention and control and resource protection in mineral resource exploitation region. *Int. J. Low-Carbon Technol.* **2019**, *14*, 142. [CrossRef]
8. Xu, Y.N. Investigation and research on the mine geological environment: Present status and outlook. *Geol. Bull. China* **2008**, *27*, 1235–1244. (In Chinese with English Abstract)
9. Chowdhury, R.; Flentje, P. Role of slope reliability analysis in landslide risk management. *Bull. Eng. Geol. Environ.* **2003**, *62*, 41–46. [CrossRef]
10. Romeo, R.W.; Floris, M.; Veneri, F. Area-scale landslide hazard and risk assessment. *Environ. Geol.* **2006**, *51*, 1–13. [CrossRef]
11. Huang, R.Q.; Li, W.L. Analysis of the geo-hazards triggered by the 12 May 2008 Wenchuan earthquake, China. *Bull. Eng. Geol. Environ.* **2009**, *68*, 363–371. [CrossRef]
12. Chen, K.T.; Wu, J.H. Simulating the failure process of the xinmo landslide using discontinuous deformation analysis. *Eng. Geol.* **2018**, *239*, 269–281. [CrossRef]
13. Wang, Q.J.; Guo, H.D.; Chen, Y.; Lin, Q.Z.; Li, H. Application of remote sensing for investigating mining geological hazards. *Int. J. Digit. Earth* **2013**, *6*, 449–468. [CrossRef]
14. Segoni, S.; Pappafico, G.; Luti, T.; Catani, F. Landslide susceptibility assessment in complex geological settings: Sensitivity to geological information and insights on its parameterization. *Landslides* **2020**, *17*, 2443–2453. [CrossRef]
15. Ahmad, H.; Chen, N.S.; Rahman, M.; Islam, M.M.; Pourghasemi, H.R.; Habumugisha, J.M. Geohazards Susceptibility Assessment along the Upper Indus Basin Using Four Machine Learning and Statistical Models. *ISPRS Int. J. Geo-Inf.* **2021**, *10*, 315. [CrossRef]
16. Westen, C.J.V.; Castellanos, E.; Kuriakose, S.L. Spatial data for landslide susceptibility, hazard, and vulnerability assessment: An overview. *Eng. Geol.* **2008**, *102*, 112–131. [CrossRef]
17. Lu, P.; Catani, F.; Tofani, V.; Casagli, N. Quantitative hazard and risk assessment for slow-moving landslides from persistent scatterer interferometry. *Landslides* **2014**, *11*, 685–696. [CrossRef]

18. Pavlova, I.; Makarigakis, A.; Depret, T.; Jomelli, V. Global overview of the geological hazard exposure and disaster risk awareness at world heritage sites. *J. Cult. Herit.* **2017**, *28*, 8445–8452. [CrossRef]
19. Vaziri, V.; Hamidi, J.K.; Sayadi, A.R. An integrated GIS-based approach for geohazards risk assessment in coal mines. *Environ. Earth Sci.* **2018**, *77*, 29. [CrossRef]
20. Mohammady, M.; Pourghasemi, H.R.; Amiri, M. Assessment of land subsidence susceptibility in Semnan plain (Iran): A comparison of support vector machine and weights of evidence data mining algorithms. *Nat. Hazards* **2019**, *99*, 951–971. [CrossRef]
21. Hemasinghe, H.; Rangali, R.S.; Deshapriya, N.L.; Samarakoon, L. Landslide susceptibility mapping using logistic regression model (a case study in Badulla District, Sri Lanka). *Procedia Eng.* **2018**, *212*, 1046–1053. [CrossRef]
22. Lee, H.; Oh, J. Establishing an ANN-Based Risk Model for Ground Subsidence Along Railways. *Appl. Sci.* **2018**, *8*, 1936. [CrossRef]
23. Youssef, A.M.; Pourghasemi, H.R.; Pourtaghi, Z.S.; Al-Katheeri, M.M. Landslide susceptibility mapping using random forest, boosted regression tree, classification and regression tree, and general linear models and comparison of their performance at Wadi Tayyah Basin, Asir Region, Saudi Arabia. *Landslides* **2016**, *13*, 839–856. [CrossRef]
24. Zhang, K.X.; Wu, X.L.; Niu, R.Q.; Yang, K.; Zhao, L.R. The assessment of landslide susceptibility mapping using random forest and decision tree methods in the Three Gorges Reservoir area, China. *Environ. Earth Sci.* **2017**, *76*, 405. [CrossRef]
25. Kim, K.D.; Lee, S.; Oh, H.J.; Choi, J.K.; Won, J.S. Assessment of ground subsidence hazard near an abandoned underground coal mine using GIS. *Environ. Geol.* **2006**, *50*, 1183–1191. [CrossRef]
26. Sharma, S.; Mahajan, A.K. A comparative assessment of information value, frequency ratio and analytical hierarchy process models for landslide susceptibility mapping of a Himalayan watershed, India. *Bull. Eng. Geol. Environ.* **2019**, *78*, 2431–2448. [CrossRef]
27. Zhou, X.T.; Wu, W.C.; Lin, Z.Y.; Zhang, G.L.; Chen, R.X.; Song, Y.; Wang, Z.L.; Lang, T.; Qin, Y.Z.; Ou, P.H.; et al. Zonation of Landslide Susceptibility in Ruijin, Jiangxi, China. *Int. J. Environ. Res. Public Health* **2021**, *18*, 5906. [CrossRef] [PubMed]
28. Cracknell, M.J.; Reading, A.M. Geological mapping using remote sensing data: A comparison of five machine learning algorithms, their response to variations in the spatial distribution of training data and the use of explicit spatial information. *Comput. Geosci.* **2014**, *63*, 22–33. [CrossRef]
29. Carranza, E.J.M.; Laborte, A.G. Random Forest predictive modeling of mineral prospectivity with small number of prospects and data with missing values in Abra (Philippines). *Comput. Geosci.* **2015**, *74*, 60–70. [CrossRef]
30. Li, X.H. Using "random forest" for classification and regression. *Chin. J. Appl. Entomol.* **2016**, *50*, 1190–1197.
31. Qin, Y.Z.; Liu, L.M. Quantitative 3D Association of Geological Factors and Geophysical Fields with Mineralization and Its Significance for Ore Prediction: An Example from Anqing Orefield, China. *Minerals* **2018**, *8*, 300. [CrossRef]
32. Wu, W.C.; Zucca, C.; Muhaimeed, A.S.; Alshafie, W.M.; Alquraish, A.M.F.; Nangia, V.; Zhu, M.Q.; Liu, G.P. Soil salinity prediction and mapping by machine learning regression in Central Mesopotamia. *Land Degrad. Dev.* **2018**, *29*, 4005–4014. [CrossRef]
33. Sun, T.; Li, H.; Wu, K.; Chen, F.; Zhu, Z.; Hu, Z. Data-Driven Predictive Modelling of Mineral Prospectivity Using Machine Learning and Deep Learning Methods: A Case Study from Southern Jiangxi Province, China. *Minerals* **2020**, *10*, 102. [CrossRef]
34. Cao, J.; Zhang, Z.; Du, J.; Zhang, L.; Song, Y.; Sun, G. Multi-geohazards susceptibility mapping based on machine learning—A case study in Jiuzhaigou, China. *Nat. Hazards* **2020**, *102*, 851–871. [CrossRef]
35. Magidi, J.; Nhamo, L.; Mpandeli, S.; Mabhaudhi, T. Application of the Random Forest Classifier to Map Irrigated Areas Using Google Earth Engine. *Remote Sens.* **2021**, *13*, 876. [CrossRef]
36. Choi, W.; Lee, H.; Kim, D.; Kim, S. Improving Spatial Coverage of Satellite Aerosol Classification Using a Random Forest Model. *Remote Sens.* **2021**, *13*, 1268. [CrossRef]
37. Qin, Y.Z.; Liu, L.M.; Wu, W.C. Machine Learning-Based 3D Modeling of Mineral Prospectivity Mapping in the Anqing Orefield, Eastern China. *Nat. Resour. Res.* **2021**. [CrossRef]
38. Youssef, A.M.; Pourghasemi, H.R. Landslide susceptibility mapping using machine learning algorithms and comparison of their performance at Abha Basin, Asir Region, Saudi Arabia. *Geosci. Front.* **2021**, *12*, 639–655. [CrossRef]
39. Shao, L.; Li, J. Geological hazards types induced by mining and their characteristics in Guizhou province. *Chin. J. Geol. Hazards Control.* **2011**, *22*, 56–60. (In Chinese with English Abstract)
40. Kong, F.J.; Li, X.B.; Hong, W.; Xie, D.F.; Li, X.; Bai, Y.X. Land cover classification based on fused data from gf-1 and modis ndvi time series. *Remote Sens.* **2016**, *8*, 741. [CrossRef]
41. Yang, C.; Wu, G.F.; Kai, D.; Shi, T.Z.; Li, Q.Q.; Wang, J.L. Improving land use/land cover classification by integrating pixel unmixing and decision tree methods. *Remote Sens.* **2017**, *9*, 1222. [CrossRef]
42. Akbari, E.; Boloorani, A.D.; Samany, N.N.; Hamzeh, S.; Soufizadeh, S.; Pignatti, S. Crop Mapping Using Random Forest and Particle Swarm Optimization based on Multi-Temporal Sentinel-2. *Remote Sens.* **2020**, *12*, 1449. [CrossRef]
43. Youssef, A.M. Landslide susceptibility delineation in the Ar-Rayth area, Jizan, Kingdom of Saudi Arabia, using analytical hierarchy process, frequency ratio, and logistic regression models. *Environ. Earth Sci.* **2015**, *73*, 8499–8518. [CrossRef]
44. Pachuau, L. Zonation of Landslide Susceptibility and Risk Assessment in Serchhip town, Mizoram. *J. Indian Soc. Remote Sens.* **2019**, *47*, 1587–1597. [CrossRef]
45. Arabameri, A.; Pradhan, B.; Rezaei, K.; Lee, C.W. Assessment of Landslide Susceptibility Using Statistical- and Artificial Intelligence-Based FR-RF Integrated Model and Multiresolution DEMs. *Remote Sens.* **2019**, *11*, 999. [CrossRef]

46. Zhang, Y.; Wu, W.C.; Qin, Y.Z.; Lin, Z.Y.; Zhang, G.L.; Chen, R.X.; Song, Y.; Lang, T.; Zhou, X.T.; Huangfu, W.C.; et al. Mapping Landslide Hazard Risk Using Random Forest Algorithm in Guixi, Jiangxi, China. *Int. J. Geo-Inf.* **2020**, *9*, 695. [CrossRef]
47. Ou, P.H.; Wu, W.C.; Qin, Y.Z.; Zhou, X.T.; Huangfu, W.C.; Zhang, Y.; Xie, L.F.; Fu, X.; Li, J.; Jiang, J.H.; et al. Assessment of landslide hazard in jiangxi using geo-information. *Front. Earth Sci. China* **2021**, *9*, 648342. [CrossRef]
48. Micheletti, N.; Foresti, L.; Robert, S.; Leuenberger, M.; Pedrazzini, A.; Jaboyedoff, M.; Kanevski, M. Machine Learning Feature Selection Methods for Landslide Susceptibility Mapping. *Math. Geosci.* **2014**, *46*, 33–57. [CrossRef]
49. Bai, Z.G. Technical characteristics of the Gaofen-1 Satellite. *Aerosp. China* **2013**, *8*, 5–9. (In Chinese)
50. Gao, H.L.; Gu, X.F.; Yu, T.; Sun, Y.; Xie, Y.; Liu, Q.Y. Validation of the calibration coefficient of the gaofen-1 pms sensor using the landsat 8 oli. *Remote. Sens.* **2016**, *8*, 132. [CrossRef]
51. Goetz, J.N.; Brenning, A.; Petschko, H.; Leopold, P. Evaluating machine learning and statistical prediction techniques for landslide susceptibility modeling. *Comput. Geosci.* **2015**, *81*, 1–11. [CrossRef]
52. Moosavi, V.; Niazi, Y. Development of hybrid wavelet packet-statistical models (WP-SM) for landslide susceptibility mapping. *Landslides* **2016**, *13*, 97–114. [CrossRef]
53. Dou, J.; Yunus, A.P.; Bui, D.T.; Merghadi, A.; Sahana, M.; Zhu, Z.F.; Chen, C.W.; Khosravi, K.; Yang, Y.; Pham, B.T. Assessment of advanced random forest and decision tree algorithms for modeling rainfall-induced landslide susceptibility in the Izu-Oshima Volcanic Island, Japan. *Sci. Total Environ.* **2019**, *662*, 332–346. [CrossRef]
54. Kalantar, B.; Pradhan, B.; Naghibi, S.A.; Motevalli, A.; Mansor, S. Assessment of the effects of training data selection on the landslide susceptibility mapping: A comparison between support vector machine (SVM), logistic regression (LR) and artificial neural networks (ANN). *Geomat. Nat. Hazards Risk* **2018**, *9*, 49–69. [CrossRef]
55. Schowengerdt, R.A. *Remote Sensing: Models & Methods for Image Processing*, 3rd ed.; Academic Press: Orlando, FL, USA, 2006.
56. Kaufman, Y.J.; Tanre, D. Atmospherically resistant vegetation index (ARVI) for eos-modis. *IEEE Trans. Geosci. Remote. Sens.* **1992**, *30*, 261–270. [CrossRef]
57. Huete, A.R. A soil-adjusted vegetation index (SAVI). *Remote Sens. Environ.* **1988**, *25*, 295–309. [CrossRef]
58. Escadafal, R.; Girard, M.C.; Courault, D. Munsell soil color and soil reflectance in the visible spectral bands of landsat mss and tm data. *Remote Sens. Environ.* **1989**, *27*, 37–46. [CrossRef]
59. Hotelling, H. Analysis of a Complex of Statistical Variables into Principal Components. *J. Educ. Psychol.* **1933**, *24*, 417–441, 498–520. [CrossRef]
60. Cichocki, A.; Amari, S. *Adaptive Blind Signal and Image Processing: Learning Algorithms and Applications*; Wiley: New York, NY, USA, 2002; p. 586.
61. Breiman, L. Random forests. *Mach. Learn.* **2001**, *45*, 5–32. [CrossRef]
62. Quinlan, J.R. Induction of Decision Trees. *Mach. Learn.* **1996**, *1*, 81–106. [CrossRef]
63. Breiman, L.; Friedman, J.H.; Olshen, R.A.; Stone, C.J. Classification and regression trees (cart). *Biometrics* **1984**, *40*, 358.
64. Breiman, L. Bagging predictors. *Mach. Learn.* **1996**, *24*, 123–140. [CrossRef]
65. Ho, T.K. The random subspace method for constructing decision forests. *IEEE Trans. Pattern Anal. Mach. Intell.* **1998**, *20*, 832–844.
66. Lee, D.H.; Kim, Y.T.; Lee, S.R. Shallow Landslide Susceptibility Models Based on Artificial Neural Networks Considering the Factor Selection Method and Various Non-Linear Activation Functions. *Remote Sens.* **2020**, *12*, 1194. [CrossRef]
67. Agterberg, F.P. Systematic approach to dealing with uncertainty of geoscience information in mineral exploration. In Proceedings of the 21st APCOM Symposium, Las Vegas, NV, USA, March 1989, Chapter 18, pp. 165–178.
68. Bonham-Carter, G.F.; Agterberg, F.P.; Wright, D.F. Weights of evidence modeling: A new approach to mapping mineral potential. *Geol. Surv. Can.* **1989**, *89*, 171–183.
69. Agterberg, F.P.; Bonham-Carter, G.F.; Cheng, Q.; Wright, D.F. Weights of evidence modeling and weighted logistic regression for mineral potential mapping. In *Computers in Geology—25 Years of Progress*; Oxford University Press, Inc.: Oxford, UK, 1993; pp. 13–32.
70. Waske, B.; van der Linden, S.; Oldenburg, C.; Jakimow, B.; Rabe, A.; Hostert, P. imageRF—A user-oriented implementation for remote sensing image analysis with random forests. *Environ. Modeling Softw.* **2012**, *35*, 192–193. [CrossRef]
71. Pradhan, B.; Lee, S. Landslide susceptibility assessment and factor effect analysis: Backpropagation artificial neural networks and their comparison with frequency ratio and bivariate logistic regression modelling. *Environ. Model. Softw.* **2010**, *25*, 747–759. [CrossRef]
72. Nicodemus, K.K. Letter to the Editor: On the stability and ranking of predictors from random forest variable importance measures. *Brief. Bioinform.* **2011**. [CrossRef]

remote sensing

MDPI

Article

Mining and Restoration Monitoring of Rare Earth Element (REE) Exploitation by New Remote Sensing Indicators in Southern Jiangxi, China

Lifeng Xie [1], Weicheng Wu [1,*], Xiaolan Huang [1], Penghui Ou [1], Ziyu Lin [1], Zhiling Wang [2], Yong Song [2], Tao Lang [2], Wenchao Huangfu [1], Yang Zhang [1], Xiaoting Zhou [1], Xiao Fu [1], Jie Li [1], Jingheng Jiang [1], Ming Zhang [1], Zhenjiang Zhang [1], Yaozu Qin [1], Shanling Peng [1], Chongjian Shao [1] and Yonghui Bai [1]

1 Key Laboratory of Digital Lands and Resources and Faculty of Earth Sciences, East China University of Technology, Nanchang 330013, China; 201810705007@ecut.edu.cn (L.X.); 201810705009@ecut.edu.cn (X.H.); 201810818013@ecut.edu.cn (P.O.); zylin@ecut.edu.cn (Z.L.); 201810818004@ecut.edu.cn (W.H.); 201810818002@ecut.edu.cn (Y.Z.); 201900818004@ecut.edu.cn (X.Z.); 201910705015@ecut.edu.cn (X.F.); 201910705016@ecut.edu.cn (J.L.); 201910818023@ecut.edu.cn (J.J.); 201910818003@ecut.edu.cn (M.Z.); 201910705005@ecut.edu.cn (Z.Z.); qyz60010@ecut.edu.cn (Y.Q.); pshanling@ecut.edu.cn (S.P.); scj350936@ecut.edu.cn (C.S.); 201960070@ecut.edu.cn (Y.B.)

2 264 Geological Team of Jiangxi Nuclear Industry, Ganzhou 341000, China; deva84@163.com (Z.W.); songy_6611@163.com (Y.S.); ecitlangtao@163.com (T.L.)

* Correspondence: wuwch@ecut.edu.cn or wuwc030903@sina.com; Tel.: +86-189-7084-7278

Received: 9 September 2020; Accepted: 24 October 2020; Published: 30 October 2020

Abstract: Rare earth elements (REEs) are widely used in various industries. The open-pit mining and chemical extraction of REEs in the weathered crust in southern Jiangxi, China, since the 1970s have provoked severe damages to the environment. After 2010, different restorations have been implemented by various enterprises, which seem to have a spatial variability in both management techniques and efficiency from one mine to another. A number of vegetation indices, e.g., normalized difference vegetation index (NDVI), soil-adjusted vegetation index (SAVI), enhanced vegetation index (EVI) and atmospherically resistant vegetation index (ARVI), can be used for this kind of monitoring and assessment but lack sensitivity to subtle differences. For this reason, the main objective of this study was to explore the possibility to develop new, mining-tailored remote sensing indicators to monitor the impacts of REE mining on the environment and to assess the effectiveness of its related restoration using multitemporal Landsat data from 1988 to 2019. The new indicators, termed mining and restoration assessment indicators (MRAIs), were developed based on the strong contrast of spectral reflectance, albedo, land surface temperature (LST) and tasseled cap brightness (TCB) of REE mines between mining and postmining restoration management. These indicators were tested against vegetation indices such as NDVI, EVI, SAVI and generalized difference vegetation index (GDVI), and found to be more sensitive. Of similar sensitivity to each other, one of the new indicators was employed to conduct the restoration assessment of the mined areas. Six typically managed mines with different restoration degrees and management approaches were selected as hotspots for a comparative analysis to highlight their temporal trajectories using the selected MRAI. The results show that REE mining had experienced a rapid expansion in 1988–2010 with a total mined area of about 66.29 km^2 in the observed counties. With implementation of the post-2010 restoration measures, an improvement of varying degrees in vegetation cover in most mines was distinguished and quantified. Hence, this study with the newly developed indicators provides a relevant approach for assessing the sustainable exploitation and management of REE resources in the study area.

Keywords: REE mines; mining and restoration assessment indicators (MRAIs); damage; time trajectory; effectiveness of management

1. Introduction

As a strategic resource, rare earth elements (REEs) are essential for high-tech applications and development across various industries [1–5]. With the increased demand for REE resources in the national and international markets, the scale of REE mining has been expanding since 1970s, and it has become one of the leading industries in southern Jiangxi, the origin of two important rivers, the Ganjiang and the Dongjiang (Figure 1). Due to their particular metallogenic character, the ion-adsorption type REE deposits are mostly hosted in the weathered crust of granitic massifs. Their mining is hence different from that of most other metal ore deposits, and the low-cost open-pit mining has led to severe damages to the ecosystem [6]. Actually, 4 decades of disordered exploitation and mining without relevant environmental protection have caused not only serious land degradation, e.g., vegetation stripping, soil erosion, nutrients losses, damage to the cultivated land and river courses and groundwater pollution, but also waste of REE resources in the mining areas. It is thus of pressing importance to implement effective measures to restore the damaged land cover and protect the environment of the mined areas in southern Jiangxi.

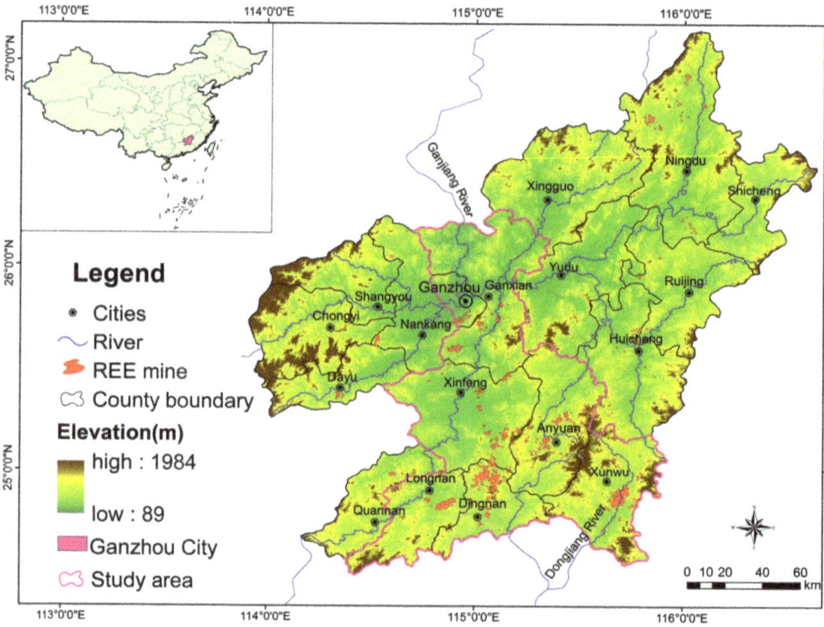

Figure 1. Location of the study area in southern Jiangxi (Ganzhou City) and distribution of the rare earth element (REE) mines.

The purpose of restoration is to recover the normal functions of the devastated ecosystems as a consequence of mining [7–10]. For this reason, the Chinese central and local governments have promulgated orders since 2009 to mitigate the degradation and restore ecosystem functions of the REE mining areas. With this effort, the environment has been improved to a certain degree in some mined areas, while degradation continues in others. However, neither systematic monitoring of the recovery status nor assessment on the effectiveness of different management approaches or techniques has been conducted.

Remote sensing has been widely applied in environmental monitoring, including in land degradation analysis, assessment of the mitigation effectiveness and, particularly, assessment of the policy impacts on the environment [11–16]. Since vegetation degradation and recovery are related to change in surface albedo [17], it is possible to use vegetation greenness, surface brightness and their change vector [18] to assess the effectiveness of such management of the mining areas [19,20]. As a topic increasingly recognized as important, research on the human–environment linkages has become a hotspot [21–26], and our study on assessment of mining, degradation and restoration is such a case.

Multitemporal remote sensing data are an important geo-information source which allow us to monitor and determine the dynamic trends of vegetation cover [27–36]. In recent years, a number of studies have been focused on monitoring of land cover change in the mining areas using vegetation indices, e.g., normalized difference vegetation index (NDVI) [37], soil-adjusted vegetation index (SAVI) [38] and enhanced vegetation index (EVI) [32,39,40]. Other studies have used thermal information to monitor coal fires [41] or radar data to examining land subsidence of coal mines [42]. It is worthwhile to mention that several authors have employed remote sensing to assess the rehabilitation or revegetation states of the mined areas [43] and have applied unmanned aerial vehicles (UAVs) for this purpose [44–46]. It is good to see that a limited number of studies were devoted to monitoring of REE mines [6,47,48]; however, none of them focused specifically on a comprehensive assessment of spatiotemporal characteristics of land degradation and restoration of the mining areas in southern Jiangxi. For this reason, the main objective of our study was to conduct monitoring and assessment research to fill this gap and to provide relevant advice for local governments for restoration management.

While examining the recovery degree of different REE mines using the above-mentioned vegetation indices, some difficulty was encountered in revealing the subtle difference of restoration, even with the generalized difference vegetation index (GDVI), which is more sensitive and of wider dynamic range than other indices in low-vegetation areas [49,50]. Nevertheless, it might be possible to incorporate GDVI with other biophysical indicators such as surface albedo [17], land surface temperature (LST) [25,51–53] and tasseled cap brightness (TCB) [54] to derive a more sensitive indicator to characterize the subtle differences of the REE mines and their restoration. Therefore, one specific objective of this research was to develop an integrated remote sensing indicator for achieving the monitoring and assessment of REE mines while taking multiple biophysical indicators into account.

2. Materials and Methods

2.1. Study Area

The study area, Ganzhou City, which encompasses 18 counties with 39,363 km^2 in surface area, is located in southern Jiangxi (Figure 1) and climatically belongs to the subtropical monsoon climatic zone. This area is mainly covered with forests which are partially conifers (*Cunninghamia lanceolata, Pinus massoniana*) and mainly mixed forests with the following dominant species: *Schima superba, Cinnamomum camphora, Acer palmatum,* Nothofagaceae Kuprian, *Liquidambar formosana, Pinus elliottii, Pinus massoniana* and *Cunninghamia lanceolata.* Other land cover types include woodlands, croplands, orchards, water-bodies and artificial land, where landscape is mainly mountains and hills interbedded with basins. The main river is Ganjiang River running north and its tributaries and subtributaries. The study area is also the origin of the Dongjiang River running south. The average annual precipitation is around 1318 mm and the annual mean temperature is 19.8 °C. The elevation ranges from 89 to 1971 m with an average of 300–500 m. With concentrated rainfall in spring and summer (March–July), soil erosion and water loss are a severe problem, especially in the red soil areas. Through this study, 1281 REE mines were identified with a total mined area of about 79 km^2 in the whole southern Jiangxi. However, our research was mainly focused in seven counties (Figure 1) including 1158 mines covering an area of 66.29 km^2.

In terms of natural resource endowment, apart from the tungsten ore reserve, southern Jiangxi has abundant ion-absorption type REE deposits and is known as the "Rare Earth Kingdom". However, due to long-term disordered mining in this area, the native vegetation cover has been severely damaged and the original geomorphologic feature has greatly changed since the 1970s. The early mining techniques included the "pool-leaching" or "heap-leaching" techniques in 1981–1995, which have caused serious vegetation degradation, soil erosion and pollution. The processing adopted in the later stage is the "in situ leaching" technique that has been widely applied since 1996 [55]. With this technique, the surface vegetation was not seriously damaged in a large area, but a great number of liquid injection holes were excavated in the ore-bearing crust, causing damage to topsoil locally and subsoil largely. A part of vegetation cover in the mining area was also immediately destroyed, and the chemical liquid would remain for a long time in soils, causing soil acidification and leading to an increased risk of landslides as such liquid serves as a lubricator. No matter which kind of technique was used, the REE exploitation has caused severe damage to the environment.

2.2. Data and Preprocessing

2.2.1. Satellite Data and Processing

Multitemporal Landsat TM, ETM+ and OLI images from 1988 to 2019 were obtained from the USGS data server (https://glovis.usgs.gov). Only cloud-free images or those with less than 5% of cloud coverage were selected. Due to the abundant rainfall and cloudy weather throughout spring and summer in the study area, October and early November are the period when it is possible to acquire images with low cloud cover, good quality and acceptable sun-elevation angle (≥40°). For this reason, only October and early November images were employed for this study.

Already orthorectified by the provider, NASA, before Landsat images were made publicly accessible, only atmospheric correction was conducted for these images using the COST model developed by Chavez [56] in combination with the dark-object subtraction (DOS) [57] approach to remove the haze effect. The band minimum was applied to estimate the haze value in each band of Landsat images, and spectral radiance was then converted into surface reflectance for each band after atmospheric correction [13,56].

Very high resolution imagery, such as QuickBird, GeoEye and SPOT, available on Google Earth was used as a complementary source for definition of the boundaries and visual verification of mines.

Elevation data (ASTER GDEMV2) with 30 m resolution were obtained from the Geospatial Data Cloud Platform of the Chinese Academy of Sciences (http://www.gscloud.cn/) and used as the background information of the study area.

2.2.2. Field Data

The first-hand data were acquired from field investigations conducted in December 2016 to December 2019 to understand the damages caused by mining to environment and the situation of restoration. Our partner, the 264 Geological Team of Jiangxi Nuclear Industry has conducted rehabilitation management in 11 abandoned REE mines in the study area. Their knowledge and experience were important to our study, especially for assessing the results of our analysis on management effectiveness, and their field pictures were direct proof for validation of our analysis.

2.3. Biophysical Features of the Mining and Restored Areas

The typical biophysical change after open-pit mining, i.e., deforestation through the peeling-off of the topsoil and vegetation cover, is the decrease or loss of vegetation greenness and increase in bareness. The major restoration efforts are to recover the mined areas with vegetation plantation such as grasses, shrubs and trees, but other management options are either to convert the mined areas into factories or solar power stations or simply to cover them with plastic mulch to prevent further soil erosion. Thus, restoration and mitigation measures and procedures are different from mine to mine, and so are the restored greenness and extent in the study areas.

Vegetation indices, surface albedo, brightness and other indicators that are able to capture the bareness information may all be useful for assessment of the mining impacts and restoration effectiveness. For this purpose, remote-sensing-based biophysical indicators such as NDVI, GDVI, EVI, SAVI, atmospherically resistant vegetation index (ARVI) and soil-adjusted and atmospherically resistant vegetation index (SARVI) [58], surface albedo [17], the first principal component (PC1), tasseled cap brightness (TCB) [54,59,60] and LST were taken into account (Table 1). Calculation of vegetation indices is known, but other indicators, e.g., albedo and LST, are less frequently applied. We present their calculation formulas here.

Table 1. Biophysical indicators used in this study.

Index	Full Name	Formula	Sources
GDVI	Generalized Difference Vegetation Index	$\left(\rho_{NIR}^2-\rho_R^2\right)/\left(\rho_{NIR}^2+\rho_R^2\right)$	[49]
NDVI	Normalized Difference Vegetation Index	$(\rho_{NIR}-\rho_R)/(\rho_{NIR}+\rho_R)$	[37]
EVI	Enhanced Vegetation Index	$2.5\times\frac{\rho_{NIR}-\rho_R}{\rho_{NIR}+6.0\rho_R-7.5\rho_B+1}$	[39]
SAVI	Soil Adjusted Vegetation Index	$(1+L)(\rho_{NIR}-\rho_R)/(\rho_{NIR}+\rho_R+L)$ Low vegetation, L = 1; intermediate, 0.5; high, 0.25	[38]
ARVI	Atmospherically Resistant Vegetation Index	$(\rho_{NIR}-\rho_{RB})/(\rho_{NIR}+\rho_{RB})$ $\rho_{RB}=(\rho_R-\gamma\times\rho_R-\rho_B)$ γ = 1, ρ_B = reflectance of blue band	[58]
SARVI	Soil Adjusted and Atmospherically Resistant Vegetation Index	$(1+L)(\rho_{NIR}-\rho_{RB})/(\rho_{NIR}+\rho_{RB}+L)$	[58]
α	Albedo		[17]
T_N	Normalized Land Surface Temperature	T_N = $(\mathrm{LST}-\mathrm{LST_{min}})/(\mathrm{LST_{max}}-\mathrm{LST_{min}})$	[25]
TCB	Tasseled Cap Brightness		[54,59,60]
PC1	The First Principal Component		[25]

Albedo (α), originally meaning "whiteness", is a measure of the diffuse reflection of solar radiation out of the total solar radiation on land surface and measured on a scale from 0 to 1. Courel et al. (1984) revealed that increase in albedo in African savanna was a result of land degradation caused by drought [17]. In our case, land degradation by mining may also lead to increase in surface albedo. Liang (2000) developed a series of algorithms for calculating albedo for various satellite sensors [61]. The one for Landsat data, which was further normalized by Smith (2010) [62], is expressed in Equation (1):

$$\alpha_{short}=\frac{0.356\rho_1+0.130\rho_3+0.373\rho_4+0.085\rho_5+0.072\rho_7-0.0018}{0.356+0.130+0.373+0.085+0.072} \quad (1)$$

where ρ_i represents the reflectance of band i; here, i is 1, 3, 4, 5 and 7.

LST was produced in terms of Chander et al. (2009) [63] and USGS (2019) [64] as shown in Equation (2):

$$\mathrm{LST}=\frac{k_2}{\ln\left(\frac{k_1}{L_\lambda}\right)+1} \quad (2)$$

where L_λ is the at-the-sensor spectral radiance (W/(m2sr μm)); k_1 and k_2 are calibration coefficients. LST in kelvin (>273.15 K) is big unit in comparison with other vegetation indices. It was thus further normalized as:

$$T_N=(\mathrm{LST}-\mathrm{LST_{min}})/(\mathrm{LST_{max}}-\mathrm{LST_{min}}) \quad (3)$$

where T_N is the normalized LST; $\mathrm{LST_{min}}$ and $\mathrm{LST_{max}}$ are the minimum and maximum LST of the observed area, i.e., the mines in our case. T_N also takes a value between 0 and 1.

The surface characteristics of REE mines at different stages such as mining disturbance, reclamation and vegetation restoration show great differences. The time trajectories of the biophysical indicators of a typical mine with restoration intervention were projected for the period 1988–2019. We noted that vegetation indices all showed a U-type curve in the mined areas, while other biophysical indicators such as albedo (α), TCB and LST demonstrated an inverse curvilinear character showing, more concretely, a mirror effect of the vegetation indices, i.e., a peak in the mined areas before restoration. An example is shown in Figure 2.

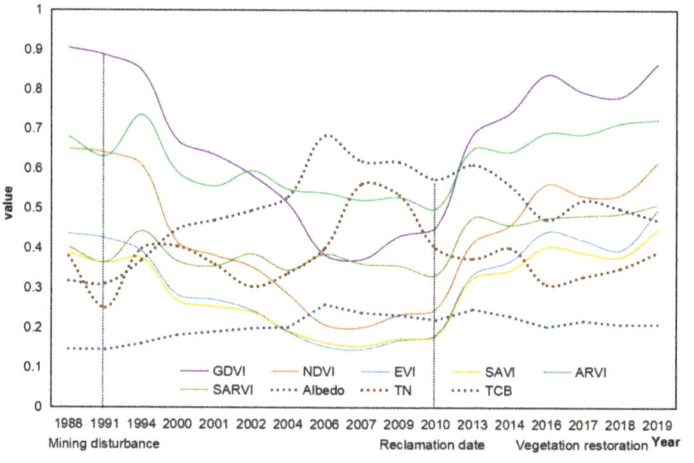

Figure 2. Time trajectories of the biophysical indicators of an REE mine (115°06′E, 24°99′N).

2.4. Pearson Correlation Analysis, Development of the New Mine-Tailored Indicators and Sensitivity Analysis

Apart from the time trajectories, a Pearson correlation analysis at the confidence level of 0.01 was applied to all biophysical indicators (i.e., NDVI, GDVI, SAVI, EVI, SARVI, ARVI, α, T_N, TCB and PC1) to uncover their correlation (Table 2).

In terms of the contrasted difference in correlation together with the mirror effect of vegetation indices versus other biophysical indicators (Table 2 and Figure 2), it seems possible to develop a mine-tailored remote sensing indicator or a set of such indicators for assessing the impacts of mining activities on the environment and the postmining mitigation effectiveness. This indicator will allow identifying the subtle difference in land cover change between premining and mining states, and between mined and pos-mining restoration intervention, and to highlight it.

Among the vegetation indices, those with soil adjustment and atmospheric correction were less sensitive to canopy cover than NDVI [13], and all these indices were less sensitive to multibiome features and of less dynamic range than GDVI in dryland systems [49,51]. For mining and mined areas, we randomly selected six typical mines with different mining periods, REE extraction approaches and restorations (see Table 6 for details) to project the time trajectories of GDVI and NDVI, and the former shows again wider dynamic range and higher sensitivity than NDVI (Figure 3). Hence, based on the results shown in both Figures 2 and 3, GDVI seems more suitable for highlighting the vegetation-related information of the mining and mined areas than other indices.

Table 2. Pearson correlation coefficients among the biophysical indicators.

	GDVI	NDVI	EVI	SAVI	ARVI	SARVI	α	T_N	TCB	PC1
GDVI	1	0.966 **	0.964 **	0.960 **	0.788 **	0.620 **	−0.713 **	−0.529 **	−0.809 **	0.356 **
NDVI		1	0.949 **	0.948 **	0.805 **	0.642 **	−0.683 **	−0.464 **	−0.776 **	0.348 **
EVI			1	0.992 **	0.787 **	0.719 **	−0.547 **	−0.477 **	−0.670 **	0.500 **
SAVI				1	0.843 **	0.788 **	−0.532 **	−0.470 **	−0.649 **	0.544 **
ARVI					1	0.844 **	−0.475 **	−0.383 **	−0.542 **	0.507 **
SARVI						1	−0.072	−0.266 **	−0.170	0.766 **
α							1	0.434 **	0.976 **	0.198 *
T_N								1	0.485 **	−0.208 *
TCB									1	0.107
PC1										1

* and ** indicate that the confidence level is 0.01 and 0.05, respectively.

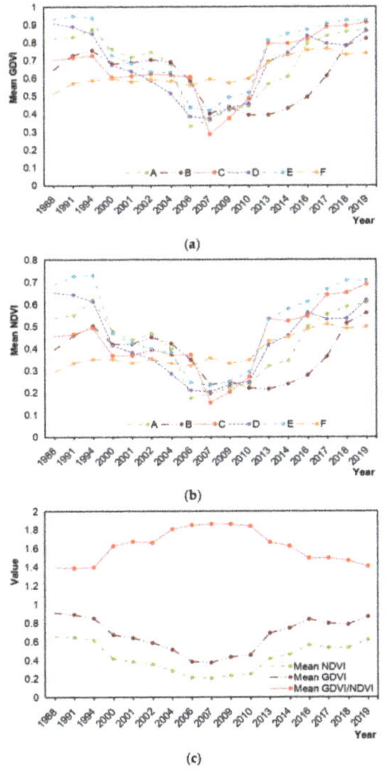

Figure 3. Time trajectories of the generalized difference vegetation index (GDVI) (**a**) and normalized difference vegetation index (NDVI) (**b**) of six typical mines. Time trajectory comparison of the mean GDVI and the mean NDVI (**c**), showing GDVI to have a wider dynamic range and higher sensitivity than NDVI in the mining areas (see Figure 5 for location of these mines and Table 6 for selection details).

To justify this selection, an importance ranking based on the correlation coefficients in Table 2 using the approach of Orloci (1978) [65] was conducted, and the weighted importance is reported in Table 3.

Table 3. Importance ranking of the biophysical indicators based on Table 2.

Biophysical Indicators	Specific Sum of Squares	Specific Sum of Squares as a Proportion (%)	Cumulative (%)
GDVI	6.359	63.588	63.588
SARVI	1.798	17.984	81.572
T_N	0.727	7.267	88.839
α	0.558	5.585	94.424
PC1	0.329	3.295	97.720
ARVI	0.113	1.130	98.849
NDVI	0.064	0.639	99.488
EVI	0.034	0.343	99.831
TCB	0.016	0.165	99.996
SAVI	0.000	0.004	100.000

Ranking was conducted following the approach proposed by Orloci (1978).

The nonvegetal part can be revealed by α, TCB and T_N, which are negatively correlated with GDVI. In view of this, the new remote sensing indicators were established as follows:

$$\text{MRAI1} = \text{GDVI}/(\alpha + T_N) \tag{4}$$

$$\text{MRAI2} = \text{GDVI}/(\alpha + \text{TCB}) \tag{5}$$

$$\text{MRAI3} = \text{GDVI}/(\alpha + T_N + \text{TCB}) \tag{6}$$

$$\text{MRAI4} = \text{GDVI}/(T_N + \text{TCB}) \tag{7}$$

The indicators in this set were termed mining and restoration assessment indicators (MRAIs). To evaluate their sensitivity, the approach proposed by Gitelson (2004) [66] was followed. Wu (2014) used this approach to assess the sensitivity of GDVI and confirmed the efficiency of this approach [49]. The sensitivity is expressed in Equation (8):

$$S_r(\text{MRAI}) = \left[\frac{d(\text{MRAI})}{d(\text{VI})}\right] \times \left[\frac{\Delta(\text{MRAI})}{\Delta(\text{VI})}\right]^{-1} \tag{8}$$

where $S_r(\text{MRAI})$ is the relative sensitivity of MRAI versus the referenced vegetation index (VI); $\frac{d(\text{MRAI})}{d(\text{VI})}$ is the first derivative of MRAI against VI or the infinitesimal change in MRAI in response to that of VI, or rather, the differences between two adjacent pixels of MRAI and VI, showing the tiny change; and $\Delta(\text{MRAI}) = \text{MRAI}_{MAX} - \text{MRAI}_{MIN}$ and $\Delta(\text{VI}) = \text{VI}_{MAX} - \text{VI}_{MIN}$ are the ranges of MRAI and VI of the observed mines in our case. If Sr > 1, MRAI is more sensitive than the referenced VI; if Sr = 1, they have the same sensitivity; if Sr < 1, MRAI is less sensitive than the referenced VI.

2.5. Application of MRAI for Mining and Restoration Monitoring

After sensitivity analysis, one MRAI, i.e., MRAI1, was selected for mining and restoration monitoring.

To compare the spatiotemporal changes and the management effectiveness at different stages of the REE mines, four observation points in time were selected, i.e., 1988 as premining, 1989–2000 as mining, 2010 as mining and mined state and 2019 as restoration state. Please be aware of the fact that those mines that had been exploited before 1988 were largely considered "no change" in the 1988–2010 period before restoration was conducted. Four concentrated REE mining areas in the study area were chosen for a detailed analysis of mining activities and vegetation greenness changes using MRAI1 as remote sensing indicator. The approach used was a differencing technique followed by a thresholding processing realized by density slicing or classification in order to identify the degradation and restoration and their subtle change degree in space and time [12,14].

3. Results

The intermediate and final results obtained following the above approaches and processing procedure are presented in this section.

3.1. Mirror Effect and Correlation among the Biophysical Indicators

It was revealed that vegetation indices have U-type time trajectories while nonvegetation biophysical indicators show anti-U curvilinear features, constituting a mirror effect of these two groups of indicators in the mining and mined areas (Figure 2).

Pearson correlation analysis illustrated that while the vegetation indices are positively correlated with each other, all of them are negatively correlated with α, T_N, TCB and PC1 (Table 2). GDVI has the best negative correlation with these nonvegetation indicators, and TCB seems the best approximation of α, followed by T_N.

Table 3 presents the result of ranking analysis based on the correlation coefficients of biophysical indicators (Table 2) and demonstrates that GDVI accounts for the most important weight (63.58%); hence, it is appropriate to select it as vegetation index to constitute the MRAIs.

3.2. Sensitivity and Dynamic Ranges of MRAIs

As presented in Figure 4, the sensitivity of all these MRAIs, i.e., *Sr*(MRAIs), in unmanaged and different vegetation restoration periods of the REE mines is higher than 1, indicating that all of the indicators in the set have higher sensitivity than vegetation indices. Thus, this set of MRAIs can be used for assessing the effects of mining and restoration management.

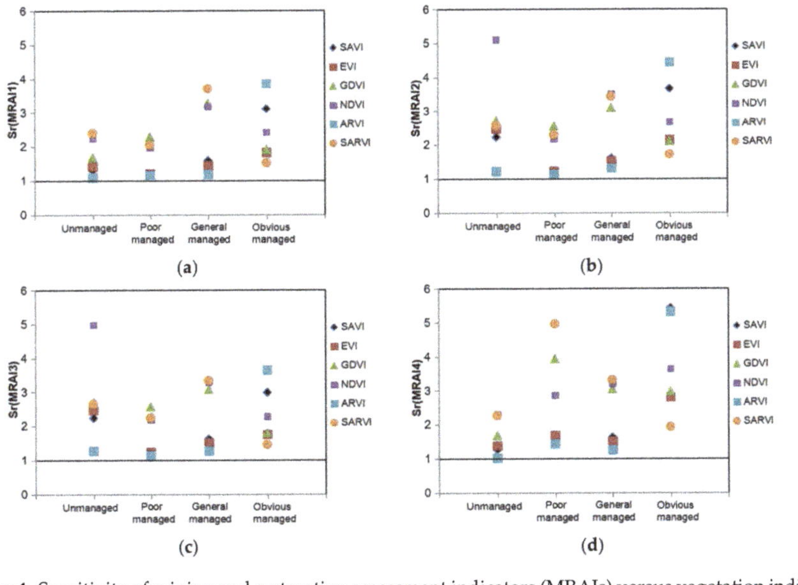

Figure 4. Sensitivity of mining and restoration assessment indicators (MRAIs) versus vegetation indices (VIs): (**a**) *Sr*(MRAI1); (**b**) *Sr*(MRAI2); (**c**) *Sr*(MRAI3); (**d**) *Sr*(MRAI4). Each sensitivity was the average of ten measurements in their corresponding mines at different stages (see Supplementary Materials).

The dynamic ranges of these indicators of REE mines for Landsat images of 2019 are shown in Table 4; their indications, taking MRAI1 as an example, are listed as follows: when MRAI1 < 1.0, it represents an unmanaged state; when 1.0 < MRAI1 ≤ 1.5, it represents a poor restoration; when 1.5 < MRAI1 ≤ 2.0, it implies a general case of restoration; and when MRAI1 > 2.0, it means a good restoration. Actually, thanks to the wide dynamic range and high sensitivity, the ranges can be much more finely divided.

Table 4. The ranges of MRAIs in REE mines.

	Range	Mean	StDev
MRAI1	[−2.86, 3.95]	1.43	0.55
MRAI2	[−6.87, 5.64]	1.21	0.51
MRAI3	[−2.18, 2.05]	0.78	0.31
MRAI4	[−2.37, 2.50]	0.97	0.39

3.3. Spatiotemporal Change of Vegetation Restoration

As previously mentioned, 1158 REE mines with a total surface area of about 66.29 km^2 in the seven counties were identified by multitemporal satellite images though the mining period of these mines was different from one to another. The spatiotemporal changes in land degradation (e.g., vegetation loss and soil erosion) and restoration of mines are shown in Table 5 and Figure 5, taking the four most serious counties where REE mines are dominantly distributed (circles a, b, c and d) as an example.

Table 5. Results of vegetation restoration in the seven counties detected by using MRAI1 indicator.

Period	Severe Degradation		Slight Degradation		No Change		Slight Recovery		Significant Recovery	
	Area (ha)	%	Area (ha)	%	Area (ha)	%	Area (ha)	%	Area (ha)	%
1988–2000	629.62	9.50	658.39	9.93	5264.43	79.42	46.72	0.72	29.83	0.45
2000–2010	1786.93	26.96	1206.82	18.21	3278.35	49.45	227.32	3.43	129.58	1.95
2010–2019	435.87	6.58	462.23	6.97	3594.99	54.23	1421.62	21.45	714.29	10.78

Here, "No Change" includes the mined area or degradation in the pre-1988 period. Before restoration, such area remained unchanged.

From 1988 to 2010, land degradation, or rather, mining area, had increased by 4281.76 ha (red and brown color in Figure 5), but from 2010 to 2019, newly mined area was reduced to 898.1 ha and rehabilitated green areas had a significant increase by 2135.9 ha, especially in Xunwu, Dingnan, Longnan, and Anyuan (green color in Figure 5). In addition, mining activity and restoration showed a spatial heterogeneity and variability in different counties.

3.4. Verification of the Detected Changes

To ensure the accuracy of the MRAI1 analysis, eight typical REE mines, namely A–F, V1 and V2 (as shown in Figure 5), were selected for verification in reference to field observation and Google Earth (Figures 6–8). The geographical information, REE extraction technique and the restoration management are all presented for each mine in Table 6. The comparison shows that the assessment results of the management effectiveness of the mined areas were well consistent with the actual situation observed in the field and on Google Earth (Figures 7 and 8), indicating that the new remote sensing indicator, MRAI1, was able to achieve the monitoring and assessment of the mining and restoration effects on the environment in space and time. On the other hand, from the local detailed analysis, we can see that in the same mining area, the new indicator MRAI1 can well reflect the tiny differences in management effectiveness.

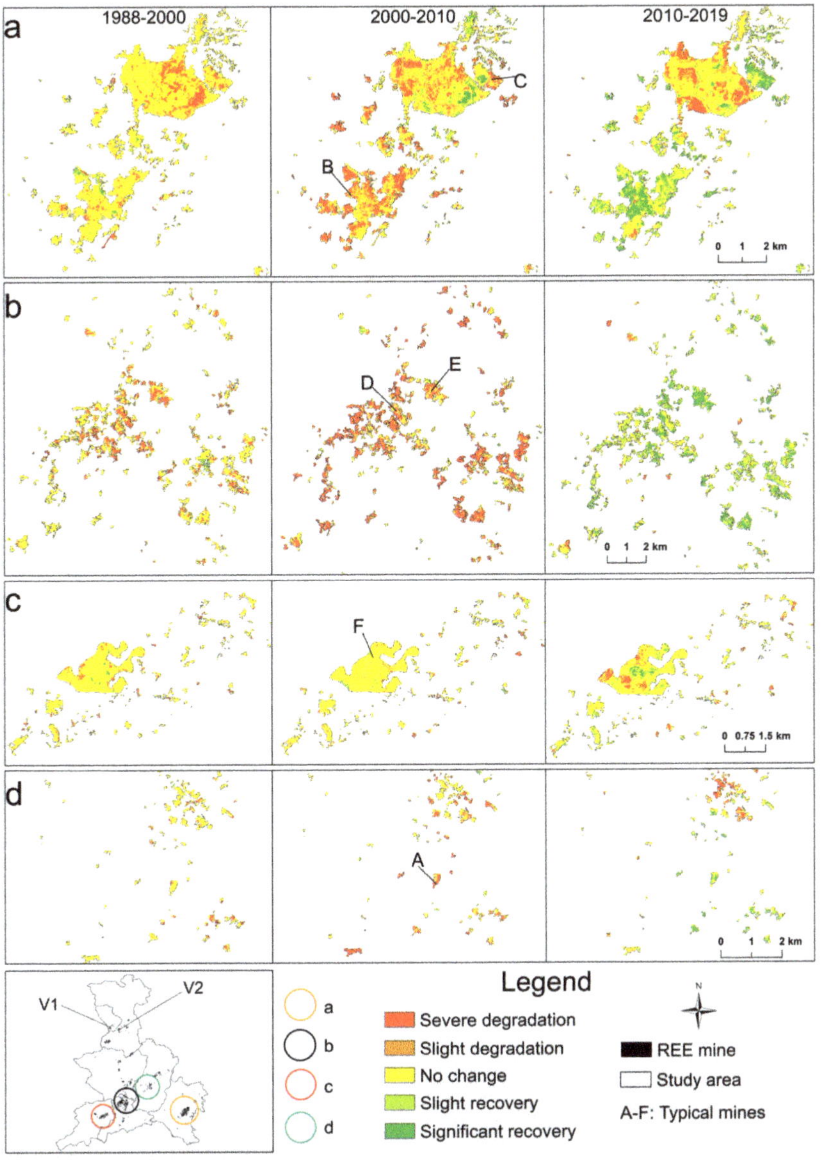

Figure 5. Spatiotemporal changes of vegetation restoration in four areas (**a**–**d**) with high mining concentration. A–F are typical mines selected for detailed analysis; V1 and V2 are example sites for verification managed by the 264 Geological Team (see Figures 7 and 8).

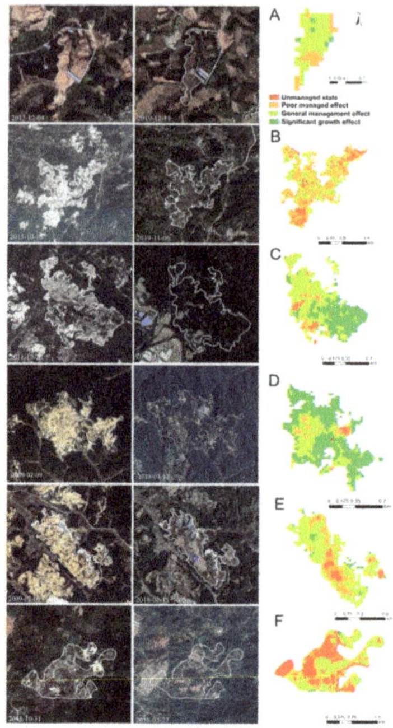

Figure 6. Comparison of our results from the six mining sites with Google Earth images from the mining to restoration stages ((**A–F**) is six typical mines, see details in Table 6) (high-resolution images: ©Google).

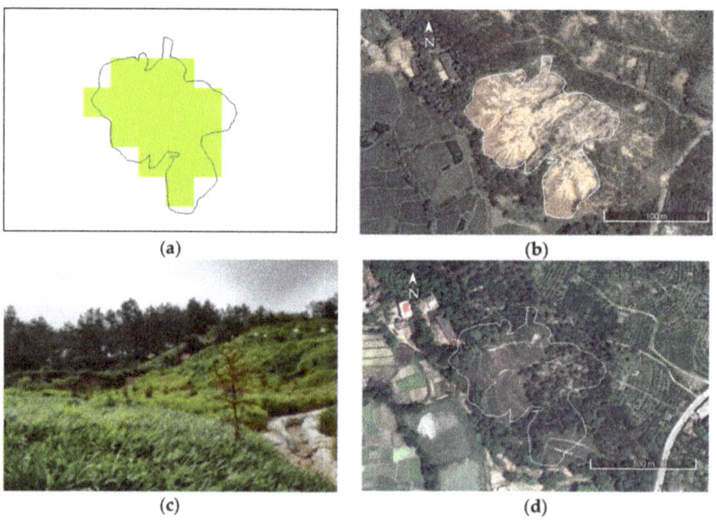

Figure 7. REE mine management in Ganlin Village in Ganzhou City (V1: 114°55′42.73″E, 25°44′4.28″N). (**a**) "Slight recovery" detected by Landsat data from 2010 to 2019; (**b**) eroded landscape after exploitation (10 August 2004, ©Google); (**c**) landscape observed in the field after management (9 July 2019), i.e., cultivated with grasses and trees; and (**d**) effect of restoration management seen on 26 April 2020 (©Google), i.e., 14 months after the management was conducted in January 2019.

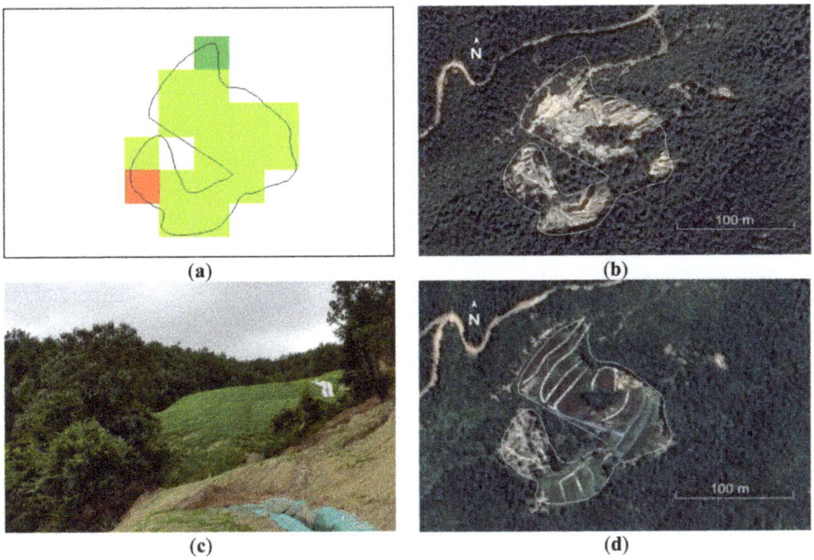

Figure 8. Restoration of REE mine close to Fengshan Village (V2: 115°0′35.97″E, 25°42′49.90″N). (**a**) Slight recovery detected by Landsat data because of the restoration engineering conducted in November 2018; (**b**) damaged landscape before management (29 February 2016, ©Google); (**c**) observed landscape in the field in July 2019; and (**d**) restoration seen on Google Earth (26 August 2019, ©Google).

Table 6. The features of the six typical managed REE mines.

Site	Longitude	Latitude	Surface Area (ha)	County	Major Features
A	115.339	25.13	8.03	Anyuan	"Heap leaching" technique, land leveling and grass plantation in 2014
B	115.679	24.85	442.76	Xunwu	"Pool leaching" technique, land leveling in 2016, solar power plant in the south, west and northeast in 2017, and large areas of grass plantation in others
C	115.725	24.90	92.55	Xunwu	"Pool leaching" technique, land leveling and tree plantation in 2009
D	115.067	24.99	87.48	Dingnan	"In situ mineral leaching" technique, a combination of grass and tree plantation in 2012
E	115.049	24.98	54.94	Dingnan	"Pool leaching" technique, land leveling in 2010, grass plantation in 2013, factory construction in some areas in 2018
F	114.846	24.83	320.55	Longnan	"Pool leaching" technique, a combination of tree and grass plantation in 2010

The forest types A–F are coniferous and mixed forests (see Section 2.1).

4. Discussion

4.1. Recovery Process

The land surface of REE mines has experienced significant changes from premining and mining to recovery in the past decades, and such changes were mainly associated with implementation of different policies. Driven by the policy of "invigorating the domestic economy and opening to the outside world" of Deng Xiaoping in 1987–1988 [14], innumerous local private companies were set up nationwide. As a result, REE mining experienced a disorderly expansion for economic reasons by a great number of private enterprises in southern Jiangxi. Our change detection using the MRAI1 indicator illustrated that the vegetation cover in the four counties considered decreased dramatically, showing a clear degradation by 4281 ha in which previous vegetation cover had become bare soil with an increased albedo and brightness in 1988–2010. This is indicative of an expanding mining process including the striping off of topsoil and forests to exploit REE deposits in the weathered crust and, at the same time, the setting up of a series of leaching ponds for application of the "pool leaching" or "heap leaching" technology, or digging/drilling of holes and ditches elsewhere to apply the "in situ leaching" technology for REE extraction after 1996 [1,55,67,68]. Vegetation degradation, soil erosion and groundwater pollution as a consequence of mining activities resulted in negative impacts on the ecosystems [68,69]. These are local phenomena in response to the macroscopic policy of the state.

With an increasing awareness of the hazardous damage to the environment caused by mining, the central government promulgated the "Regulations on the Protection of the Geological Environment of Mines" (MoLR 2009) [70], with an order of "who mined who governs". Hence, restoration management of the abandoned mines became an obligation for mining enterprises in southern Jiangxi in 2010. For this reason, a number of pilot projects have been implemented [67] and different treatments, including chemical and biological approaches, have been tested and proposed [68,71–75]; for the time being, biological rehabilitation is dominant because of the purpose of ecological restoration.

The common vegetation recovery management is plantation of conifers (e.g., *Cunninghamia lanceolata*, *Pinus massoniana*), grasses (e.g., *Paspalum wettsteinii* Hack) and orchards (e.g., navel orange and other local specialty fruits) in the mined areas located in hills and mountains that showed spatial differences in levels of recovery within our detection results (Figure 5). Thanks to these efforts, 2135 ha of abandoned mines have been managed and have, to a certain degree, recovered in the period of 2010–2019 (Table 5). However, in 2019, about 67% of the total area of the observed REE mines was still in a state of degradation or remained unmanaged (Table 5). This is probably the results of the following cases: (1) restoration of some mined areas was still at the stage of land leveling and vegetation was not yet planted; (2) for large mined areas surrounding the cities, the management action was to build factories and solar power plants, which was not regarded as an ecological restoration and led such areas to be classified as unmanaged mines by MRAIs; (3) restoration management had not yet been conducted for the newly mined area after 2010. Clearly, much reclamation work remains to be conducted in this region.

4.2. Advantages and Disadvantages of Our Approaches

It is clear that with their sensitivity and wide dynamic ranges, the integrated MRAIs can help distinguish the subtle differences between premining and mining and between mined and postmining management, of which the latter was actually graded into two levels for facilitating visualization in maps, though it may be divided into many more levels. Actually, Li et al. (2018) and Zhang et al. (2015) used NDVI to identify the REE mines locally [6,48], and Peng et al. (2016) employed ecological quality index to analyze these areas in southern Jiangxi [47]. Largely speaking, their studies were relevant but had less sensitivity and accuracy than our approach, because some of them even failed to distinguish clouds from mines or required more indicators to compose the quality index. From this viewpoint, our approach based on MRAIs seems more robust and practical, requiring less data and having superior applicability. As a matter of fact, our new indicators contain both vegetation and nonvegetation characteristics and are capable of catching more features of land surface change than other single indicators or indices. As such, aside from REE mines, it will be possible to apply them in studies of other kinds of open-pit mining areas such as copper, limestone, iron and coal mines.

However, as with other remote sensing approaches, MRAIs were unable to identify completely mined areas where "in situ leaching" technique had been applied due to the fact that drilling operations kept the surface vegetation cover to certain extent. This problem can be resolved if UAV technology is employed, as it is able to provide higher resolution and hyperspectral data [44]. The limitation of our approach lies in that most of the MRAIs developed, except for MRAI2, may not be applicable if remote sensing data do not possess the thermal band that allows us to derive LST. In this case, MRAI2 is the only choice.

5. Conclusions

This paper presented the development of a set of mine-tailored remote sensing indicators (MRAIs) and their application to detection of the spatial and temporal changes of the REE mining and mined areas and assessment of the effectiveness of restoration management in southern Jiangxi. Our study showed that these new indicators are applicable for achieving such research goals with reliability. We believe that these indicators not only are suitable for research on REE mines, but also have potential for assessment of other kinds of open-pit mines.

This study considered the state of surface vegetation cover, albedo, brightness and surface temperature to assess mining and the effectiveness of the postmining management after different policy implementations. The research still remains at the surface layer, and a more holistic and comprehensive assessment integrating MRAIs with soil and groundwater samples will be required as a next step, as the multidimensional assessment may provide more relevant advice for local authorities for sustainable exploitation and restoration of REE mines.

Supplementary Materials: The following are available online at http://www.mdpi.com/2072-4292/12/21/3558/s1, Table S1: List of the images sensor and acquisition dates selected for this study and Annex 1: Sensitivity analysis.

Author Contributions: Conceptualization, L.X. and W.W.; Data curation, L.X., X.H., P.O., W.H., Y.Z., X.F., J.L., J.J., M.Z., Z.Z. and Y.B.; Formal analysis, L.X.; Funding acquisition, W.W.; Investigation, L.X., Z.L., W.Z., Y.S., T.L. and X.Z.; Methodology, L.X., Z.L., Y.Q., S.P. and C.S.; Project administration, W.W.; Resources, P.O.; Software, X.H.; Writing—review & editing, W.W. All authors have read and agreed to the published version of the manuscript.

Funding: This research was supported by the Start-up Fund for Scientific Research of the East China University of Technology (DHTP2018001) and the fund of the Jiangxi Talent Program (900/2120800004), both granted to Dr Weicheng Wu.

Acknowledgments: The field trips were received by the 264 Geological Team of Jiangxi Nuclear Industry. Landsat images were freely obtained from the USGS data server (https://glovis.usgs.gov).

Conflicts of Interest: The authors declare no conflict of interest.

References

1. Cheng, J.; Che, L. Current mining situation and potential development of rare earth in China. *Chin. Rare Earths* **2010**, *31*, 65–69. (In Chinese)
2. Dushyantha, N.; Batapola, N.; Ilankoon, I.M.S.K.; Rohitha, S.; Premasiri, R.; Abeysinghe, B.; Ratnayake, N.; Dissanayake, K. The story of rare earth elements (REEs): Occurrences, global distribution, genesis, geology, mineralogy and global production. *Ore Geol. Rev.* **2020**, 122. [CrossRef]
3. Wübbeke, J. Rare earth elements in China: Policies and narratives of reinventing an industry. *Resour. Policy* **2013**, *38*, 384–394. [CrossRef]
4. Wang, J.; Guo, M.; Liu, M.; Wei, X. Long-term outlook for global rare earth production. *Resour. Policy* **2020**, *65*, 101569. [CrossRef]
5. Kamenopoulos, S.N.; Agioutantis, Z. Geopolitical Risk Assessment of Countries with Rare Earth Element Deposits. *Min. Metall. Explor.* **2020**, *37*, 51–63. [CrossRef]
6. Li, H.; Lei, J.; Wu, J. Analysis of land damage and recovery process in rare earth mining area based on multi-source sequential NDVI. *Trans. Chin. Soc. Agric. Eng.* **2018**, *34*, 232–240. (In Chinese) [CrossRef]

7. Li, T.; Lü, Y.; Fu, B.; Comber, A.J.; Harris, P.; Wu, L. Gauging policy-driven large-scale vegetation restoration programmes under a changing environment: Their effectiveness and socio-economic relationships. *Sci. Total Environ.* **2017**, *607*, 911–919. [CrossRef] [PubMed]
8. Ngugi, M.R.; Neldner, V.J.; Doley, D.; Kusy, B.; Moore, D.; Richter, C. Soil moisture dynamics and restoration of self-sustaining native vegetation ecosystem on an open-cut coal mine. *Restor. Ecol.* **2015**, *23*, 615–624. [CrossRef]
9. Pei, J.; Wang, L.; Wang, X.; Niu, Z.; Kelly, M.; Song, X.P.; Huang, N.; Geng, J.; Tian, H.; Yu, Y.; et al. Time Series of Landsat Imagery Shows Vegetation Recovery in Two Fragile Karst Watersheds in Southwest China from 1988 to 2016. *Remote Sens.* **2019**, *11*, 2044. [CrossRef]
10. Tang, Y.; Shao, Q.; Liu, J.; Zhang, H.; Gong, G. Did Ecological Restoration Hit Its Mark? Monitoring and Assessing Ecological Changes in the Grain for Green Program Region Using Multi-source Satellite Images. *Remote Sens.* **2019**, *11*, 358. [CrossRef]
11. Tucker, C.J.; Dregne, H.; Newcomb, W.W. Expansion and Contraction of the Sahara Desert from 1980 to 1990. *Science* **1991**, *253*, 299–301. [CrossRef] [PubMed]
12. Wu, W. *Monitoring Land Degradation in Drylands by Remote Sensing*; Marini, A., Talbi, M., Eds.; Springer: Berlin/Heidelberg, Germany, 2009. [CrossRef]
13. Wu, W.; De Pauw, E.; Helldén, U. Assessing woody biomass in African tropical savannahs by multiscale remote sensing. *Int. J. Remote Sens.* **2013**, *34*, 4525–4549. [CrossRef]
14. Wu, W.; De Pauw, E.; Zucca, C. Using remote sensing to assess impacts of land management policies in the Ordos rangelands in China. *Int. J. Digit. Earth* **2013**, *6*, 81–102. [CrossRef]
15. Zucca, C.; Wu, W.; Dessena, L.; Mulas, M. Assessing the Effectiveness of Land Restoration Interventions in Dry Lands by Multitemporal Remote Sensing—A Case Study in Ouled DLIM (Marrakech, Morocco). *Land Degrad. Dev.* **2015**, *26*, 80–91. [CrossRef]
16. Li, J.; Pei, Y.; Zhao, S.; Xiao, R.; Zhang, C. A Review of Remote Sensing for Environmental Monitoring in China. *Remote Sens.* **2020**, *12*, 1130. [CrossRef]
17. Courel, M.F.; Kandel, R.S.; Rasool, S.I. Surface albedo and the Sahel drought. *Nature* **1984**, *307*, 528–531. [CrossRef]
18. Lambin, E.F.; Strahler, A.H. Indicators of land-cover change for change-vector analysis in multitemporal space at coarse spatial scales. *Int. J. Remote Sens.* **1994**, *15*, 2099–2119. [CrossRef]
19. Zhang, J.; Wang, T.; Ge, J. Assessing Vegetation Cover Dynamics Induced by Policy-Driven Ecological Restoration and Implication to Soil Erosion in Southern China. *PLoS ONE* **2015**, *10*, e0131352. [CrossRef]
20. Carabassa, V.; Ortiz, O.; Alcañiz, J.M. RESTOQUARRY: Indicators for self-evaluation of ecological restoration in open-pit mines. *Ecol. Indic.* **2019**, *102*, 437–445. [CrossRef]
21. Geist, H.J.; Lambin, E.F. Proximate Causes and Underlying Driving Forces of Tropical Deforestation. *BioScience* **2002**, *52*, 143–150. [CrossRef]
22. Coppin, P.; Jonckheere, I.; Nackaerts, K.; Muys, B.; Lambin, E. Digital change detection methods in ecosystem monitoring: A review. *Int. J. Remote Sens.* **2004**, *25*, 1565–1596. [CrossRef]
23. Verburg, P.H.; Soepboer, W.; Veldkamp, A.; Limpiada, R.; Espaldon, V.; Mastura, S.S. Modeling the spatial dynamics of regional land use: The CLUE-S model. *Environ. Manag.* **2002**, *30*, 391–405. [CrossRef]
24. Veldkamp, A.; Verburg, P.H.; Kok, K.; De Koning, G.H.J.; Priess, J.; Bergsma, A.R. The Need for Scale Sensitive Approaches in Spatially Explicit Land Use Change Modeling. *Environ. Modeling Assess.* **2001**, *6*, 111–121. [CrossRef]
25. Wu, W.; Mhaimeed, A.S.; Al-Shafie, W.M.; Ziadat, F.; Dhehibi, B.; Nangia, V.; Pauw, E.D. Mapping soil salinity changes using remote sensing in Central Iraq. *Geoderma Reg.* **2014**, *2*, 21–31. [CrossRef]
26. Aspinall, R. Modelling land use change with generalized linear models—A multi-model analysis of change between 1860 and 2000 in Gallatin Valley, Montana. *J. Environ. Manag.* **2004**, *72*, 91–103. [CrossRef] [PubMed]
27. Wulder, M.A.; White, J.C.; Goward, S.N.; Masek, J.G.; Irons, J.R.; Herold, M.; Cohen, W.B.; Loveland, T.R.; Woodcock, C.E. Landsat continuity: Issues and opportunities for land cover monitoring. *Remote Sens. Environ.* **2008**, *112*, 955–969. [CrossRef]
28. George, M.; Manal, N.; Irna, V.D.M.; Jonathan, L. Evaluating exposure to land degradation in association with repetitive armed conflicts in North Lebanon using multi-temporal satellite data. *Environ. Monit. Assess.* **2014**, *186*, 7655–7672. [CrossRef]

29. Maynard, J.J.; Karl, J.W.; Browning, D.M. Effect of spatial image support in detecting long-term vegetation change from satellite time-series. *Landsc. Ecol.* **2016**, *31*, 2045–2062. [CrossRef]
30. Jana, A.; Maiti, S.; Biswas, A. Seasonal change monitoring and mapping of coastal vegetation types along Midnapur-Balasore Coast, Bay of Bengal using multi-temporal landsat data. *Modeling Earth Syst. Environ.* **2015**, *2*, 7. [CrossRef]
31. Töyrä, J.; Pietroniro, A. Towards operational monitoring of a northern wetland using geomatics-based techniques. *Remote Sens. Environ.* **2005**, *97*, 174–191. [CrossRef]
32. Wu, W.; Lambin, E.F.; Courel, M.F. Land use and cover change detection and modelling for North Ningxia, China. In Proceedings of the Mapasia 2002, Bangkok, Thailand, 7–9 August 2002.
33. Lyu, X.; Li, X.; Dang, D.; Dou, H.; Gong, J. A new method for grassland degradation monitoring by vegetation species composition using hyperspectral remote sensing. *Ecol. Indic.* **2020**, *114*, 106310. [CrossRef]
34. Alharthi, A.; El-Sheikh, M.A.; Elhag, M.; Alatar, A.A.; Eid, E.M. Remote sensing of 10 years changes in the vegetation cover of the northwestern coastal land of Red Sea, Saudi Arabia. *Saudi J. Biol. Sci.* **2020**. [CrossRef]
35. Huylenbroeck, L.; Laslier, M.; Dufour, S.; Georges, B.; Michez, A. Using remote sensing to characterize riparian vegetation: A review of available tools and perspectives for managers. *J. Environ. Manag.* **2020**, *267*, 110652. [CrossRef] [PubMed]
36. Yonatan, T.; María, M.L. Monitoring tropical forest degradation using remote sensing. Challenges and opportunities in the Madre de Dios region, Peru. *Remote Sens. Appl. Soc. Environ.* **2020**, *19*. [CrossRef]
37. Tucker, C.J. Red and photographic infrared linear combinations for monitoring vegetation. *Remote Sens. Environ.* **1979**, *8*, 127–150. [CrossRef]
38. Huete, A.R. A soil-adjusted vegetation index (SAVI). *Remote Sens. Environ.* **1988**, *25*, 295–309. [CrossRef]
39. Huete, A.R.; Liu, H.; Batchily, K.; Leeuwen, W.V.A. A comparison of vegetation indices over a global set of TM images for EOS-MODIS. *Remote Sens. Environ.* **1997**, *59*, 440–451. [CrossRef]
40. Demirel, N.; Düzgün, E.; Emil, M.K. Landuse change detection in a surface coal mine area using multi-temporal high-resolution satellite images. *Int. J. Min. Reclam. Environ.* **2011**, *25*, 342–349. [CrossRef]
41. Kuenzer, C.; Hecker, C.; Zhang, J.; Wessling, S.; Wagner, W. The potential of multidiurnal MODIS thermal band data for coal fire detection. *Int. J. Remote Sens.* **2008**, 29. [CrossRef]
42. Zheng, M.; Deng, K.; Fan, H.; Du, S. Monitoring and Analysis of Surface Deformation in Mining Area Based on InSAR and GRACE. *Remote Sens.* **2018**, *10*, 1392. [CrossRef]
43. Vidal-Macua, J.J.; Nicolau, J.M.; Vicente, E.; Heras, M.M.D.L. Assessing vegetation recovery in reclaimed opencast mines of the Teruel coalfield (Spain) using Landsat time series and boosted regression trees. *Sci. Total Environ.* **2020**, *717*, 137250. [CrossRef]
44. Ren, H.; Zhao, Y.; Xiao, W.; Hu, Z. A review of UAV monitoring in mining areas: Current status and future perspectives. *Int. J. Coal Sci. Technol.* **2019**, *6*, 320–333. [CrossRef]
45. Johansen, K.; Erskine, P.D.; McCabe, M.F. Using Unmanned Aerial Vehicles to assess the rehabilitation performance of open cut coal mines. *J. Clean. Prod.* **2019**, *209*, 819–833. [CrossRef]
46. Padró, J.C.; Carabassa, V.; Balagué, J.; Brotons, L.; Alcaniz, J.M.; Pons, X. Monitoring opencast mine restorations using Unmanned Aerial System (UAS) imagery. *Sci. Total Environ.* **2019**, *657*, 1602–1614. [CrossRef] [PubMed]
47. Peng, Y.; He, G.; Zhang, Z.; Jiang, W.; Ouyang, Z.; Wang, G. Eco-environmental dynamic monitoring and assessment of rare earth mining area in Southern Ganzhou using remote sensing. *Acta Ecol. Sin.* **2016**, *36*, 1676–1685.
48. Zhang, H.; Zhong, B.; Hong, Y.; Wan, H.; Liu, Q. Dynamic Monitoring of Rare Earths Mining Area in Ganzhou during Recent 20 Years based on Remote Sensing. *Remote Sens. Technol. Appl.* **2015**, *30*, 376–382. [CrossRef]
49. Wu, W. The Generalized Difference Vegetation Index (GDVI) for Dryland Characterization. *Remote Sens. Environ.* **2014**, *6*, 1211–1233. [CrossRef]
50. Wu, W.; Al-Shafie, W.M.; Mhaimeed, A.S.; Ziadat, F.; Nangia, V.; Payne, W.B. Soil Salinity Mapping by Multiscale Remote Sensing in Mesopotamia, Iraq. *IEEE J. Sel. Top. Appl. Earth Obs.* **2017**, *7*, 4442–4452. [CrossRef]
51. Gunasekara, N.K.; Al-Wardy, M.M.; Al-Rawas, G.A.; Charabi, Y. Applicability of VI in arid vegetation delineation using shadow-affected SPOT imagery. *Environ. Monit. Assess.* **2015**, *187*, 454. [CrossRef]

52. Wu, W.; Zucca, C.; Muhaimeed, A.S.; Al-Shafie, W.M.; Al-Quraishi, A.F.; Nangia, V.; Zhu, M.; Liu, G. Soil salinity prediction and mapping by machine learning regression in Central Mesopotamia, Iraq. *Land Degrad. Dev.* **2018**, *29*, 4005–4014. [CrossRef]
53. Wu, W.; Zucca, C.; Karam, F.; Liu, G. Enhancing the performance of regional land cover mapping. *Int. J. Appl. Earth Obs. Geoinf.* **2016**, *52*, 422–432. [CrossRef]
54. Crist, E.P.; Cicone, R.C. Application of the tasseled cap concept to simulated thematic mapper data. *Photogramm. Eng. Remote Sens.* **1984**, *50*, 343–352. [CrossRef]
55. Ding, J. Historical review of the ionic rare earth mining: In honor of the 60 anniversary of GNMRI. *Nonferrrous Met. Sci. Eng.* **2012**, *3*, 14–19. [CrossRef]
56. Chavez, P.S.J. Image-Based Atmospheric Corrections—Revisited and Improved. *Photogramm. Eng. Remote Sens.* **1996**, *62*, 1025–1036. [CrossRef]
57. Chavez, P.S. An improved dark-object subtraction technique for atmospheric scattering correction of multispectral data. *Remote Sens. Environ.* **1988**, *24*, 459–479. [CrossRef]
58. Kaufman, Y.J.; Tanre, D. Atmospherically resistant vegetation index (ARVI) for EOS-MODIS. *IEEE Trans. Geosci. Remote Sens.* **1992**, *30*, 261–270. [CrossRef]
59. Baig, M.H.A.; Zhang, L.; Shuai, T.; Tong, Q. Derivation of a tasselled cap transformation based on Landsat 8 at-satellite reflectance. *Remote Sens. Lett.* **2014**, *5*, 423–431. [CrossRef]
60. Huang, C.; Wylie, B.; Yang, L.; Homer, C.; Zylstra, G. Derivation of a tasselled cap transformation based on Landsat 7 at-satellite reflectance. *Int. J. Remote Sens.* **2002**, *23*, 1741–1748. [CrossRef]
61. Liang, S. Narrowband to broadband conversions of land surface albedo I. *Remote Sens. Environ.* **2001**, *76*, 213–238. [CrossRef]
62. Smith, R.B. The Heat Budget of the Earth's Surface Deduced from Space. Available online: https://yceo.yale.edu/faq-page#t3n88 (accessed on 10 June 2020).
63. Chander, G.; Markham, B.L.; Helder, D.L. Summary of current radiometric calibration coefficients for Landsat MSS, TM, ETM+, and EO-1 ALI sensors. *Remote Sens. Environ.* **2009**, *113*, 893–903. [CrossRef]
64. United States Geological Survey (USGS). Landsat 8 (L8) Data Users Handbook, Version 5.0. 2019. Available online: https://digitalcommons.unl.edu/usgspubs/110 (accessed on 10 June 2020).
65. Orlóci, L. *Multivariate Analysis in Vegetation Research*; Springer: Dordrecht, The Netherlands, 2013.
66. Gitelson, A.A. Wide Dynamic Range Vegetation Index for Remote Quantification of Biophysical Characteristics of Vegetation. *J. Plant Physiol.* **2004**, *161*, 165–173. [CrossRef] [PubMed]
67. Guo, Z.; Zhao, K.; Jin, J.; Wang, G.; Zhu, Y. Reviews on Environmental Assessment and Pollution Prevention of Ion Adsorption Type Rare Earth Ores. *Chin. Rare Earths* **2019**, *40*, 115–126. (In Chinese) [CrossRef]
68. Chen, X.; Cai, Q.Y.; Yu, X.S.; Liu, Y.Z.; Wang, X.F. Vertical distributions of soil environmental factors in ion-type rare earth mining of southern Jiangxi—A case study in Longnan rare earth mining area. *Chin. Rare Earths* **2015**, *36*, 23–28. (In Chinese)
69. Li, M.S. Ecological restoration of mineland with particular reference to the metalliferous mine wasteland in China: A review of research and practice. *Sci. Total Environ.* **2006**, *357*, 38–53. [CrossRef] [PubMed]
70. MoLR (Ministry of Land and Resources of the People's Republic of China). Regulations on the Protection of the Geological Environment of Mines. 3 March 2009. Available online: http://www.gov.cn/flfg/2009-03/05/content_1251130.htm (accessed on 3 March 2020).
71. Wei, X.; Zhang, S.; Shimko, J.; Dengler, R.W. Mine drainage: Treatment technologies and rare earth elements. *Water Environ. Res.* **2019**, *91*, 1061–1068. [CrossRef]
72. Zeng, G.; Ling, B.; Li, Z.; Luo, S.; Sui, X.; Guan, Q. Fluorine removal and calcium fluoride recovery from rare-earth smelting wastewater using fluidized bed crystallization process. *J. Hazard. Mater.* **2019**, *373*, 313–320. [CrossRef]
73. Hedin, B.C.; Capo, R.C.; Stewart, B.W.; Hedin, R.S.; Lopano, C.L.; Stuckman, M.Y. The evaluation of critical rare earth element (REE) enriched treatment solids from coal mine drainage passive treatment systems. *Int. J. Coal Geol.* **2019**, *208*, 54–64. [CrossRef]

74. Deng, Z.; Qin, L.; Wang, G.; Luo, S.; Peng, C. Ammonia Nitrogen Pollution and Progress in Its Treatment of Ionic Rare Earth Mines. *Chin. Rare Earths* **2019**, *40*, 120–129. (In Chinese) [CrossRef]
75. Wei, Z.; Hao, Z.; Li, X.; Guan, Z.; Cai, Y.; Liao, X. The effects of phytoremediation on soil bacterial communities in an abandoned mine site of rare earth elements. *Sci. Total Environ.* **2019**, *670*, 950–960. [CrossRef]

Publisher's Note: MDPI stays neutral with regard to jurisdictional claims in published maps and institutional affiliations.

remote sensing

MDPI

Article

Improved CNN Classification Method for Groups of Buildings Damaged by Earthquake, Based on High Resolution Remote Sensing Images

Haojie Ma [1,2,†], Yalan Liu [1,*], Yuhuan Ren [1,†], Dacheng Wang [1], Linjun Yu [1] and Jingxian Yu [1,2]

1 Aerospace Information Research Institute, Chinese Academy of Sciences, Beijing 100101, China; mahj@radi.ac.cn (H.M.); renyh@radi.ac.cn (Y.R.); wangdc@radi.ac.cn (D.W); yulj201831@aircas.ac.cn (L.Y.); yujx@radi.ac.cn (J.Y.)
2 University of Chinese Academy of Sciences, Beijing 100049, China
* Correspondence: liuyl@aircas.ac.cn; Tel.: +86-139-1103-2598
† These authors contributed equally to this work.

Received: 3 December 2019; Accepted: 5 January 2020; Published: 11 January 2020

Abstract: Effective extraction of disaster information of buildings from remote sensing images is of great importance to supporting disaster relief and casualty reduction. In high-resolution remote sensing images, object-oriented methods present problems such as unsatisfactory image segmentation and difficult feature selection, which makes it difficult to quickly assess the damage sustained by groups of buildings. In this context, this paper proposed an improved Convolution Neural Network (CNN) Inception V3 architecture combining remote sensing images and block vector data to evaluate the damage degree of groups of buildings in post-earthquake remote sensing images. By using CNN, the best features can be automatically selected, solving the problem of difficult feature selection. Moreover, block boundaries can form a meaningful boundary for groups of buildings, which can effectively replace image segmentation and avoid its fragmentary and unsatisfactory results. By adding Separate and Combination layers, our method improves the Inception V3 network for easier processing of large remote sensing images. The method was tested by the classification of damaged groups of buildings in 0.5 m-resolution aerial imagery after the earthquake of Yushu. The test accuracy was 90.07% with a Kappa Coefficient of 0.81, and, compared with the traditional multi-feature machine learning classifier constructed by artificial feature extraction, this represented an improvement of 18% in accuracy. Our results showed that this improved method could effectively extract the damage degree of groups of buildings in each block in post-earthquake remote sensing images.

Keywords: earthquake; damaged groups of buildings; classification; remote sensing images; Convolution Neural Network (CNN); block vector data

1. Introduction

During the rescue and recovery phases following an earthquake, damaged buildings may indicate the locations of trapped people [1]. Hence, building damage maps are key to post-earthquake rescue and reconstruction. The use of traditional manual field survey methods to obtain building damage information presents relatively high accuracy and confidence. However, there are still some shortcomings such as large workload, low efficiency, high costs and unintuitive information, meaning that these methods cannot meet the requirements for the fast acquisition of building damage information [2]. With the progress of sensors and space technology, remote sensing can now provide detailed spatial and temporal information for target areas, whilst usually requiring little field work. Therefore, remote sensing has been widely used in various post-disaster rescue operations, being particularly important for earthquake-stricken areas where it is often difficult to conduct field surveys for the first time [3]. To some extent, previous studies have proved that relatively accurate information of building damage can be obtained from remote sensing data [4].

Many methods to extract information about building damage, caused by earthquakes, from remote sensing images have been presented [5–7]. These can be divided into single-temporal and multi-temporal evaluation methods according to the number of images being used. In single-temporal evaluation methods, only one kind of post-earthquake image is used for the extraction of information. Multi-temporal evaluation methods, in turn, use at least two-temporal images. Due to the influence of data acquisition, sensor revisiting cycle, and filming angle and time, multi-temporal evaluation methods are difficult to be applied in practice [8]. Single-temporal evaluation methods are less restricted and have become an effective technical means to directly extract and evaluate the damage information of buildings using remote sensing images after earthquakes [9]. Chen [10] used object-oriented methods to segment remote sensing images, and classified image objects by Classification and Decision Tree (CART), Support Vector Machine (SVM), and Random Forest (RF) in the machine learning methods. The results showed that, among the three machine learning methods, RF was the best in extracting information about damaged buildings. Janalipour et al. [11] used high spatial resolution remote sensing images as a background to manually select and extract features based on the fuzzy genetic algorithm, establishing a semi-automatic detection system for building damage. This system has higher robustness and precision compared with machine learning methods such as the RF and SVM. In the single-temporal evaluation methods, the spectral, textural and morphological features of the image are mainly used for its classification. This process is often based on the object-oriented classification algorithm for the extraction of information about building destruction [12–14]. However, these object-oriented classification methods present problems such as difficulties in feature space selection and unsatisfactory image segmentation.

In recent years, deep learning technology has achieved great success in the image application field [15], becoming more and more popular in the applications of remote sensing [16]. Convolutional Neural Network (CNN) is a common method of deep learning. Since LeNet5 [17] has achieved satisfactory results in handwritten number recognition tasks compared with traditional methods, a large number of deeper and more complex CNN, such as AlexNet [18], VGGNet [19], Inception V3 [20], and ResNet [21], have made a great breakthrough in large-scale image classification tasks. These CNN models went deeper and deeper, reaching 152 layers by the time of ResNet. A deeper network layer means that deeper image features can be extracted, making the image classification results more accurate. Compared with traditional object-oriented classification methods, CNN-based methods can select and extract classification features automatically, presenting a strong self-learning ability and robustness [22]. Meng Chen et al. [23] proposed a method combining image segmentation with CNN to extract building damage information from remote sensing images, and effectively extracted the damage information of buildings after earthquakes. However, its robustness is largely affected by the accuracy of image segmentation, which affects the effectiveness of the successive tasks and limits practical applications.

With the continuous improvement of basic Geographic Information System (GIS) data in recent years, their applications for the extraction of richer and clearer disaster information have become more popular [24,25]. Ye et al. [8] combined the block information derived from urban road vector data in post-earthquake remote sensing images and constructed a multi-feature classification model with building blocks as its unit. The results showed that this method had high accuracy for the classification of the damage degree of groups of buildings. Therefore, GIS data can be applied to the damage assessment of buildings. In other words, we can accurately achieve the boundary segmentation of the groups of buildings by overlaying the analysis of GIS data and remote sensing images instead of using image segmentation.

To overcome the problem of feature selection and image segmentation in object-oriented classification, this paper proposes a new strategy to extract the damage information of groups of buildings via remote sensing images by combining CNN and GIS data. Using block vector data, all buildings in each block were treated as a group of buildings unit, and the Inception V3 network in CNN was used as the basic classification network to classify the damaged groups of buildings, which was compared with traditional machine learning methods. The rest of this paper is organized as follows. Section 2 describes the details of the method. Section 3 provides the descriptions of the study area. Section 4 presents the analysis of the experimental results. Finally, Section 5 concludes this paper.

2. Method

2.1. The Basic Architecture of Inception V3

The basic architecture of CNN usually includes alternating overlapping convolution and pooling layers, full connection layer, input layer, and output layer. The convolution layer usually includes convolution operation and nonlinear transformation [26]. In the Inception V3 network, besides these basic structures, there are some special operations such as the addition of the well-designed Inception Module to replace the last fully connected layer with global average pooling and the addition of the BN (Batch Normalization) [27] method.

The Inception Module in Inception V3 can improve the efficiency of parameter utilization. One of the structures is shown in Figure 1. It is similar to a small network in a large network, and its structure can be repeatedly stacked to form a larger network. In the Inception Module structure, the convolution of 1×1 can organize information across channels, improve the expression ability of the network, and raise or lower the dimensions of the output channel. Moreover, the idea of spatial asymmetric convolution [20] is introduced to split a large two-dimensional convolution into two smaller one-dimensional convolutions. This helps to reduce many parameters, accelerate the operation, and reduce the overfitting; on the other hand, it adds a layer of nonlinearity, increases the diversity of features and expands the expression ability of the model and its ability to deal with more and richer spatial features.

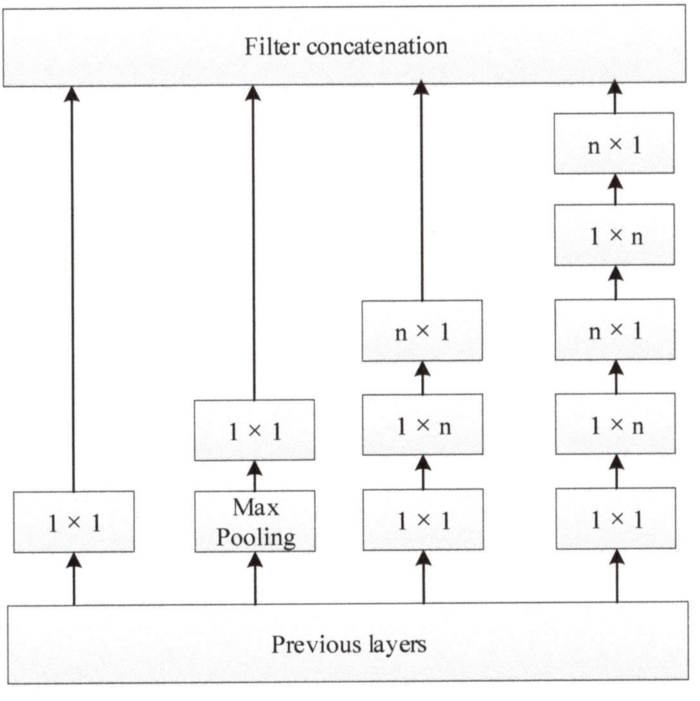

Figure 1. Inception Module structure in Inception V3.

Inception V3 removes the last fully connected layer and replaces it with a global average pooling layer. In AlexNet and VGGNet, the full connection layer accounts for almost 90% of all the parameters, which causes overfitting. Therefore, by removing the full connection layer, the model can be trained faster, and overfitting can be reduced.

The BN method added to Inception V3 is one of the most effective regularization methods. When BN is used for a certain layer of the neural network, it will standardize the internal processing of each mini-batch of data, to normalize the output to N (0,1) and reduce the changes in the distribution of internal neurons. It can considerably accelerate the training speed of large convolutional networks and greatly improve the classification accuracy after convergence.

2.2. Method Flow

In this study, to utilize Inception V3, the remote sensing image was cropped into sub-images of S × S pixels, and these sub-images with fixed size were used as an input for the Inception V3. Each sub-image was processed with Inception V3 to output the probability of each category of damage for groups of buildings. The higher the value, the greater the probability of belonging to the damaged category.

Although CNN can well predict the damage category of groups of buildings [23,28,29] in rectangular images with a fixed size, for groups of buildings in each block, it often results in a larger error to predict directly using CNN due to irregular shape and different sizes of blocks. Therefore, this paper proposed a method to avoid the bigger error when using CNN to predict the categories of blocks presenting irregular shapes and different sizes. The main idea of the method was to cut the rectangular image of fixed size in the minimum bounding rectangle of the block according to a specific step size, and the obtained rectangular image could directly predict its category by using CNN. However, many of the obtained rectangular images were outside the block range, so the threshold of the overlap ratio had to be set to filter out rectangular images outside the block range. Finally, the prediction results of all rectangular images within each block were averaged to obtain the category of the block.

By using the method described above combined with block vector data, the groups of buildings in post-earthquake remote sensing images were classified by the basic processes shown in Figure 2.

The first part was to cut the remote sensing images into rectangular sub-images with a fixed size and train the Inception V3 network to obtain the trained network weights.

The second part consisted of three steps. First, an S × S window was used to scan the images contained in the minimum bounding rectangle of each block with the step size S, and a certain number of S × S sub-images were obtained. Then, the overlap ratio between each sub-image and its block was calculated, and the sub-images with an overlap area greater than 50% were used as inputs for the Inception V3 in order to predict their category probability. Finally, the category probabilities of all effective sub-images in each block were integrated to obtain the category probability of the block.

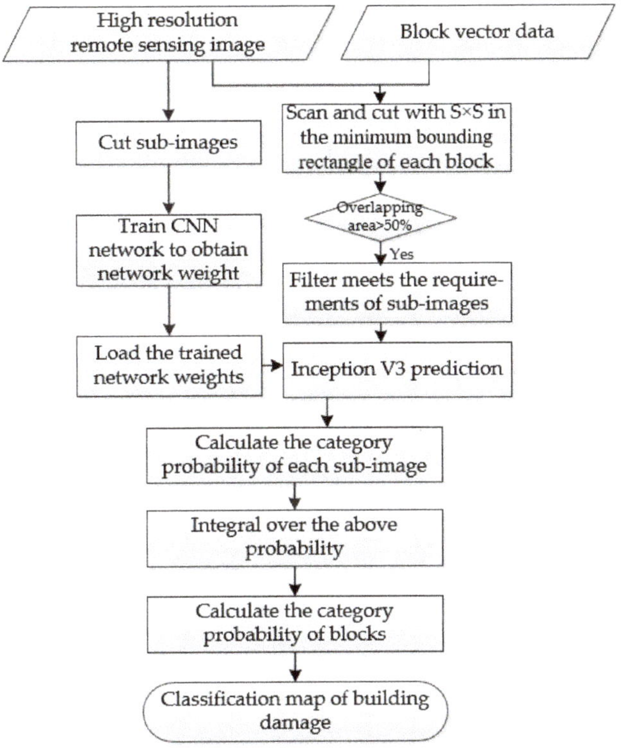

Figure 2. Classification process of damaged groups of buildings by remote sensing images combined with Convolutional Neural Network (CNN) and Geographic Information System (GIS).

2.3. *The Improved Convolutional Neural Network (CNN)*

The remote sensing image usually has a larger width than that of the natural image, so they cannot be directly input into traditional CNN for processing. In order to enable the classification network to more easily process large remote sensing images for application and considering that the method proposed in this paper took city blocks as the basic processing unit, superimposition of block vector data for processing was required. Thus, this paper added Separate and Combination layers to Inception V3, as shown in Figure 3.

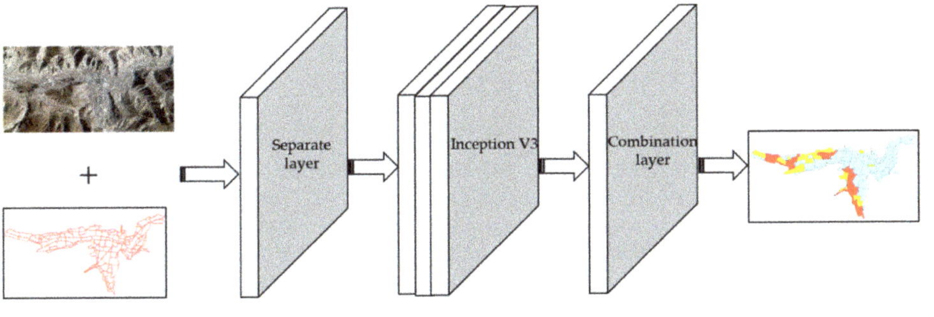

Figure 3. Improved CNN classification framework for damaged groups of buildings after earthquakes.

The role of the Separate layer here was to use a sliding window of 224 × 224 to scan and cut with a step size of 224 within the minimum bounding rectangle of each block. Taking block 1 as an example, as shown in Figure 4, 16 sub-images of 224 × 224 pixels were cut by scanning block 1. Then, the overlap area between these 16 sub-images and block 1 was calculated, and the 7 sub-images in which the overlap ratio was greater than 50% were selected as the valid sub-images, that is, the sub-images belonging to block 1. After the Separate layer, a number of 224 × 224 pixels effective sub-images were generated for each block, and these sub-images were used as the input for the Inception V3 network for classification.

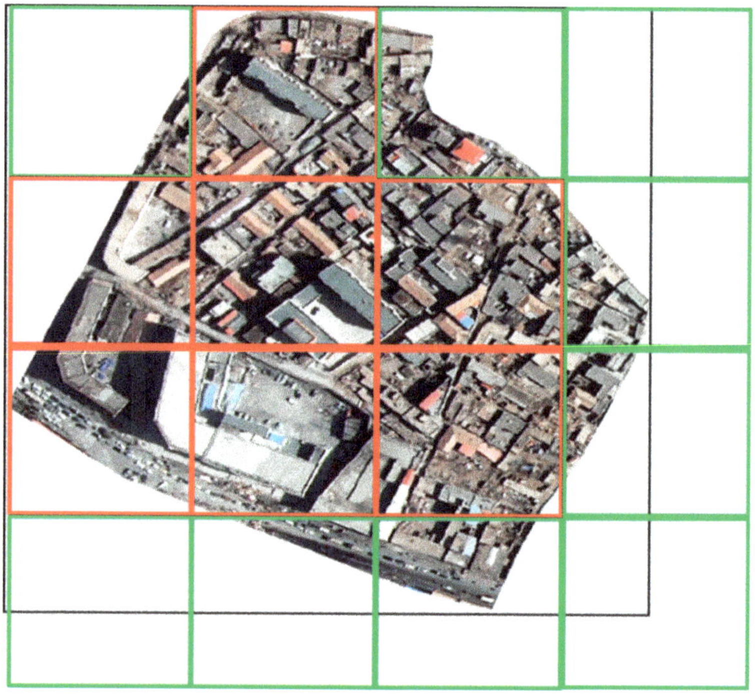

Figure 4. Sub-image cutting schematic diagram of block 1 (red boxes are valid sub-images).

The purpose of the Combination layer was to combine the classification results of all valid sub-images in each block. For example, the function to integrate the probability values of 7 sub-images in block 1 is expressed as:

$$P_{1,j} = \frac{1}{7}\sum_{i=1}^{7} A_{i,j} \quad (1)$$

where $A_{i,j}$ is the probability that the i valid sub-image in block 1 is classified into j class, and $P_{1,j}$ is the probability that block 1 is divided into j class.

3. Data

In order to test the effectiveness of this method, 0.5 m-resolution aerial remote sensing images were selected. These images were acquired on the second day after the 7.1-magnitude earthquake in the Yushu Tibetan Autonomous Prefecture of Qinghai Province on 14 April 2010. The studied area was severely affected by the disaster and there were a large number of collapsed buildings (Figure 5).

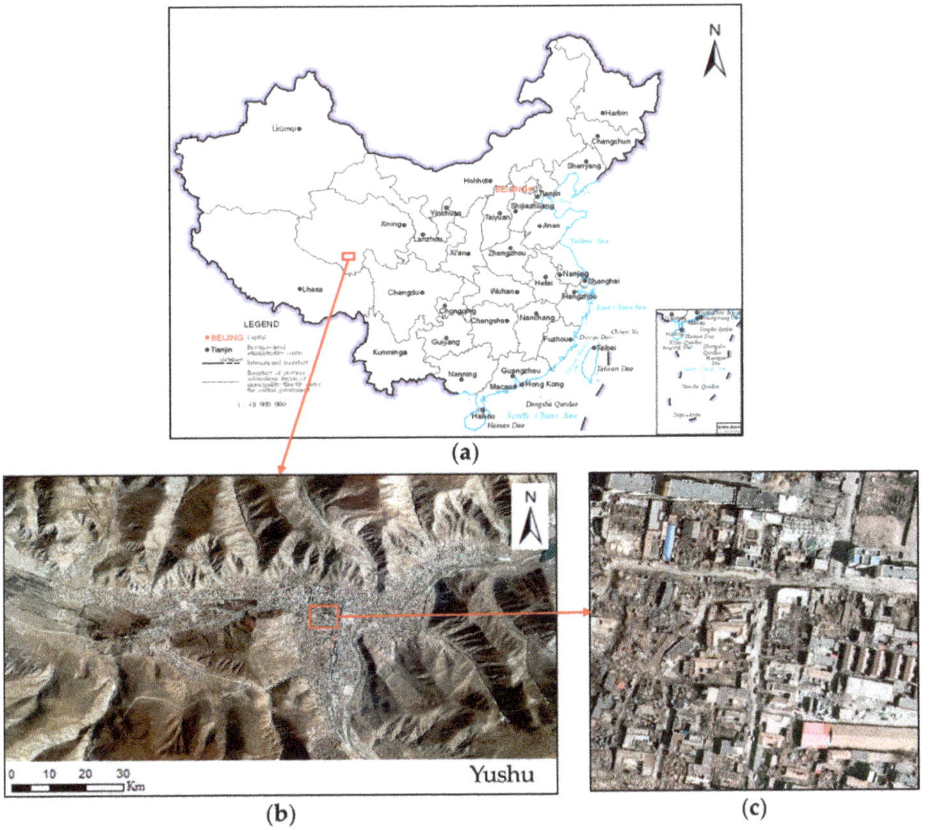

Figure 5. Location of the study area and the remote sensing image: (**a**) China map, (**b**) the image of Yushu Tibetan Autonomous Prefecture, and (**c**) part of the enlarged image.

The earthquake damage of buildings was divided into five grades according to the "Guidelines for Earthquake Damage and its Loss Assessment" formulated by the China Earthquake Administration and the "Classification Standard for Building Earthquake Damage Levels" formulated by the Ministry of Housing and Urban-Rural Development of China. In remote sensing images, classification is based mainly on the overall and detailed image characteristics of the buildings after the earthquake. Additionally, in the classification process, the sub-study area method is adopted to make a general assessment of the damage degree of all buildings in the sub-study area. In this paper, the sub-study area could be divided into a sub-image of 224 × 224 pixels, or a block. Referring to previous research results [8,30,31], and according to the post-earthquake remote sensing images, this paper divided the damage of the groups of buildings into three levels: serious damage (all destroyed or most collapsed), moderate damage (about half collapsed), and slight damage (generally intact or a small part collapsed). The specific classifications are shown in Table 1. The collapse rate, c, was the ratio between the number of collapsed buildings (or collapsed building area) and the total number of buildings (or total building area).

Table 1. Classification of damaged groups of buildings in post-earthquake remote sensing images.

Category	Character Description	Collapse Rate
Serious damage	All destroyed or most collapsed	>70%
Moderate damage	About half collapsed	30–70%
Slight damage	Generally intact or a small part collapsed	<30%

After some necessary processing, the above experimental images can be trained within the CNN. The sub-images of 224 × 224 pixels were clipped from the remote sensing images. Then, one category was assigned to each sub-image by visual interpretation. The samples for each category of Table 1 are shown in Figure 6 (where negative samples are open space and water bodies are near buildings, etc.).

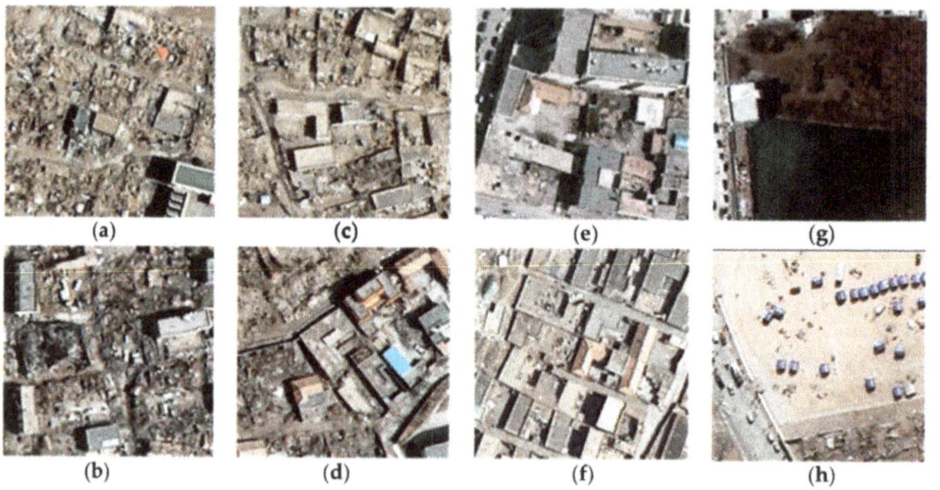

Figure 6. Typical classification samples for damaged groups of buildings: (**a**,**b**) serious damage, (**c**,**d**) moderate damage, (**e**,**f**) slight damage, (**g**,**h**) negative samples.

Data used for training usually needs to be enhanced, similar to natural images, and processes such as rotation, flipping, and color conversion can be used [32]. However, this enhancement is different from that applied to natural images. Most objects in natural images can only be rotated at very small angles, yet, the buildings can be rotated at any angle in this study. In addition, remote sensing images are often displayed after being stretched, so data enhanced by stretching could make the Inception V3 model more robust. Through the process of enhancing, a total of 16,803 samples were obtained. These were divided into three groups: 10,764 samples were used as a training set, 4599 as a validation set, and 1440 as a test set.

4. Results and Discussion

4.1. Multi-Feature Machine Learning Classification Method

To compare the accuracy of the results obtained by the method described in this paper with that of the results of traditional machine learning methods, the experiment involving the machine learning method was firstly carried out. First, various features of the image were extracted manually, and then, according to the features extracted from each block, the damage categories of the groups of buildings were classified by the machine learning method.

The features selected in this study were: contrast, dissimilarity, correlation, entropy, and homogeneity. The extracted image features are shown in Figure 7. It can be seen from the original raster image in Figure 7 that the buildings in the left half of the image were completely collapsed, while the buildings in the right half were less collapsed. Comparing the five characteristic diagrams in Figure 7, it can be seen that the difference between the collapsed buildings on the left side and the non-collapsed buildings on the right side is clearly reflected in the feature map.

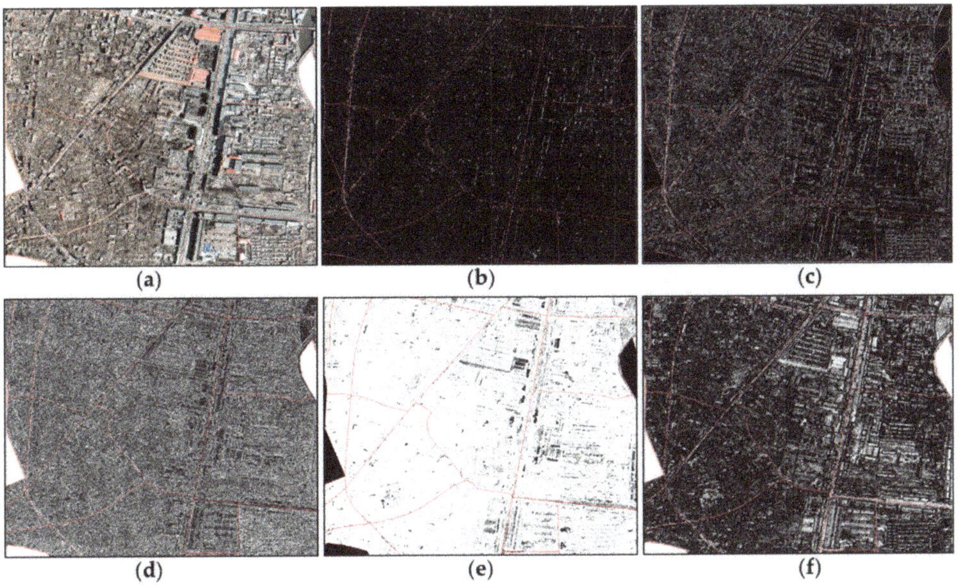

Figure 7. Feature maps: (**a**) raster images, (**b**) contrast, (**c**) dissimilarity, (**d**) correlation, (**e**) entropy, and (**f**) homogeneity.

In order to see more easily the range of values of each feature for the different damage categories, a quantitative statistical analysis was conducted for the feature maps of each category of damage, and the results are shown in Figure 8. Changes in building damage grade cause the values of the five features to change accordingly, which can reflect the damage grade of groups of buildings. Therefore, these five features can be used to classify the damage of groups of buildings.

After completing the feature extraction, the feature maps on each band were superimposed with the block vector data, and the average value of different features in each block were calculated by the ArcGIS software. The results of statistical calculations for multiple features were used as the feature vector of the block to classify the damage of groups of buildings, and the multi-feature classification model was established. According to the statistical results of the aforementioned features, SVM, CART and RF in machine learning methods were used to obtain the classification accuracy of the damage of groups of buildings.

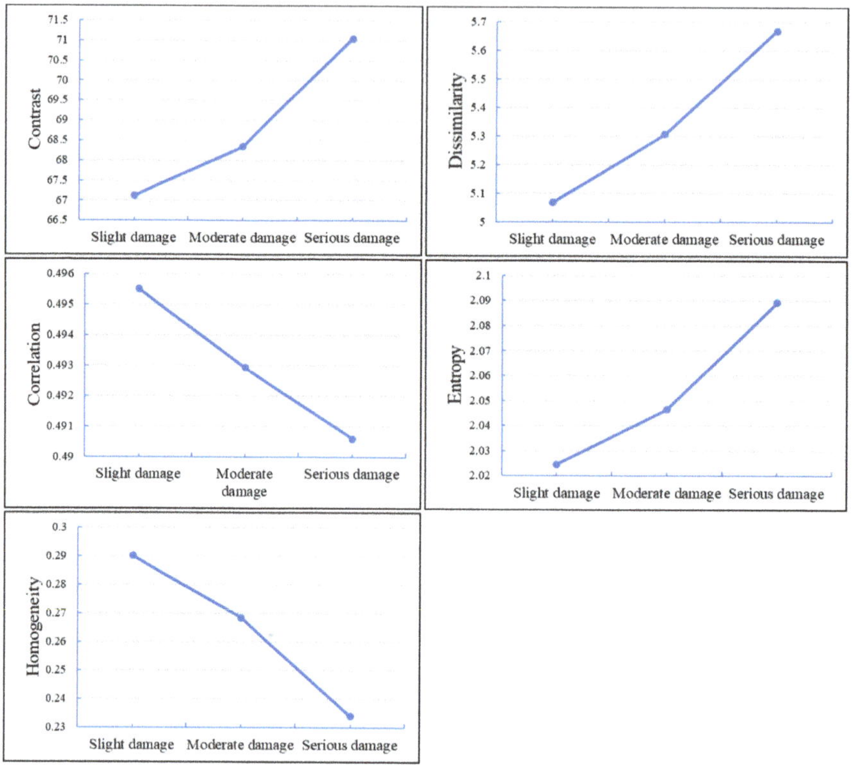

Figure 8. Quantitative statistical analysis curves for different features of each damage category.

The final classification results obtained by using the three machine learning classifiers are shown in Table 2. SVM had the highest classification accuracy, reaching 72%. The parameter C corresponding to the optimal accuracy of SVM was 1, which was the penalty coefficient for the classification error item in this classifier. The larger C corresponded to a greater punishment degree for the misclassification of samples. Therefore, the higher the accuracy in training samples, the lower the generalization ability. Conversely, the smaller C corresponded to the more misclassified samples that were allowed in the training sample then the stronger the generalization ability.

Table 2. Comparison of three machine learning classifiers for groups of buildings damage classification.

Classifier	Optimal Precision	Parameter Corresponding to the Optimal Precision
SVM	72.5%	Penalty factor for error terms: C = 1
CART	70%	Maximum depth of the tree: max_depth = 6
RF	60%	Number of trees in RF: n_estimators = 3

4.2. Improved CNN Classification Method

In this study, the deep learning library TensorFlow was used to replicate the Inception V3 network model. To facilitate the model training, the Separate and the Combination layers shown in Figure 3 were removed. During the training process, the parameters were gradually optimized and adjusted, and the final parameters were as follows: the optimizer was Stochastic Gradient Descent with mini batches, the batch size was 32, the dropout ratio was 0.5. The learning rate started from 0.01 and dropped to 0.001 when the training reached 20,000 steps, and to 0.0001 at 28,000 steps.

After the training, the accuracy of the Inception V3 model finally reached 96.39% for the verification set. The loss value and accuracy change for the verification set during this training process are shown in Figure 9.

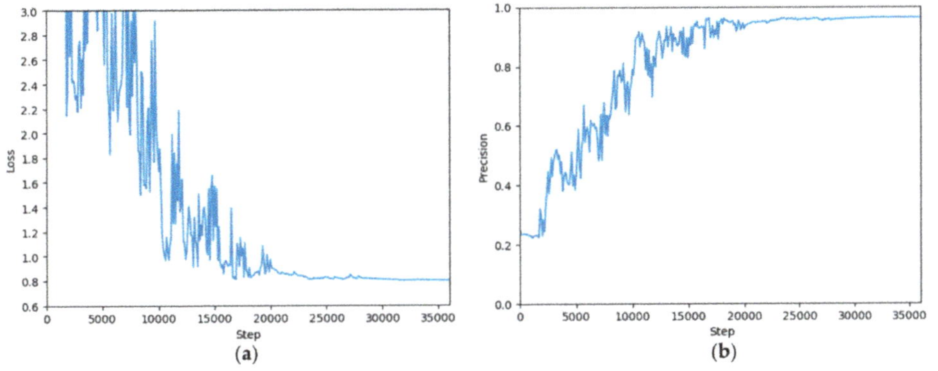

Figure 9. Loss and accuracy changes for the validation set during training: (**a**) loss curve, (**b**) accuracy curve.

The trained model was then tested on the test set of 1440 rectangular images, with an accuracy of 92.22%. The confusion matrix between the test results on the test set and the ground truth is shown in Table 3. The confusion matrix showed that the number of wrong judgements between the serious and the moderate damage was 56 and the number of wrong judgements between the moderate and the slight damage was 52, and there was almost no misjudgment between the serious and the slight damage.

Table 3. Confusion matrix of ground truth and test results of the CNN model.

Category	Ground Truth					
	Serious Damage	Moderate Damage	Slight Damage	Negative Samples	Total	User Accuracy
Serious damage	457	29	0	0	486	94.03%
Moderate damage	27	249	15	0	291	85.57%
Slight damage	1	37	531	0	569	93.32%
Negative samples	1	0	2	90	93	96.77%
Total	486	315	549	90	1440	
Producer accuracy	94.03%	79.05%	96.72%	100.00%		92.22%

Finally, the Separate and Combination layers were added to the Inception V3 network to directly input aerial images of the Yushu earthquake and the vector map of urban blocks, and to output the vector map of building damage. After completion of the process described above, the building damage information was extracted from the aerial images after the Yushu earthquake (Figure 10). To compare the results extracted by the improved CNN method with the ground truth, visual interpretation was carried out by combining the pre-earthquake and post-earthquake images. The visual interpretation results were taken as the ground truth reference results. The results obtained by visual interpretation is shown in Figure 11. Together, Figures 10 and 11 show that the extraction result was basically consistent with the visual interpretation result, although there were still some misjudgments in some blocks.

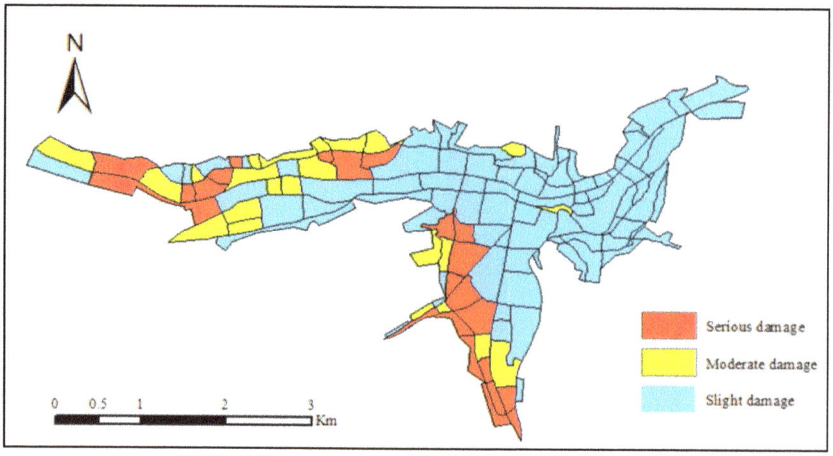

Figure 10. Damage classification map for groups of buildings based on the improved CNN method.

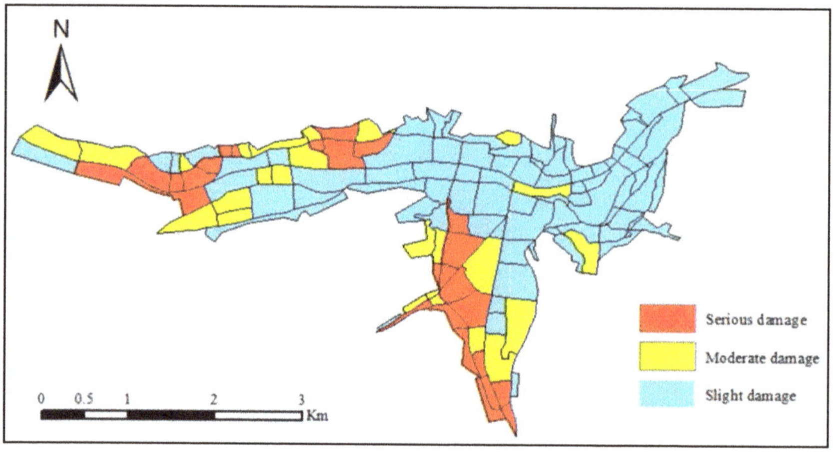

Figure 11. Damage classification map for groups of buildings based on visual interpretation.

To quantitatively compare the results of CNN extraction and visual interpretation, the confusion matrix table is shown in Table 4. The mis-detected blocks mainly resulted from wrong judgement between the moderate and the slight damage, and the number of wrong judgements between the two categories was 7. Finally, the overall accuracy was 90.07%, and Kappa Coefficient was 0.81.

Table 4. Confusion matrix of visual interpretation and extraction results of the improved CNN method.

	Visual Interpretation Results				
Category	**Serious Damage**	**Moderate Damage**	**Slight Damage**	**Total**	**User Accuracy**
Serious damage	21	1	1	23	91.3%
Moderate damage	2	19	2	23	82.61%
Slight damage	2	5	78	85	91.76%
Total	25	25	81	131	
Producer accuracy	84%	76%	96.29%		90.07%

Figure 12 shows examples of some typical blocks. Blocks (1), (2) and (3) are the correct classification blocks, while (4) and (5) are the wrong classification blocks. Table 5 shows the probability distribution of the classification results of the five blocks in Figure 12. From the probability of each category, we concluded that the probability difference between the misclassified category and the correct category was not significant, although the classification category was wrong. This indicated that the final classification category was not the only judging criterion for the blocks with difficulty to distinguish collapse types. It was also possible to determine the confidence degree of each collapse type according to the probability of each category, to evaluate the collapse type of each block more accurately.

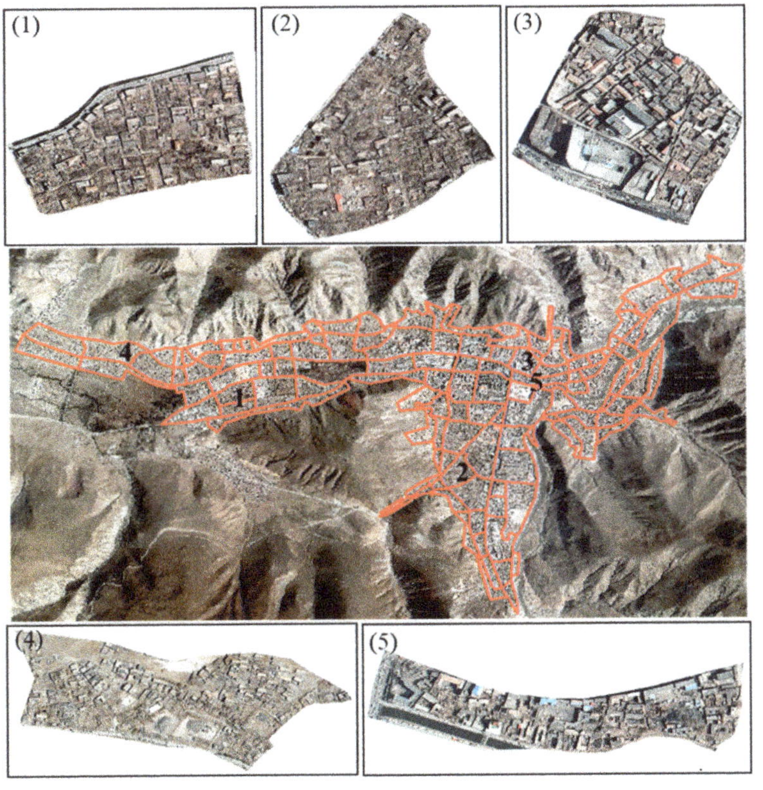

Figure 12. Block building classification example: (1) moderate damage, (2) serious damage, (3) slight damaged, (4) moderate damage, (5) moderate damage.

Table 5. Comparison of the test results of the improved CNN method with the results of visual interpretation.

Block	Categories Obtained by Visual Interpretation	Test Results of the Improved CNN Method			
		Categories	Probability of Serious Damage	Probability of Moderate Damage	Probability of Slight Damage
(1)	Moderate damage	Moderate damage	29.29%	48.35%	11.79%
(2)	Serious damage	Serious damage	79.12%	11.09%	0.01%
(3)	Slight damage	Slight damage	0.02%	0.23%	96.30%
(4)	Moderate damage	Serious damage	27.64%	23.91%	17.55%
(5)	Moderate damage	Slight damage	0.83%	25.50%	56.42%

4.3. Discussion

In this study, some measures were taken to prevent overfitting in the training of the CNN. In fact, there was not enough data on building damage to meet the requirements. When these limited samples were used to train the large CNN, it tended to over fit. For this reason, the data set was enhanced and expanded to increase the diversity of samples. Furthermore, in the training process of the CNN, when the loss function value did not decline after a certain number of steps, the training was terminated in advance to avoid excessive learning.

In this study, the classification of damaged groups of buildings achieved high accuracy through training, but there were also some wrong classifications, which were mainly attributed to the following two reasons. First, the background environment of the high-resolution remote sensing images after an earthquake are far more complex than those of the natural images, so the background environment has a greater impact on the classification results. For example, the characteristics of collapsed adobe house and bare soil are very similar, thus, it is easy to judge bare soil as collapsed adobe house. Second, because the labeling of sample categories is judged by visual interpretation and lacks the support of ground survey data, this may cause some error labels.

The improved CNN approach proposed in this study can be extended to other CNNs. With the continuous advancement of deep learning, a CNN with higher accuracy and better performance will be developed in the near future. Adding Separate and Combination layers to better network architectures may allow the achievement of a better classification effect.

5. Conclusions

By combining an improved CNN approach with GIS data, this paper proposed a new strategy to extract the damage information of groups of buildings in remote sensing images after earthquakes. From our experiment, we found that CNNs could effectively solve the problem of difficult feature selection, which is an advantage over traditional object-oriented classification methods. Compared with the traditional multi-feature machine learning classification method constructed by artificial feature extraction, accuracy is greatly improved, and a satisfactory effect can be achieved. Block vector data in GIS can form a meaningful boundary for groups of buildings, effectively replacing image segmentation and avoiding its fragmentary and unsatisfactory results. At the same time, our method was able to avoid the big error that arises when the CNN is used to predict irregular shapes and different sizes of blocks.

However, due to the limited number of samples used for training in CNN and the confusion between collapsed buildings and bare ground, classification errors for the groups of buildings can be easily caused, meaning that there are still some errors in comparison with the ground truth.

Therefore, extending the training data set, including remote sensing images of different types and resolutions, is future work to be tested for improvement. The methods combining multi-classifiers including CNN should be considered to improve the classification accuracy of groups of buildings.

Author Contributions: All authors contributed in a substantial manner to the manuscript. H.M. conceived, designed and performed research and wrote the manuscript. Y.L. and Y.R. made contributions to the design of the research and data analysis. All authors discussed the basic structure of the manuscript. All authors read and approved the submitted manuscript.

Funding: This research was funded by the National Key Research and Development Program, Project NO. 2017YFC1500902.

Conflicts of Interest: The authors declare no conflicts of interest.

References

1. Dell'Acqua, F.; Gamba, P. Remote sensing and earthquake damage assessment: Experiences, limits, and perspectives. *Proc. IEEE* **2012**, *100*, 2876–2890. [CrossRef]
2. Chen, W. *Research of Remote Sensing Application Technology Based on Earthquake Disaster Assessment*; China Earthquake Administration Lanzhou Institute of Seismology: Lanzhou, China, 2007.
3. He, M.; Zhu, Q.; Du, Z. A 3D shape descriptor based on contour clusters for damaged roof detection using airborne LiDAR point clouds. *Remote Sens.* **2016**, *8*, 189. [CrossRef]
4. Menderes, A.; Erener, A.; Sarp, G. Automatic detection of damaged buildings after earthquake hazard by using remote sensing and information technologies. *Procedia Earth Planet. Sci.* **2015**, *15*, 257–262. [CrossRef]
5. Gong, L.; Li, Q.; Zhang, J. Earthquake building damage detection with object-oriented change detection. In Proceedings of the IEEE International Geoscience and Remote Sensing Symposium-IGARSS, Melbourne, Australia, 21–26 July 2013; pp. 3674–3677.
6. Janalipour, M.; Mohammadzadeh, A. Building Damage Detection Using Object-Based Image Analysis and ANFIS from High-Resolution Image (Case Study: BAM Earthquake, Iran). *IEEE J. Sel. Top. Appl. Earth Obs. Remote Sens.* **2015**, *9*, 1937–1945. [CrossRef]
7. Nie, J.; Yang, S.; Fan, Y. Building losses assessment for Lushan earthquake utilization multisource remote sensing data and GIS. In Proceedings of the Mippr: Automatic Target Recognition and Navigation, Enshi, China, 31 October–1 November 2015; p. 98120J.
8. Ye, X.; Wang, J.; Qin, Q. Damaged building detection based on GF-1 satellite remote sensing image: A case study for Nepal MS8.1 earthquake. *Acta Seismol. Sin.* **2016**, *38*, 477–485.
9. Dong, L.; Shan, J. A comprehensive review of earthquake-induced building damage detection with remote sensing techniques. *ISPRS J. Photogramm. Remote Sens.* **2013**, *84*, 85–99. [CrossRef]
10. Chen, J. *Research on Extraction Methods of Damaged Buildings after Earthquake Based on Optical Remote Sensing*; China Earthquake Administration Lanzhou Institute of Seismology: Lanzhou, China, 2018.
11. Janalipour, M.; Mohammadzadeh, A. A fuzzy-ga based decision making system for detecting damaged buildings from high-spatial resolution optical images. *Remote Sens.* **2017**, *9*, 349. [CrossRef]
12. Zhang, L.; Zhang, J. Research on building collapse rate calculation method based on object-oriented classification. *Earthquake* **2009**, *29*, 139–145.
13. Li, F.; Ma, C.; Zhang, G. Rapid assessment of earthquake damage to buildings based on object-oriented classification. *J. Henan Polytech. Univ. Sci.* **2011**, *30*, 55–60.
14. Wang, Y.; Wang, X.; Dou, A. Building damage detection of the 2008 Wenchuan, China earthquake based on object-oriented classification method. *Earthquake* **2009**, *29*, 54–60.
15. Nogueira, K.; Penatti, O.A.B.; Santos, J.A.D. Towards better exploiting convolutional neural networks for remote sensing scene classification. *Pattern Recognit.* **2017**, *61*, 539–556. [CrossRef]
16. Zhu, X.X.; Tuia, D.; Mou, L. Deep learning in remote sensing: A comprehensive review and list of resources. *IEEE Geosci. Remote Sens. Mag.* **2017**, *5*, 8–36. [CrossRef]
17. Lecun, Y.; Bottou, L.; Bengio, Y. Gradient-based learning applied to document recognition. *Proc. IEEE* **1998**, *86*, 2278–2324. [CrossRef]
18. Krizhevsky, A.; Sutskever, I.; Hinton, G.E. Imagenet classification with deep convolutional neural networks. In Proceedings of the Advances in Neural Information Processing Systems, Lake Tahoe, CA, USA, 3–6 December 2012; pp. 1097–1105.
19. Simonyan, K.; Zisserman, A. Very deep convolutional networks for large-scale image recognition. *arXiv* **2014**, arXiv:1409.1556.
20. Szegedy, C.; Vanhoucke, V.; Ioffe, S. Rethinking the inception architecture for computer vision. In Proceedings of the IEEE Conference on Computer Vision and Pattern Recognition, Seattle, WA, USA, 27–30 June 2016; pp. 2818–2826.
21. He, K.; Zhang, X.; Ren, S. Deep residual learning for image recognition. In Proceedings of the IEEE Conference on Computer Vision and Pattern Recognition, Seattle, WA, USA, 27–30 June 2016; pp. 770–778.
22. Haykin, S. *Neural Networks and Learning Machines, 3/E*; Pearson Education: Upper Saddle River, NJ, USA, 2010.
23. Chen, M.; Wang, X.; Dou, A. The extraction of post-earthquake building damage information based on convolutional neural network. *Int. Arch. Photogramm. Remote Sens. Spat. Inf. Sci.* **2018**, *42*, 161–165. [CrossRef]

24. Wang, L.; Wang, X.; Ding, X. Study on loss assessment of construction earthquake damage based on remote sensing and GIS. *Earthquake* **2007**, *27*, 77–83.
25. Samadzadegan, F.; Rastiveisi, H. Automatic detection and classification of damaged buildings, using high resolution satellite imagery and vector data. *Int. Arch. Photogramm. Remote Sens. Spat. Inf. Sci.* **2008**, *37*, 415–420.
26. Li, Y.; Hao, Z.; Lei, H. Overview of convolutional neural networks. *J. Comput. Appl.* **2016**, *36*, 2508–2515.
27. Ioffe, S.; Szegedy, C. Batch normalization: Accelerating deep network training by reducing internal covariate shift. *arXiv* **2015**, arXiv:1502.03167.
28. Duarte, D.; Nex, F.; Kerle, N. Satellite image classification of building damages using airborne and satellite image samples in a deep learning approach. *ISPRS Ann. Photogramm. Remote Sens. Spat. Inf. Sci.* **2018**, *4*, 89–96. [CrossRef]
29. Ji, M.; Liu, L.; Buchroithner, M. Identifying collapsed buildings using post-earthquake satellite imagery and convolutional neural networks: A case study of the 2010 Haiti earthquake. *Remote Sens.* **2018**, *10*, 1689. [CrossRef]
30. Cao, D.; Shi, X.; Zhang, J. Study on the statistical characteristics of seismic disaster information of building in remote sensing image. *Remote Sens. Land Resour.* **2001**, *13*, 42–46.
31. Wei, Z.; Huanfeng, S.; Chunlin, H. Building earthquake damage information extraction from a single post-earthquake PolSAR image. *Remote Sens.* **2016**, *8*, 171.
32. Perez, L.; Wang, J. The effectiveness of data augmentation in image classification using deep learning. *arXiv* **2017**, arXiv:1712.04621.

remote sensing

MDPI

Letter

Improved Anchor-Free Instance Segmentation for Building Extraction from High-Resolution Remote Sensing Images

Tong Wu [1,2], Yuan Hu [1,2], Ling Peng [1,*] and Ruonan Chen [1,2]

1 Aerospace Information Research Institute, Chinese Academy of Sciences, Beijing 100094, China; wutong@aircas.ac.cn (T.W.); huyuan@radi.ac.cn (Y.H.); chenrn@aircas.ac.cn (R.C.)
2 University of Chinese Academy of Sciences, Beijing 100049, China
* Correspondence: pengling@aircas.ac.cn

Received: 21 July 2020; Accepted: 4 September 2020; Published: 8 September 2020

Abstract: Building extraction from high-resolution remote sensing images plays a vital part in urban planning, safety supervision, geographic databases updates, and some other applications. Several researches are devoted to using convolutional neural network (CNN) to extract buildings from high-resolution satellite/aerial images. There are two major methods, one is the CNN-based semantic segmentation methods, which can not distinguish different objects of the same category and may lead to edge connection. The other one is CNN-based instance segmentation methods, which rely heavily on pre-defined anchors, and result in the highly sensitive , high computation/storage cost and imbalance between positive and negative samples. Therefore, in this paper, we propose an improved anchor-free instance segmentation method based on CenterMask with spatial and channel attention-guided mechanisms and improved effective backbone network for accurate extraction of buildings in high-resolution remote sensing images. Then we analyze the influence of different parameters and network structure on the performance of the model, and compare the performance for building extraction of Mask R-CNN , Mask Scoring R-CNN , CenterMask, and the improved CenterMask in this paper. Experimental results show that our improved CenterMask method can successfully well-balanced performance in terms of speed and accuracy, which achieves state-of-the-art performance at real-time speed.

Keywords: building extraction; improved anchor-free instance segmentation; high-resolution remote sensing images; deep learning

---

1. Introduction

In pace with the high-speed development of high-resolution remote sensing data in both China and International Community, the spatial information, geometric structures, textural features and intensity information contained in remote sensing images and point cloud data are becoming clearer, which makes it possible to identify and detect terrestrial objects. Among them, building is an important feature of city and the most important place for human production and life [1]. The geometry, area, or the dimensions of buildings gained from two-dimensional information-rich optical remote sensing images and three-dimensional information-containing point cloud data [2] are the relevant urban metrics. They can effectively represent the urban spatial structure, and then quantify the morphology of city [3], reflect the processes that occur during a city's development [4], and monitor urban management and planning strategies [5]. Thus, the identification and extraction of the individual building will play a vital role in a wide range of applications such as urban planning, safety supervision, real-estate management and Geo-database updates.

For decades, researchers have made considerable efforts to extract buildings from remote sensing data. Huertas and Nevatia [6] assumed that the building is composed of rectangular components (such as "T" shape, "L" shape and "E" shape), and then designed a building model to detect buildings. Irvin and McKeown [7] used the spatial constraint relationship between shadows and buildings to extract buildings. Inglada [8] achieved automatic recognition of buildings in high resolution optical remote sensing images by support vector machine (SVM) classification of geometric image features. Meng et al. [9] proposed a novel object-oriented building extraction method based on fuzzy SVM. Awrangjeb et al. [10], designed an innovative image line guided segmentation technique to extract the roof planes based on light detection and ranging (LIDAR) and orthoimage, and applied a newly proposed rule-based procedure to removing planes constructed on trees. In Reference [11], to represent an individual building or tree, researchers clustered the non-ground LIDAR points, and then the planar roof segments were extracted from each cluster of points and refined using rules, such as the coplanarity of points and their locality. Gilani et al. [12] used LIDAR data to present a non-manifold points creation methods that provides a better interpolation of roof regions, these geometric features were preserved to achieve automated identification and segmentation of the building roof. Besides, they also used features from point cloud and orthoimagery to extract and regularise the buildings, so as to overcome the limitations of shadow and partly occlusion [13]. As can be seen from above, scholars have tried many methods to extract buildings from optical remote sensing images, point cloud data, or the fusion of optical images and point cloud data. However, the features of buildings used in all the above methods are artificially designed shallow features, which are time-consuming and labor-intensive, sparse for feature distribution, and cannot express higher-level semantic information.

In the past few years, deep learning algorithms have made breakthrough progress in the field of image processing. And the convolutional neural network (CNN) has been used to abstract multi-level and metaphysical features from original images, which undoubtedly provides a huge advantage for obtaining the complex spectrum, texture, and geometric features contained in remote sensing images. Based on that, currently, most researchers have used semantic segmentation frameworks (such as U-Net [14], SegNet [15], and DeepLab [16]), to achieve efficient and automatic buildings extraction from high-resolution remote sensing images. For instance, Xu et al. [17] designed a segmentation method with deep residual networks and a guided filter to gain buildings from remote sensing images. In Reference [18], to balance high accuracy with low network complexity, Shrestha and Vanneschi proposed an enhanced fully convolutional network (FCN) structure by adding conditional random fields (CRFs) and successfully obtained buildings. Li et al. [19] integrated latent high-order structural features learned by the adversarial network into semantic segmentation network during network training, and can effectively rectify spatial inconsistency on aerial images.

Although the CNN-based semantic segmentation methods have achieved promising results, they still have drawbacks.The main problem lies in a large number of closely adjacent buildings existing in remote sensing images. But semantic segmentation can not distinguish different objects of the same category. Under complex and fluid geographical environment, it may lead to edge connection and is unfavorable for the application and research which focused on single building extraction. Compared with the task of image semantic segmentation, instance segmentation can not only identify the individual buildings on the image, but also give the pixel-levels semantic categories of each target on this basis.

At present, most state-of-the-art approaches to instance segmentation are based on the two-stage object detection model, and Mask R-CNN [20] is one of the classic models and standard frameworks for them, which has achieved the best result of a single model in COCO instance segmentation challenge. It designed RoIAlign instead of RoI pooling layer used in Faster R-CNN [21], and plugged an FCN branch into the original classification and regression network branch to predict the mask. As scholars continue to study, there have been many works [22–25] on improving the Mask R-CNN, but few considered the speed of instance segmentation. Inspired by SSD [26] and YOLO [27], some scholars have designed instance segmentation model based on one-stage object detection methods. For example, YOLACT [28] used RetinaNet as the basic network and added two parallel branches to complete the mask prediction task: the first branch applied FCN to generating a series of prototype masks independent of a single instance; the second branch added mask coefficients to prediction head to encode the representation of an instance in the prototype mask space. Finally, after Non-Maximum Suppression(NMS) operation, the output results of the two branches were linearly combined to get the final prediction results. Based on state-of-the-art instance segmentation approaches, some researchers designed novel building extraction methods from high-resolution remote sensing images. Potlapally et al. [29] extracted various types of remote sensing features including buildings by employing Mask R-CNN. Ji et al. [30] utilized building extraction network implemented with a Mask R-CNN branch for object-based instance segmentation, and a multi-scale FCN branch for pixel-based semantic segmentation to locate changed buildings as well as the changed pixels from aerial remote sensing images. Li et al. [31] improved Mask R-CNN by adding key points map, and completed well preservation of geometric details for buildings. Su et al. [32] proposed an advanced Cascade Mask R-CNN which named HQ-ISNet and made the predicted instance masks more accurate.

As can be seen from above, CNN-based instance segmentation methods for building extraction from high-resolution remote sensing images are generally via two-stage, just like Mask R-CNN, which focus primarily on extraction performance, few take the speed and real-time ability into account. Although there are some instance segmentation methods, such as YOLACT, are built on one-stage detector that directly predicts boxes without proposal step. They still rely heavily on pre-defined anchors, which are sensitive to data sets and hyperparameters (e.g., aspect ratio, ratio, input size, etc.). In addition, to ensure sufficient overlap with most ground truth boxes, excessively anchor setting results in the matters of higher computation/storage cost and imbalance between positive and negative samples. In order to address these problem, recently, many kinds of research [33–37] tended to replace the anchors with anchor-free by using corner/center points. Compared with anchor-based detectors, anchor-free will contribute to more efficient computation and better performance.

Therefore, in this paper, we propose an improved anchor-free instance segmentation method based on CenterMask [37] with spatial and channel attention-guided mechanisms for accurate extraction of buildings in high-resolution remote sensing images. It maintains good performance yet realize efficient. The main contributions of our work can be summarized as follows:

- An improved anchor-free instance segmentation architecture is proposed for building extraction from high-resolution remote sensing images, which is composed of an efficient one-stage anchor-free object detector FCOS [33] and a novel spatial and channel attention-guided mask branch.
- Besides, we design a more effective backbone network by improving VoVNetV2 that is designed in Reference [37], and receives better performance than ResNet and VoVNetV2.
- In order to improve the segmentation performance of CenterMask, we also develop the mask branch in CenterMask. A spatial and channel attention-guided mask (SCAG-Mask) branch is designed in this paper to effectively optimize the building extraction behavior.

The next content of the paper is organized as follows—Section 2 introduces our instance segmentation network architecture and the details for building extraction. Section 3 details the dataset and evaluation approach used in the experiment. Section 4 describes the experimental results and analysis in detail. Then Section 5 draws a conclusion and looks forward to the follow-up work.

2. Methods

The overall architecture of the improved CenterMask is shown in Figure 1. It consists of four parts, and (a) in Figure 1 represents the feature extraction network. This paper uses the improved VoVNetV2-57 to complete the convolution feature extraction of the input image. It consists of three convolutional layers and four stages (C2-C5) with different numbers of one-shot aggregation (OSA) modules. Each OSA module has five convolutional layers. In order to realize the fusion of the shallow position information and deep semantic information of the convolutional neural network, (b) in Figure 1 is connected to feature pyramid network (FPN) [38]. In this paper, the output of C3–C5 for improved VoVNetV2-57 network is operated through upsampling and horizontal connection operations to generate P3–P5, and P6 and P7 are obtained through convolution operations on the basis of P5, thus rich feature information extraction from a single-resolution input image is completed. After passing through FPN network, the FCOS bounding box prediction network is connected to each scale feature map to generate the region of interests(RoIs), as shown in (c) of Figure 1. By judging the relative size of the generated RoI and input image, the feature map matching the scale of the RoI is selected and operated by RoIAlign [20]. After processing, the RoI and the feature map are fed into the spatial and channel attention-guided mask branch in Figure 1d for segmenting the instance in this RoI. Finally, the exported instance segmentation extraction results are obtained.

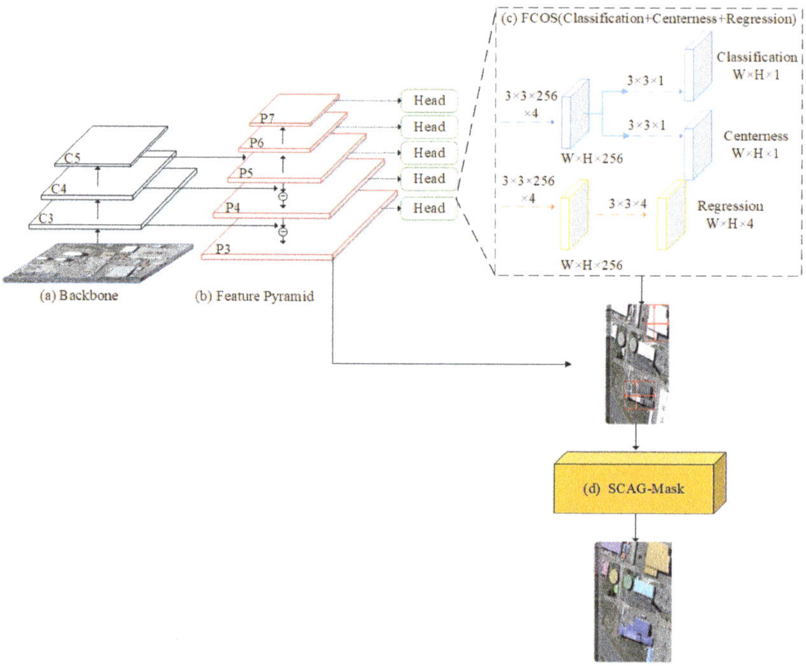

Figure 1. The network architecture of Improved CenterMask, where C3, C4, and C5 denote the feature maps of the backbone network, and P3 to P7 are the feature levels used for the final prediction. W, H separately represent the width and height of feature maps.

2.1. Improved VoVNetV2

VoVNet [39] is an efficient feature extraction network. It uses one-time aggregation module. In this module, each convolution layer generates two kinds of connections. One is connected with the next layer through convolution to gain larger receptive fields. The other one is connected to the final output layer to aggregate features.

However, with the deepening of the network and the stacking of OSA modules, the accuracy of the model will be saturated. According to the ResNet [40] literature, this is because the deepening of the network causes the problem of gradient explosion and gradient disappearance. In Reference [37], to boost the performance of VoVNet, the identity mapping and an effective Squeeze-Excitation(eSE) module are added. In this paper, we let the C4 and C5 stages retain the added identity mapping as shown in Figure 2a. At the same time, we improve the eSE module, and the improved eSE module (im-eSE) is proposed and added to the OSA module at each stage of C2-C5 to learn the correlation between feature maps to filter out the more powerful feature maps, as shown in Figure 2b.

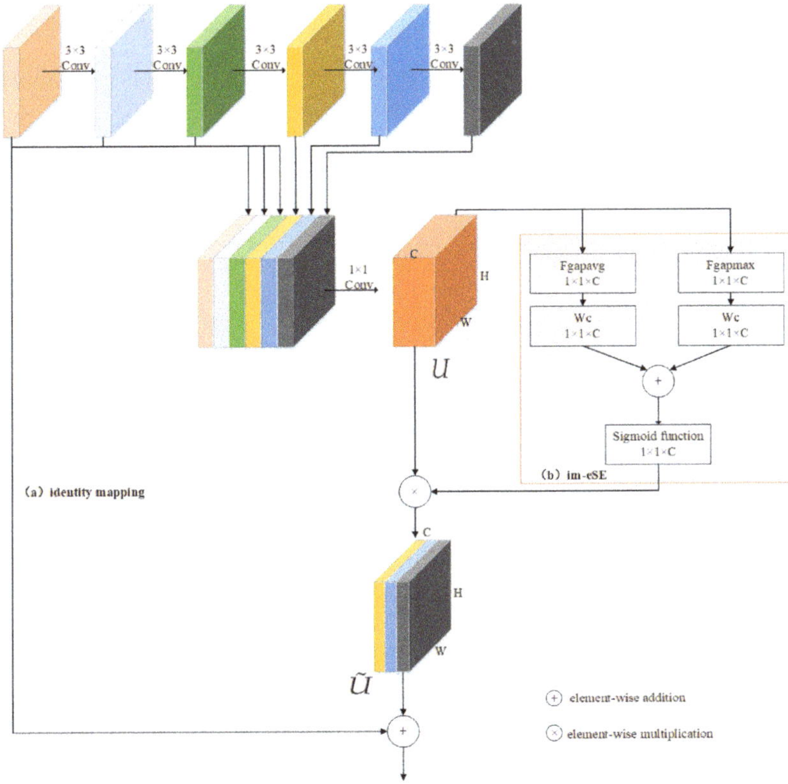

Figure 2. The structure of feature extraction network, where Fgapavg is global average pooling, Fgapmax is global max pooling, WC is fully-connected layer. W, H, C separately represent the width, height, and the number of channels for feature maps.

In order to obtain the global statistics while retaining the degree of the most significant part of each feature map, im-eSE module in this paper uses a global average pooling and a global max pooling. Suppose that after passing through the original OSA module, a feature map U of $W \times H \times C$ is obtained. Then the gating unit in im-eSE module can be expressed as:

$$A_{im-eSE}(U) = \sigma(Wc(Fgapavg(U)) \oplus Wc(Fgapmax(U))), \quad (1)$$

where, σ is sigmoid function. $\oplus$ denotes element-wise addition. Fgapavg and Fgapmax are defined as:

$$Fgapavg(U) = \frac{1}{WH} \sum_{i,j=1}^{W,H} U_{i,j}, \quad (2)$$

$$Fgapmax(U) = max_{i \in W, j \in H} U_{i,j}. \quad (3)$$

Finally, output $\tilde{U}$ of the im-eSE module is obtained by multiplying the gating unit $A_{im-eSE}(U)$ with the feature map U, see formula (4), where $\otimes$ denotes element-wise multiplication. After this, for the C4 and C5 stages, identity mapping is added to $\tilde{U}$, so as to obtain the output feature maps of this OSA module.

$$\tilde{U} = A_{im-eSE}(U) \otimes U. \tag{4}$$

2.2. FCOS

FCOS uses pixel-by-pixel idea for object detection. For each point (x, y) on the feature map, it will be mapped back to the coordinates of the input image as (x_s, y_s). formula (5) shows the conversion relationship between these, where s represents the downsampling rate of the current feature map relative to the input image.

$$x_s = \left\lfloor \frac{s}{2} \right\rfloor + x \times s, y_s = \left\lfloor \frac{s}{2} \right\rfloor + y \times s. \tag{5}$$

For any ground truth box B_i given in this paper, is defined as $(x_0^{(i)}, y_0^{(i)}, x_1^{(i)}, y_1^{(i)}, C)$, where $(x_0^{(i)}, y_0^{(i)})$ and $(x_1^{(i)}, y_1^{(i)})$ are the coordinates of the upper left and lower right corners of the ground truth box, C is the building category. If (x_s, y_s) calculated by formula (5) falls into any ground truth box, it is temporarily considered to be a positive sample, and its regression target is (l^*, t^*, r^*, b^*). Here l^*, t^*, r^*, b^* are the distances from the location to the four sides of the ground truth box, as shown in (6). Otherwise, it will be considered as a negative sample and its category will be set to 0. If the point is located in multiple ground truth boxes, it is considered an ambiguous sample. And then we select the ground truth box with the minimal area as the regression target. In addition, for multi-level prediction with FPN, if the regression target of positive sample at level i satisfies $max(l^*, t^*, r^*, b^*) \geq m_i$ or $max(l^*, t^*, r^*, b^*) \leq m_{i-1}$, it will be set as a negative sample, and can not regress a bounding box anymore. Here m_i is the maximum distance that feature level i can regress. In this paper, we set m_2, m_3, m_4, m_5, m_6 and m_7 as 0, 64, 128, 256, 512 and ∞, respectively.

$$l^* = x_s - x_0^{(i)}, t^* = y_s - y_0^{(i)}, r^* = x_1^{(i)} - x_s, b^* = y_1^{(i)} - y_s. \tag{6}$$

As can be seen from Figure 1, FCOS contains classification subnet and regression subnet. In classification subnet, we firstly use four $3 \times 3 \times 256$ convolutional layers, and then through a 3×3 convolutional layer with the channel number of 1 to predict the probability that each bounding box belongs to a building in the range of $W \times H$. The design of the regression subnet is same to classification subnet, the only difference is that the number of channels in the last layer is 4, that is, each bounding box will regress a four-dimensional vector, indicating the offset of the bounding box from the related ground truth box. In other words, without the anchor-box, FCOS directly predicts a 4D vector plus a class label at each spatial location on a level of feature maps.

Besides, in order to suppress the low-quality bounding boxes generated at the locations that are far from the center of a target object. FCOS adds a centerness branch to the classification subnet to predict the deviation of a pixel to the center of its corresponding bounding box. Given the regression targets l^*, t^*, r^*, b^* for a location, the centerness target can be defined as,

$$centerness = \sqrt{\frac{min(l^*, r^*)}{max(l^*, r^*)} \times \frac{min(t^*, b^*)}{max(t^*, b^*)}}, \tag{7}$$

When $l^* == r^*, t^* == b^*$, it is the most ideal state, and the centerness value is 1. In this paper, the binary cross-entropy(BCE) loss is used to calculate the loss caused by different centerness values. The smaller the centerness value is, the greater BCE loss will be, so that the predicted bounding box will be close to the center point during training. When testing, the value of centerness is equivalent to a weight, which is used to multiply the classification score of each RoI. Thus the centerness can down weight the scores of bounding boxes far from the center of an object. Finally, the detection performance can be effectively improved by filtering out the bounding boxes with low scores by NMS operation.

2.3. Spatial and Channel Attention-Guided Mask

We judge the feature map matched with the RoIs obtained by FCOS according to formula (8), where k_{max} is the maximum level of the feature map. A_{input} and A_{RoI} represent the area of the input image and the area of RoI, respectively. If the area of the RoI is greater than half of the input image area, the RoI is assigned to the feature map with the largest scale. In this paper, we set k_{max}=7, if k is lower than the minimum level (in this paper, we set the minimum level to P3), then k is forcibly assigned to the minimum scale feature map P3.

$$k = k_{max} - \log_2 \frac{A_{input}}{A_{RoI}}. \tag{8}$$

Inspired by the spatial attention mechanism, this paper uses a SCAG-Mask mechanism to guide the mask subnet to predict class-specific masks by focusing on meaningful pixels and suppressing non-information pixels. The details are shown in Figure 3 and formula (9).

$$\begin{aligned} X = F_{1\times1}(deconv_{2\times2}((\sigma(F_{3\times3}(P_{max} \circ P_{avg})) \otimes x_i) \oplus (\sigma(\\ Wc(Fgapavg(x_i)) \oplus Wc(Fgapmax(x_i))) \otimes x_i))), \end{aligned} \tag{9}$$

where x_i denotes the input feature map obtained through RoIAlign operation and four convolutional layers. P_{max} is the max pooling of all feature maps for each position. P_{avg} is the average pooling for all feature maps for each position. $\circ$ represents concatenate operation. $F_{1\times1}$ and $F_{3\times3}$ separately denote 1×1 conv and 3×3 conv. $\oplus$ and $\otimes$ respectively represent element-wise addition and element-wise multiplication.

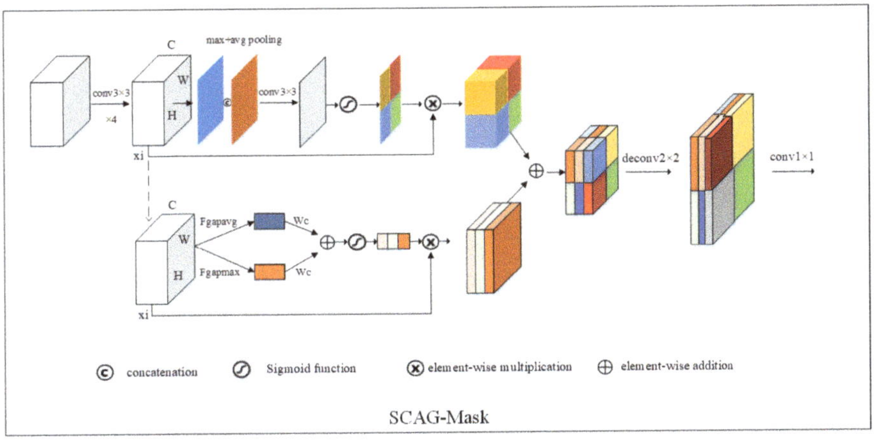

Figure 3. The structure of the spatial and channel attention-guided mask (SCAG-Mask), where Fgapavg and Fgapmax are global average pooling and global max pooling same as Figure 2, WC is fully-connected layer. W, H, C separately represent the width, height, and the number of channels for feature maps.

2.4. Multi-Task Loss

During training time, we compute a multi-task loss on each RoI as:

$$Loss = L_{cls} + L_{reg} + L_{centerness} + L_{mask} + L_{maskiou}, \tag{10}$$

where the classification loss L_{cls} is α-Balance focal loss defined in Reference [41], the regression loss L_{reg} is GIoU loss identical as in Reference [42], the centerness loss $L_{centerness}$ and mask loss L_{mask} are same as those in Reference [33]. In addition, this paper uses L_2 loss to calculate MaskIoU loss $L_{maskiou}$.

3. Dataset and Evaluation Metrics

3.1. Dataset Description

In this paper, we gain high-resolution remote sensing building semantic segmentation database data of Wuhan University (WHU building dataset, http://study.rsgis.whu.edu.cn/pages/download/building_dataset.html). The original remote sensing images in the database were from Christchurch, New Zealand, which cover 220,000 diverse buildings. After the downsampling, labeling and segmentation operations of the Jishunping team [43] of Wuhan University, more than 8000 images of 512-pixel × 512-pixel, 0.3 m spatial resolution were produced. In this paper, through removing the abnormal images in the building dataset manually and converting the semantic segmentation dataset to COCO instance segmentation task format by using the Suzuki [44] contours detection algorithm, we obtain 4268 images as training set data and 719 images as test set.

3.2. Evaluation Metrics

We adopt average precision (AP) and average detection time to quantitatively assess the performance of the instance segmentation method. A certain number of indexes are needed, namely, true positive (*TP*),false negative (*FN*), false positive (*FP*), and precision–recall curve (PRC). *TP* denotes the number of correct detections; *FN* denotes the number of missing detections; and *FP* denotes the number of mistaken detections. PRC is a curve drawn by the precision metric and the recall metric. The precision metric measures the proportion of correct detections to all detections. The recall metric measures the proportion of correct detections to all ground-truth boxes. The precision metric and recall metric are defined as,

$$Precision = \frac{TP}{TP + FP}, \tag{11}$$

$$Recall = \frac{TP}{TP + FN}. \tag{12}$$

AP—the AP metric computes the area under the PRC. A higher AP value indicates better performance, and vice versa. The AP used here includes AP of box detection AP^{box} and AP of mask extraction AP^{mask}. Besides, the AP^{box} and AP^{mask} of small, medium and large size objects are further refined. Among them, small-sized objects are objects with a pixel area of less than 32^2. Medium-sized objects are objects with a pixel area greater than 32^2 but less than 96^2. Large size objects are objects with a pixel area greater than 96^2.

4. Results and Discussion

4.1. Parameter Setting Experiments

Table 1 shows the detailed hyperparameter settings. In the experiment, we set weight decay as 0.0001, the momentum as 0.9, and the pooler_scales as (0.125, 0.0625, 0.03125), then, we use ResNet101 and VoVNetV2-57 with different learning rates commit several experiments. Note that, we utilize the training set data described in Section 3 to fit the model when changing the backbone and learning rate; and we use test set to evaluate the different performances of models generated by different backbone and learning rates. The performances of different models on the test set are shown in Table 2.

Table 1. Summarization of hyperparameters fixed in the paper.

Hyperparameters	Setting Details
No. of FILTERS	$3 \times 3conv, 64$ $3 \times 3conv, 64$ $3 \times 3conv, 128$ $[3 \times 3conv, 128, \times 5, concat, \&1 \times 1conv, 256] \times 1$ $[3 \times 3conv, 160, \times 5, concat, \&1 \times 1conv, 512] \times 1$ $[3 \times 3conv, 192, \times 5, concat, \&1 \times 1conv, 768] \times 4$ $[3 \times 3conv, 224, \times 5, concat, \&1 \times 1conv, 1024] \times 3$
BATCH SIZE	8
EPOCHS	200
WEIGHT DECAY	0.0001
MOMENTUM	0.9
WARMUP_FACTOR	0.333
WARMUP_ITERS	500
POOLER_ScALES	(0.125, 0.0625, 0.03125)
POOLER_SAMPLING_RATIO	2.0
POOLER_RESOLUTION	14
FPN_STRIDES	[8, 16, 32, 64, 128]
FCOS_CENTER_SAMPLE_POS_RADIUS	1.5
FOCAL_LOSS_ALPHA	0.25
FOCAL_LOSS_GAMMA	2.0

In this paper, we consider the impact of different learning rates on model performance. As shown in Table 2, when using ResNet101 as the feature extraction network, the model performance is optimal when the learning rate is 0.005. And when using VoVNetV2-57 as the feature extraction network, the model performance is optimal when the learning rate is 0.001. Comparing the AP values for performance of two best-case models, we can find that the VoVNetV2-57 has significant advantages as a feature extraction network, especially when extracting small and medium-scale buildings.

Table 2. Comparison of average precision (AP) on the test set under different parameter configurations.

Feature Extraction Network	Learning Rate	AP^{box}	AP_s^{box}	AP_m^{box}	AP_l^{box}	AP^{mask}	AP_s^{mask}	AP_m^{mask}	AP_l^{mask}
ResNet101	0.01	0.6444	0.4483	0.8142	0.7344	0.6085	0.4009	0.7758	0.7672
ResNet101	0.005	0.6544	0.4552	0.8285	0.747	0.6116	0.3972	0.7836	0.7725
ResNet101	0.0001	0.629	0.4287	0.8047	0.6741	0.5981	0.3884	0.7694	0.7311
VoVNetV2-57	0.0005	0.6773	0.4869	0.8436	0.7205	0.6272	0.4234	0.7933	0.7323
VoVNetV2-57	0.001	0.6799	0.4915	0.8451	0.7208	0.6296	0.4274	0.7959	0.7437

Besides, we examine the impact of our improvements in this paper on model performance. Since our backbone is improved on VoVNetV2-57, we control the learning rate at 0.001 in the following experiments. The performances of our improved models on test set are shown in Table 3.

As can be seen from Table 3, when use optimized backbone without changing other parts of the model, the AP value(the second row in Table 3) of the obtained model on buildings detection and segmentation has been improved to a certain extent. Especially when extracting large buildings, the AP value increased significantly. When only use SCAG-Mask without changing other parts of the model. Compared with the AP value gained by original CenterMask (the first row in Table 3), the AP value of bounding box detection of small and medium-sized buildings has decreased. But due to our SCAG-Mask focus on spatial and channel attention guiding, the AP value of the building segmentation has been significantly improved. Compared with the results obtained by optimizing the backbone (the second row in Table 3), although the AP value acquired by using SCAG-Mask in the segmentation of medium and large buildings is significantly improved, the AP of bounding box detection reduces heavily. Thus, it seems that optimize the backbone only is better than only use SCAG-Mask due to its comprehensive performance improvement. At last, we synthetically use the improved backbone and SCAG-Mask. The obtained AP value is shown in the fourth row of Table 3.

And we can see that the AP value for large building detection and the AP value for small, medium, and large building instance object segmentation are both significantly improved. Although compared with the backbone optimizing only (the second row in Table 3), the detection AP of small buildings and medium-sized buildings have slight decreases, but compared with other performance improvements, these slight decreases can be ignored. So the combination of backbone and SCAG-Mask is effective.

Table 3. Comparison of AP on the test set under different improved models.

Model	AP^{box}	AP_s^{box}	AP_m^{box}	AP_l^{box}	AP^{mask}	AP_s^{mask}	AP_m^{mask}	AP_l^{mask}
CenterMask with VoVNetV2-57	0.6799	0.4915	0.8451	0.7208	0.6296	0.4274	0.7959	0.7437
CenterMask with improved VoVNetV2-57(ours)	0.6847	0.4923	0.846	0.7365	0.6307	0.4279	0.7969	0.7542
CenterMask with SCAG-Mask(ours)	0.6778	0.488	0.8434	0.7296	0.631	0.4269	0.7985	0.7561
CenterMask with improved VoVNetV2-57+SCAG-Mask(ours)	0.6841	0.492	0.8444	0.7377	0.6342	0.4309	0.7992	0.7574

After the above experiments, here, we visualize and compare the building extraction results on the test set, which from the optimal model obtained by using ResNet101 as feature extraction network, the optimal model gained by applying VoVNetV2-57 as backbone, and the three models acquired by using the improved methods in our work, as Figure 4 shown. And we can find that the performances of the segmentation of target bounding box (Figure 4a), the prediction of dense buildings (Figure 4b), the identification of confusing object (Figure 4c) and occlusion buildings (Figure 4d) by synthetically using the improved VoVNetV2-57 and SCAG-Mask are better than the others, especially the use of ResNet101 and original VoVNetV2-57. It further illustrates the effectiveness of the network structure and the trained model used in this paper.

4.2. Comparison with State-of-the-Art Methods

This paper compares and analyzes the method effects of Mask R-CNN, Mask Scoring R-CNN, CenterMask and improved CenterMask on the same test set under the same experimental environment. Same as parameter setting experiments, we use the training set data described in Section 3 to fit the model when adopting different algorithms; and we use test set to evaluate the different performances of models under different methods. The corresponding AP values and average detection time of four methods are shown in Table 4. As can be seen from Table 4, CenterMask and improved CenterMask by not setting the anchor mechanism perform more efficiently than Mask R-CNN and Mask Scoring R-CNN which belong to two-stage instance segmentation method. And the average detection time is only 49.88 ms. In addition, the improved CenterMask used in this paper, while ensuring efficiency, demonstrates that it is not inferior to the performance of Mask R-CNN and Mask Scoring R-CNN. Especially in the extraction of medium and large scale buildings, it can even achieve higher AP than the two methods. Although the anchor mechanism is discarded, the AP of improved CenterMask is slightly lower than Mask R-CNN and Mask Scoring R-CNN when extracting small-scale buildings. However, in terms of comprehensive AP and average detection efficiency, improved CenterMask used in this paper still has great advantages.

Table 4. Comparison of AP on the test set with state-of-the-art methods.

Method	AP^{box}	AP_s^{box}	AP_m^{box}	AP_l^{box}	AP^{mask}	AP_s^{mask}	AP_m^{mask}	AP_l^{mask}	Average Detection Time (ms)	GPU
Mask R-CNN	0.6585	0.5049	0.789	0.7121	0.6384	0.4663	0.771	0.7468	69.49	Xp
Mask scoring R-CNN	0.6604	0.5058	0.7881	0.7124	0.6451	0.4705	0.7806	0.756	72.83	Xp
CenterMask	0.6799	0.4915	0.8451	0.7208	0.6296	0.4274	0.7959	0.7437	49.88	Xp
Improved CenterMask	0.6841	0.492	0.8444	0.7377	0.6342	0.4309	0.7992	0.7574	49.88	Xp

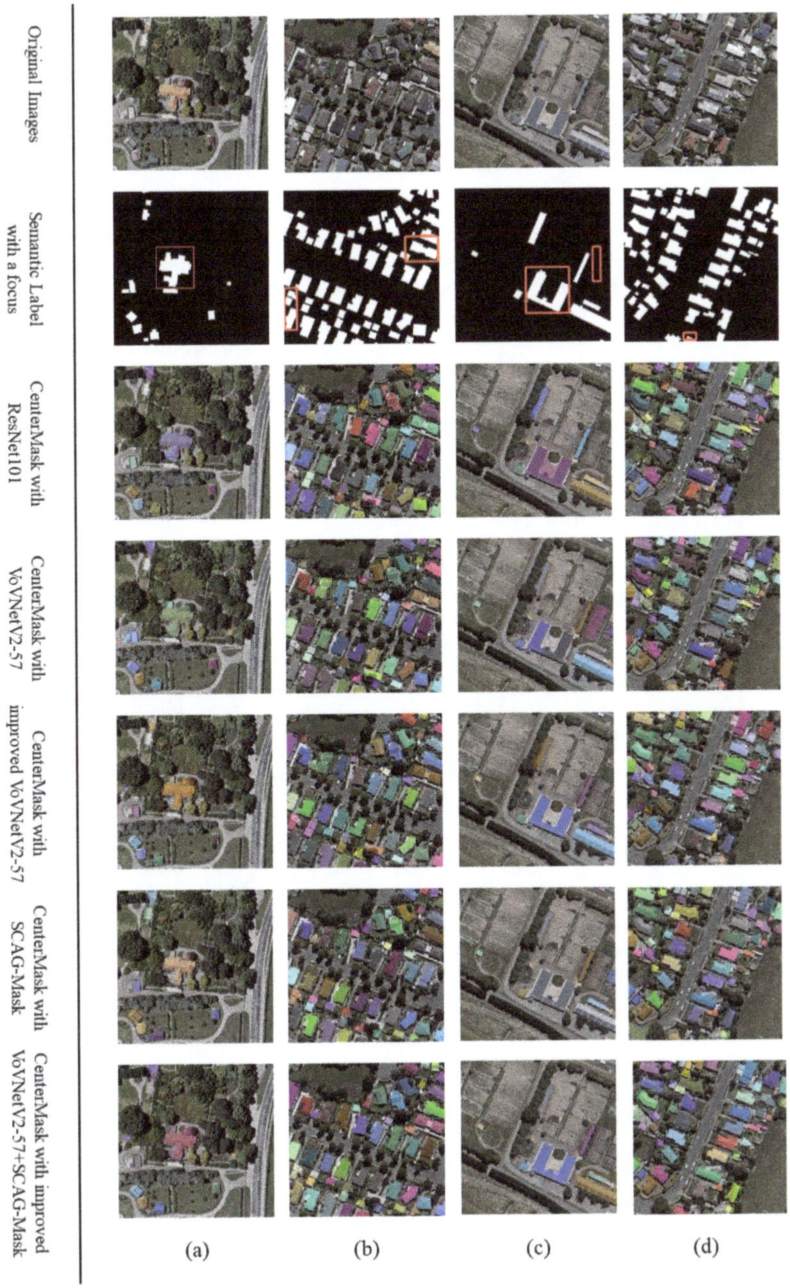

Figure 4. Comparison of building extraction results on test set for five models. (**a**) Performance of segmentation for target bounding box; (**b**) Performance of dense buildings prediction; (**c**) Performance of confusing object identification; (**d**) Performance of occlusion buildings extraction.

4.3. Usability Analysis in Practical Application

In order to further illustrate the effectiveness of the method proposed in this paper, we take building extraction in China's urban villages as example to verify the usability of our method in practical application. This paper collects the building samples in China's urban villages from Google Earth imagery with spatial resolution of 0.11 m. To realize data augmentation, these samples are vertical and horizontal flipped, and 90 degrees, 180 degrees and 270 degrees counterclockwise rotated. Finally, by using professional labeling software, 2726 instance segmentation samples of China's urban villages buildings are obtained. They are further divided into 2326 training samples and 400 test samples. The training samples are fed into the method proposed in this paper to fit a model suitable for the extraction of urban villages buildings. And the test samples are used for verifying the performance of the proposed method in the extraction of China's urban villages buildings.

Figure 5 shows the extraction results of China's urban villages buildings on test samples with proposed method in this paper. We can find out that despite the high density of buildings and narrow streets and lanes in urban villages, the method in this paper can still effectively extract buildings in China's urban villages. In terms of quantitative indicators, the AP^{box} of the model on the urban village buildings test set is 0.799, and the AP^{mask} is 0.728. The precision rate of building extraction can be as high as 0.91 and the recall rate can be as high as 0.95 for single sample. This fully demonstrates that the improved anchor-free instance segmentation method proposed in this paper can achieve good extraction performance on China's urban villages building samples. And the method can achieve real-time building extraction, the prediction of entire test samples only takes 12 s, which further proves the practicability of this method.

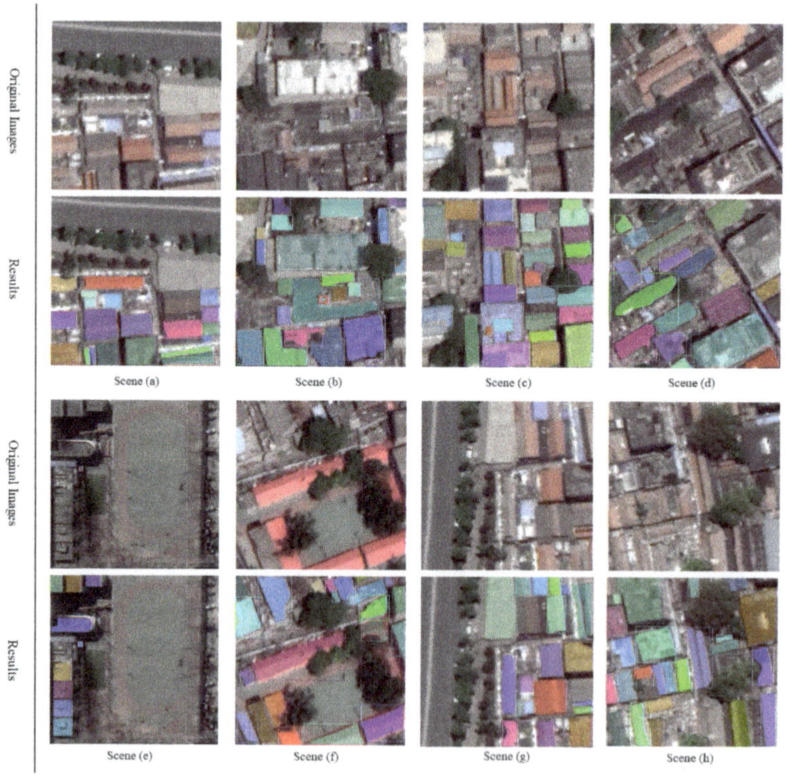

Figure 5. Building extraction results on building test samples in China's urban villages.

5. Conclusions

We have proposed an improved real-time anchor-free one-stage instance segmentation method and a more effective backbone network. By adding the SCAG-Mask to the anchor-free one stage instance detection, our improved CenterMask achieves state-of-the-art performance at real-time speed. Although Mask R-CNN and Mask Scoring R-CNN show better performances than our improved CenterMask on AP_s^{box} and AP_s^{mask}, we are still full of confidence in our improved CenterMask due to its well-balanced performance in terms of speed and accuracy. In addition, we take buildings extraction in China's urban villages as example to verify the usability of our method in practical applications and get good extraction results and satisfactory speed. In the future, we will further improve the performance of the model in small object detection and segmentation. And some post-processing techniques may be explored and compared to obtain the best framework for building extraction.

Funding: This research was funded by the Beijing Municipal Science and Technology Project under grant No.Z191100001419002.

Conflicts of Interest: The authors declare no conflict of interest.

References

1. Ding, Z.; Wang, X.; Li, Y.; Zhang, S. Study on Building Extraction from High-Resolution Images Using Mbi. *Int. Arch. Photogramm. Remote Sens. Spat. Inf. Sci* **2018**, *42*, 283–287. [CrossRef]
2. Shinohara, T.; Xiu, H.; Matsuoka, M. FWNet: Semantic Segmentation for Full-Waveform LiDAR Data Using Deep Learning. *Sensors* **2020**, *20*, 3568. [CrossRef]
3. Colaninno, N.; Roca, J.; Pfeffer, K. An automatic classification of urban texture: Form and compactness of morphological homogeneous structures in Barcelona. In Proceedings of the 51st Congress of the European Regional Science Association: New Challenges for European Regions and Urban Areas in a Globalised World, Barcelona, Spain, 30 August–3 September 2011.
4. Hermosilla, T.; Palomar-Vázquez, J.; Balaguer-Beser, Á.; Balsa-Barreiro, J.; Ruiz, L.A. Using street based metrics to characterize urban typologies. *Comput. Environ. Urban Syst.* **2014**, *44*, 68–79. [CrossRef]
5. Van de Voorde, T.; Jacquet, W.; Canters, F. Mapping form and function in urban areas: An approach based on urban metrics and continuous impervious surface data. *Landsc. Urban Plan.* **2011**, *102*, 143–155. [CrossRef]
6. Huertas, A.; Nevatia, R. Detecting buildings in aerial images. *Comput. Vision Graph. Image Process.* **1988**, *41*, 131–152. [CrossRef]
7. Irvin, R.B.; McKeown, D.M. Methods for exploiting the relationship between buildings and their shadows in aerial imagery. *IEEE Trans. Syst. Man Cybern.* **1989**, *19*, 1564–1575. [CrossRef]
8. Inglada, J. Automatic recognition of man-made objects in high resolution optical remote sensing images by SVM classification of geometric image features. *ISPRS J. Photogramm. Remote. Sens.* **2007**, *62*, 236–248. [CrossRef]
9. Meng, Y.; Peng, S. Object-oriented building extraction from high-resolution imagery based on fuzzy SVM. In Proceedings of the 2009 International Conference on Information Engineering and Computer Science, Wuhan, China, 19–20 December 2009; IEEE: Piscataway, NJ, USA, 2009; pp. 1–6.
10. Awrangjeb, M.; Zhang, C.; Fraser, C.S. Automatic extraction of building roofs using LIDAR data and multispectral imagery. *ISPRS J. Photogramm. Remote. Sens.* **2013**, *83*, 1–18. [CrossRef]
11. Awrangjeb, M.; Fraser, C.S. Automatic segmentation of raw LiDAR data for extraction of building roofs. *Remote Sens.* **2014**, *6*, 3716–3751. [CrossRef]
12. Gilani, S.A.N.; Awrangjeb, M.; Lu, G. Segmentation of airborne point cloud data for automatic building roof extraction. *Gisci. Remote Sens.* **2018**, *55*, 63–89. [CrossRef]
13. Gilani, S.A.N.; Awrangjeb, M.; Lu, G. An automatic building extraction and regularisation technique using lidar point cloud data and orthoimage. *Remote Sens.* **2016**, *8*, 258. [CrossRef]
14. Ronneberger, O.; Fischer, P.; Brox, T. U-net: Convolutional networks for biomedical image segmentation. In *International Conference on Medical Image Computing and Computer-Assisted Intervention*; Springer: Berlin/Heidelberg, Germany, 2015; pp. 234–241.

15. Badrinarayanan, V.; Kendall, A.; Cipolla, R. SegNet: A Deep Convolutional Encoder-Decoder Architecture for Image Segmentation. *IEEE Trans. Pattern Anal. Mach. Intell.* **2017**, *39*, 2481–2495. [CrossRef]
16. Chen, L.C.; Papandreou, G.; Kokkinos, I.; Murphy, K.; Yuille, A.L. DeepLab: Semantic Image Segmentation with Deep Convolutional Nets, Atrous Convolution, and Fully Connected CRFs. *IEEE Trans. Pattern Anal. Mach. Intell.* **2018**, *40*, 834–848. [CrossRef]
17. Xu, Y.; Wu, L.; Xie, Z.; Chen, Z. Building Extraction in Very High Resolution Remote Sensing Imagery Using Deep Learning and Guided Filters. *Remote Sens.* **2018**, *10*, 144. [CrossRef]
18. Shrestha, S.; Vanneschi, L. Improved fully convolutional network with conditional random fields for building extraction. *Remote Sens.* **2018**, *10*, 1135. [CrossRef]
19. Li, X.; Yao, X.; Fang, Y. Building-A-Nets: Robust Building Extraction from High-Resolution Remote Sensing Images with Adversarial Networks. *IEEE J. Sel. Top. Appl. Earth Obs. Remote Sens.* **2018**, *11*, 3680–3687. [CrossRef]
20. He, K.; Gkioxari, G.; Dollar, P.; Girshick, R. Mask R-CNN. *IEEE Trans. Pattern Anal. Mach. Intell.* **2020**, *42*, 386–397. [CrossRef]
21. Ren, S.; He, K.; Girshick, R.; Sun, J. Faster R-CNN: Towards Real-Time Object Detection with Region Proposal Networks. *IEEE Trans. Pattern Anal. Mach. Intell.* **2017**, *39*, 1137–1149. [CrossRef]
22. Huang, Z.; Huang, L.; Gong, Y.; Huang, C.; Wang, X. Mask Scoring R-CNN. In Proceedings of the 2019 IEEE/CVF Conference on Computer Vision and Pattern Recognition (CVPR), Long Beach City, CA, USA, 16–20 June 2019; pp. 6409–6418.
23. Cai, Z.; Vasconcelos, N. Cascade R-CNN: Delving Into High Quality Object Detection. In Proceedings of the 2018 IEEE/CVF Conference on Computer Vision and Pattern Recognition, Salt Lake City, UT, USA, 18–22 June 2018; pp. 6154–6162.
24. Liu, S.; Qi, L.; Qin, H.; Shi, J.; Jia, J. Path Aggregation Network for Instance Segmentation. In Proceedings of the 2018 IEEE/CVF Conference on Computer Vision and Pattern Recognition, Salt Lake City, UT, USA, 18–23 June 2018; pp. 8759–8768.
25. Li, Y.; Chen, Y.; Wang, N.; Zhang, Z.X. Scale-Aware Trident Networks for Object Detection. In Proceedings of the 2019 IEEE/CVF International Conference on Computer Vision (ICCV), Seoul, Korea, 27 October–2 November 2019; pp. 6053–6062.
26. Liu, W.; Anguelov, D.; Erhan, D.; Szegedy, C.; Reed, S.E.; Fu, C.Y.; Berg, A.C. SSD: Single Shot MultiBox Detector. In *European Conference on Computer Vision*; Springer: Berlin/Heidelberg, Germany, 2016; pp. 21–37.
27. Redmon, J.; Divvala, S.; Girshick, R.; Farhadi, A. You Only Look Once: Unified, Real-Time Object Detection. In Proceedings of the 2016 IEEE Conference on Computer Vision and Pattern Recognition (CVPR), Las Vegas, NV, USA, 27–30 June 2016; pp. 779–788.
28. Bolya, D.; Zhou, C.; Xiao, F.; Lee, Y.J. YOLACT: Real-Time Instance Segmentation. In Proceedings of the 2019 IEEE/CVF International Conference on Computer Vision (ICCV), Seoul, Korea, 27 October–2 November 2019; pp. 9156–9165.
29. Potlapally, A.; Chowdary, P.S.R.; Shekhar, S.R.; Mishra, N.; Madhuri, C.S.V.D.; Prasad, A. Instance Segmentation in Remote Sensing Imagery using Deep Convolutional Neural Networks. In Proceedings of the 2019 International Conference on contemporary Computing and Informatics (IC3I), Singapore, 12–14 December 2019; pp. 117–120.
30. Ji, S.; Shen, Y.; Lu, M.; Zhang, Y. Building instance change detection from large-scale aerial images using convolutional neural networks and simulated samples. *Remote Sens.* **2019**, *11*, 1343. [CrossRef]
31. Li, Q.; Mou, L.; Hua, Y.; Sun, Y.; Jin, P.; Shi, Y.; Zhu, X.X. Instance segmentation of buildings using keypoints. *arXiv* **2020**, arXiv:2006.03858.
32. Su, H.; Wei, S.; Liu, S.; Liang, J.; Wang, C.; Shi, J.; Zhang, X. HQ-ISNet: High-Quality Instance Segmentation for Remote Sensing Imagery. *Remote Sens.* **2020**, *12*, 989. [CrossRef]
33. Tian, Z.; Shen, C.; Chen, H.; He, T. FCOS: Fully Convolutional One-Stage Object Detection. In Proceedings of the 2019 IEEE/CVF International Conference on Computer Vision (ICCV), Seoul, Korea, 27 October–2 November 2019; pp. 9626–9635.
34. Duan, K.; Bai, S.; Xie, L.; Qi, H.; Huang, Q.; Tian, Q. CenterNet: Keypoint Triplets for Object Detection. In Proceedings of the 2019 IEEE/CVF International Conference on Computer Vision (ICCV), Seoul, Korea, 27 October–2 November 2019; pp. 6568–6577.

35. Law, H.; Deng, J. CornerNet: Detecting Objects as Paired Keypoints. *Int. J. Comput. Vis.* **2020**, *128*, 642–656. [CrossRef]
36. Chen, H.; Sun, K.; Tian, Z.; Shen, C.; Huang, Y.; Yan, Y. BlendMask: Top-Down Meets Bottom-Up for Instance Segmentation. In Proceedings of the CVPR 2020: Computer Vision and Pattern Recognition, Virtual, Seattle, WA, USA, 14–19 June 2020; pp. 8573–8581.
37. Lee, Y.; Park, J. CenterMask: Real-Time Anchor-Free Instance Segmentation. In Proceedings of the CVPR 2020: Computer Vision and Pattern Recognition, Virtual, Seattle, WA, USA, 14–19 June 2020; pp. 13906–13915.
38. Lin, T.Y.; Dollár, P.; Girshick, R.; He, K.; Hariharan, B.; Belongie, S. Feature pyramid networks for object detection. In Proceedings of the IEEE Conference on Computer Vision and Pattern Recognition, Honolulu, HI, USA, 21–26 July 2017; pp. 2117–2125.
39. Lee, Y.; Hwang, J.W.; Lee, S.; Bae, Y.; Park, J. An energy and gpu-computation efficient backbone network for real-time object detection. In Proceedings of the IEEE Conference on Computer Vision and Pattern Recognition Workshops, Long Beach City, CA, USA, 16–20 June 2019.
40. He, K.; Zhang, X.; Ren, S.; Sun, J. Deep residual learning for image recognition. In Proceedings of the IEEE conference on computer vision and pattern recognition, Las Vegas, NV, USA, 27–30 June 2016; pp. 770–778.
41. Lin, T.Y.; Goyal, P.; Girshick, R.; He, K.; Dollar, P. Focal Loss for Dense Object Detection. *IEEE Trans. Pattern Anal. Mach. Intell.* **2020**, *42*, 318–327. [CrossRef] [PubMed]
42. Rezatofighi, H.; Tsoi, N.; Gwak, J.; Sadeghian, A.; Reid, I.; Savarese, S. Generalized Intersection Over Union: A Metric and a Loss for Bounding Box Regression. In Proceedings of the 2019 IEEE/CVF Conference on Computer Vision and Pattern Recognition (CVPR), Long Beach, CA, USA, 15–20 June 2019; pp. 658–666.
43. Ji, S.; Wei, S.; Lu, M. Fully Convolutional Networks for Multisource Building Extraction From an Open Aerial and Satellite Imagery Data Set. *IEEE Trans. Geosci. Remote Sens.* **2019**, *57*, 574–586. [CrossRef]
44. Suzuki, S.; Abe, K. Topological Structural Analysis of Digitized Binary Images by Border Following. *Graph. Model. Graph. Model. Image Process. Comput. Vis. Graph. Image Process.* **1985**, *30*, 32–46. [CrossRef]

remote sensing

MDPI

Article

Application Study on Double-Constrained Change Detection for Land Use/Land Cover Based on GF-6 WFV Imageries

Jingxian Yu [1,2], Yalan Liu [1,*], Yuhuan Ren [1], Haojie Ma [1,2], Dacheng Wang [1], Yafei Jing [1,2] and Linjun Yu [1]

1 Aerospace Information Research Institute, Chinese Academy of Sciences, Beijing 100101, China; yujx@radi.ac.cn (J.Y.); renyh@radi.ac.cn (Y.R.); mahj@radi.ac.cn (H.M.); wangdc@aircas.ac.cn (D.W.); jingyafei19@mails.ucas.edu.cn (Y.J.); yulj201831@aircas.ac.cn (L.Y.)

2 University of Chinese Academy of Sciences, Beijing 100049, China

* Correspondence: liuyl@aircas.ac.cn; Tel.: +86-139-1103-2598

Received: 28 July 2020; Accepted: 7 September 2020; Published: 11 September 2020

Abstract: As a new satellite sensor of the GaoFen (GF) series, GF-6 Wide Field of View (WFV) with the resolution of 16 m has the characteristics of wide coverage, high-frequency imaging and has four new bands of two red-edge, yellow, and purple compared with GF-1 WFV. In order to test the validity of the supplementary bands of GF-6WFV data for change detection of land use/land cover (LULC), this study applied the Double-constrained Change Detection Method (DCDM) that uses the double constraints (change vector intensity and correlation coefficient) for change detection on object-level. According to two GF-6WFV imageries acquired in the Xiong'an New Area in June of 2018 and 2019, feature analysis was performed to determine whether the new bands are helpful to detect the change of LULC first. Then, by coupling these selected features, the intensity of change vector and correlation coefficient were used as the double constraints to perform the change detection. The study demonstrates that the relevant features of the two red-edge bands can achieve the overall accuracy of 89% for change detection of LULC and improved by 2% comparing with using the corresponding temporal GF-1WFV data, while the purple and yellow bands cannot provide enough effective information for this detection. This study can provide theoretical support for the in-depth applications of GF-6 WFV data products in the change detection fields and has explored its applicability and potential in resource and environment monitoring, it is helpful to the further applications.

Keywords: land use/land cover (LULC); GF-6 WFV; object-oriented; change detection; double constraints

1. Introduction

Land use/land cover (LULC) is a combination of surface elements covered by natural land and artificial construction structures. LULC change is a widely concerned issue in global environmental change and sustainable development presently. The urbanization in China has promoted the increase in land use/cover change during recent 10 years due to the national economic and social development, while the growing needs of implementation of land planning, comprehensive land management, cadastral management, farmland protection, law enforcement, and supervision for the LULC updating information is more and more urgent. Due to the short revisiting period, macroscopic and wide acquisition coverage, remote sensing technology is widely applied to the field of environmental monitoring by the related departments of China [1,2].

Change detection is the approach to detect the surface changes over time by remote sensing images acquired at different times for the same interesting area. The premise is that these changes will lead to Digital Number (DN) changes in the corresponding area between different images, and the

DN changes caused by real changes in ground truth is far greater than those caused by other factors, such as spectral changes caused by the observation angle, solar altitude angle, and atmospheric conditions at different times [3]. The basic approaches for change detection are direct comparison and post-classification comparison [4]. The former is to process the preprocessed original image through image enhancement, image transformation, or to use specific operators to extract image features, and compare the features of each image at the corresponding position to obtain the result of change detection [5]. The latter is based on the classifications for different imageries. It excessively depends on the accuracy of classification, therefore, its results for change detection are often unstable. In contrast, the results of the direct comparison method are more stable [6]. Bruzzone and Prieto first proposed an unsupervised change detection method based on the difference graph in 2002 [7]. Many subsequent researches on change detection, such as Change Vector Analysis (CVA) [8–12], have been based on their studies. As an extension of the difference method, CVA combines the difference value of bands between images into a change vector and uses the change vector to measure the change between bi-temporal images. The change intensity is obtained by determining the Euclidean distance between two data points in the N-dimensional space. Yang et al. conducted a comparative study on CVA and other commonly used methods; their results show that CVA has the lowest probability of omission [13]. Actually, CVA has been widely used in change detection of LULC, the main reason is that it is more suitable for change detection based on remote sensing images in terms of that it can use more or even all the band information to detect changing pixels [14]. However, this method also has shortcomings, that is, it is difficult to determine the change threshold.

According to this, Wang et al. of our team proposed the object-oriented Double-Constrained Change Detection (DCDM) model that combines CVA with correlation analysis between different objects to conduct change detection in 2018 [15]. Correlation analysis studies the degree of correlation between objects in multi-temporal imageries by calculating their correlation coefficient. Obviously, the greater the degree of correlation between an object in images of different time phases, the less likely it is to change. Therefore, using the dual constraints of the correlation coefficient and CVA can better determine the change threshold and reduce more errors and omissions.

At the same time, based on an object-level, DCDM can reduce the "salt and pepper" phenomenon compared to the pixel-level method and improves the detection accuracy [16–20]. However, these researches only based on the images of GaoFen (GF)-1 Wide Field of View (WFV) [21,22].

The GF-6 satellite, also known as the "High-resolution Land Emergency Monitoring Satellite", was launched on 2 June 2018. It is equipped with two cameras, one is the panchromatic/multi-spectral camera with high-resolutions of 2 m and 8 m, and the other is the WFV camera with 16 m multi-spectral medium-resolution and 800 km wide-coverage of observation. The WFV can provide great support for the investigations of agriculture, forestry, land and resources, disaster emergency, and other industries in China [23]. The parameters of its sensors include 4 more bands of purple, yellow, and two red-edge bands than those of GF-1WFV.

The above researches show that use object-level CVA and correlation analysis as double constraints to conduct change detection can obtain higher detection accuracy, so in order to study whether the 4 new added bands of GF-6WFV data are suitable for change detection, we used the object-oriented DCDM to do research on GF-6WFV data. The aim of the study is to provide a theoretical basis for the in-depth application of GF-6 WFV data products in change detection fields and explore the applicability and the potential of GF-6 in resource and environment monitoring [24]. We used the significance analysis of spectral, texture, and shape features for each band of GF-6WFV. Based on optimal features selection, the change vector intensity of objects after image segmentation for bi-temporal imageries and the correlation coefficient between the objects were used for coupling features to realize the automatic extraction of the change information. Finally, the comparison test was conducted with a pair of same time-phrase of GF-1WFV data to evaluate the result of GF-6 data.

2. Materials and Methods

2.1. Analysis of Band Spectral Characteristics

When detect changes for ground features from satellite remote sensing data, the values of features and the spectral characteristic curves are based on the multi-spectral information. Consequently, to make full use of this information for subsequent analysis, one needs to perform a holistic statistical analysis of the information to derive the spectral statistical characteristic values of each band, such as the maximum, mean, standard deviation, entropy, etc., for providing the basis for the subsequent feature selection and various types of analysis [25].

The remote sensing images can be expressed in three dimensions of x, y, and z, where x and y are used to represent plane space, and z is band information, as shown in Expression (1).

$$I = f\ (x, y, z) \tag{1}$$

According to the needs of different analyses, it can be expressed by following three means:

(1) The image space $I = f\ (x, y, 0)$, simply and directly describes the spatial distribution of ground objects in the image and the relationship between spectral response and spatial position, but it is not sensitive to the relationship between bands.

(2) The spectral space $I = f\ (x, 0, z)$, it mainly reflects a spectral curve corresponding to the average radiation value of each pixel and depicts the characteristics of the electromagnetic wave energy with the wavelength change, that is, the spectral characteristic. The spectral response is the function of wavelength. Because of the difference in brightness between different objects as well as the same objects in different bands, these constitute the spectral feature information of the ground objects and make it possible to use the spectral space for image analysis and interpretation.

(3) The feature space $I = f\ (0, y, z)$, is a two-dimensional feature space in which the radiation intensity values of different features in two bands are plotted on a two-dimensional plane.

Statistical analysis was conducted on the 8 bands of GF-6WFV images to derive the related statistical characteristic values such as mean, maximum, and standard deviation.

2.2. Feature Analysis and Optimization of LULC

The process of the change detection for multi-temporal remote sensing images is to extract change information for the ground objects from the region of interests by the change detection algorithm and to obtain the changes through analysis and description. The core of change detection is to determine whether the features in the smallest detection unit have changed by the quantification of the difference between the features of this unit in different temporal images. Therefore, the selection of features is crucial [26]. Feature analysis was used to verify the influence of the new additional bands of GF-6WFV on the accuracy of change detection of LULC.

The surface features of remote sensing imagery can be described by spectral features, texture features, and spatial features. In this paper, these three types of features are selected for the quantitative analysis of objects in different temporal remote sensing imageries. The spectral features, often refer to the reflectance spectral characteristics, can be expressed by single-band grayscale mean, standard deviation, and band-specific feature index, such as Normalized Difference Vegetation Index (NDVI) and Normalized Difference Water Index (NDWI). The textual features describe the repeated local patterns in the image and their arrangement and include Angular Second Moment (ASM), dissimilarity, and correlation [27]. The shape features describe the contour features of objects in the image, which nclude aspect ratio, shape index, and area index. Its feature analysis can be carried out according to the classes of construction land, vegetation, bare land, and water body.

$$\mathrm{NDVI} = (\mathrm{NIR} - \mathrm{R})/(\mathrm{NIR} + \mathrm{R}) \tag{2}$$

$$\mathrm{NDWI} = (\mathrm{Green} - \mathrm{NIR})/(\mathrm{Green} + \mathrm{NIR}) \tag{3}$$

Green, R, NIR are B2, B3, and B4 bands in GF-6WFV data, respectively (See the next section for details).

The spectral feature curves are used to represent the different features of objects in spectral space. Therefore, it is necessary to select the valid features and remove the redundant and invalid features to improve the efficiency for change detection.

Feature selection is to select N ($N \leq M$) features from the given M features to optimize the features, which is the key step for the pre-processing of change detection [25,26]. The purpose of this study is to select the features that are sensitive to LULC changes from the spectral and textural features of the images and to apply the selected features to the change detection model for reducing the dimension of data and to keep the original useful information to effectively describe the characteristic of changes for all the classes to the maximum extent.

In order to test the significance of the difference between the features for selected categories in each band, a test of significance approach was used to realize the optimization of features. The test of significance approach is used to test whether there is a difference between the effects of the experimental group and the control group or two different treatments and whether this difference is significant. The more significant the difference between classes in a certain feature, the more effective the feature can distinguish these classes, then it can be as the optimal feature. In this paper, Variance Analysis is adopted for this significance test. Its formula is as follows:

$$F = MS_b / MS_w \tag{4}$$

where F is the statistic calculated by Analysis of Variance (ANOVA), MS_b is the mean square between groups, and MS_w is the mean square within a group [28].

When performing ANOVA on a feature, each F value has its corresponding p-value. The larger the F value and the smaller the p-value, the smaller the possibility of accepting the null hypothesis, that is, the difference in feature between groups is large. However, when the result of ANOVA is significant, the feature has a significant difference between the classes, but it does not indicate whether there are significant differences among all the classes under this feature or only among some classes.

In this case, multiple comparisons are required, that is, pairwise comparisons to determine the significance of each feature level. In multiple comparisons, if the calculated p-value is less than 0.05 at a given confidence level, the null hypothesis is rejected, and the two feature types are significantly different under this feature. Otherwise, the difference is not significant.

The confidence level does not indicate the size of the difference between groups, but it can show whether the difference between the two test groups is statistically significant. According to the needs of feature selection, one-way ANOVA with single-factor and multiple comparisons are taken to analyze the significant differences of various features between different classes, and the optimal features are selected after comprehensive analysis. In this paper, we used a 95% confidence interval (CI), that is, the p-value is equal to 0.05.

According to the above principles, the most suitable features for change detection in GF-6WFV data were selected.

2.3. *Double-Constrained Change Detection*

The "double-constrained" here refers to obtaining the change information by two thresholds through the change vector intensity and the correlation coefficient of objects in two-phase images. The double-constrained change detection is performed after image segmentation and feature optimization. The process is shown in Figure 1, and the specific methods are described as follows.

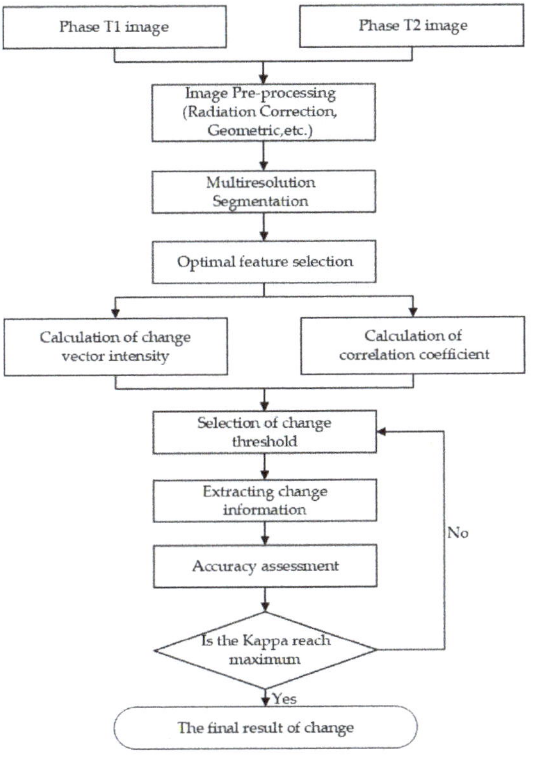

Figure 1. Flow chart of object-level double-constrained change detection (DCDM).

CVA is based on the radiation changes between images of different time phases, focusing on the analysis of the differences in each band to determine the intensity and direction characteristics of the change. Each pixel can generate a change vector with two characteristics of change direction and change intensity. The intensity of change is the Euclidean distance between two data points in n-dimensional space [5,7,10], its calculation is shown in Formula (5).

$$||\Delta G|| = \sqrt{\sum_{k=1}^{n} (g_k - h_k)^2} \quad (5)$$

$||\Delta G||$ is the intensity of the change vector. The larger the value is, the greater the possibility of the change is. Conversely, the less the probability of the change is. Under normal circumstances, the initial threshold of the intensity of the change vector is set to 0.8, which needs to be adjusted appropriately according to different regions. g_k and h_k represent the features in the images of T1 and T2 phases, respectively, and n is the number of selected features.

Generally, it is not accurate enough to use CVA alone to express the features of changes, so the correlation coefficient was used to further measure the possibility of change together with CVA in this study. The correlation coefficient is the result of correlation analysis in change detection and can be used to evaluate the correlation between the same object in different time phases images. The larger the correlation coefficient, the greater the correlation of the object between two images and the less likely for them to change. The addition of the correlation coefficient can better determine the change threshold and reduce more errors and omissions. The correlation coefficient can be calculated by Formula (6):

$$R = \frac{\sum_{k=1}^{n}\left\{\left[x_{i1}^{k} - \overline{x}_{i1}\right]\cdot\left[x_{i2}^{k} - \overline{x}_{i2}\right]\right\}}{\sqrt{\sum_{k=1}^{n}\left[x_{i1}^{k} - \overline{x}_{i1}\right]^{2} \times \sum_{k=1}^{n}\left[x_{i2}^{k} - \overline{x}_{i2}\right]^{2}}} \tag{6}$$

where R is the correlation coefficient, the objects with small correlation coefficients are less likely to change. n is the number of selected features, which x_i^k represents the average DN of the i-th object in the k-band in a certain time-phase image and $\overline{x}_i$ represents the average DN of the n-band of the i-th object in a certain-phase image.

Through the above data analysis, the results are applied to the Double-constrained Change Detection Model. The first process is the multi-scale segmentation for the input images, the second is the optimal features selection for each object, the third is the calculation of the correlation coefficient and the intensity of change vector, and the last is to extract the change areas in the image using the double constraints.

2.4. Accuracy Assessment

A confusion matrix was used to evaluate the detection accuracy of the above change detection method. The results are as shown in Table 1. Each column of the confusion matrix represents the predicted categories, and the total number of each column represents the number of samples predicted to be in the appointed category. Each row represents the actual category, and the total number in each row represents the number of instances in a certain category.

Table 1. Confusion matrix of change detection for land use/land cover (LULC).

Evaluation Data / Detection Result	Unchanged	Changed	Total
Unchanged	N_{nn}	N_{cn}	N_{tn}
Changed	N_{nc}	N_{cc}	N_{tc}
Total	N_{nt}	N_{ct}	N

Some parameters, such as commission errors, omission errors, precision, and overall accuracy, are usually applied to assess the detection accuracy.

Commission Errors = N_{cn}/N_{tn} represents the proportion of samples in which unchanged categories are detected as changed categories;

Omission Errors = N_{nc}/N_{tc} represents the proportion of the undetected samples that are the actual changes;

Precision = N_{cn}/N_{ct} represents the proportion of the real change samples in all detected changes;

Overall Accuracy = $(N_{nn} + N_{cc})/N$ represents the proportion of samples correctly detected;

Kappa is used to measure classification accuracy; the result of performing a KAPPA analysis is a KHAT statistic (an estimate of KAPPA), its formula is as follows:

$$k_{hat} = \frac{N\cdot(N_{nn} + N_{cc}) - (N_{tn}\cdot N_{nt} + N_{tc}\cdot N_{ct})}{N^2 - (N_{tn}\cdot N_{nt} + N_{tc}\cdot N_{ct})} \tag{7}$$

Among them, N_{nn} represents the number of samples that have not changed both in the test results and the reality, N_{cn} represents the number of samples that have been erroneously detected as unchanged, N_{tn} represents the total number of samples that have not changed in the test results, and N_{nc} represents the number of unchanged samples that have been erroneously detected as changed, N_{cc} represents the number of samples whose detection result is changed and is in line with the actual situation, N_{tc} represents the total number of changed samples of the detection result, N_{nt} represents the total number of actually unchanged samples, N_{ct} represents the total number of actually changed samples, and N is the total number of samples.

The specific change detection process is shown in Figure 1.

3. Study Area and Data

The Xiong'an New Area is a state-level new area established in Hebei, China on 1 April 2017, and covers Xiongxian, Rongcheng, and Anxin counties. It serves to further promote the coordinated development of the Beijing-Tianjin-Hebei region and relief the nonessential functions of Beijing in an orderly fashion [29]. After the Xiong'an New Area was started following the Shenzhen Special Economic Zone and Shanghai Pudong New Area, the urban-rural, industrial and mining, and residential land in this new area has increased substantially and has derived its land use and cover being undergone obvious changes. Therefore, this is very suitable as a study area for change detection of LULC. The studied area for this study is located between 38°54′22′′–39°9′11′′ N and 115°42′50′′–116°13′26′′ E and occupies a land area of 1183.6 square kilometers, shown in Figure 2. In the selected area, there are mainly four types of ground objects: vegetation, construction-land, bare land, and water bodies. Among them, vegetation covers the largest proportion, followed by water bodies, built-up areas, and bare land.

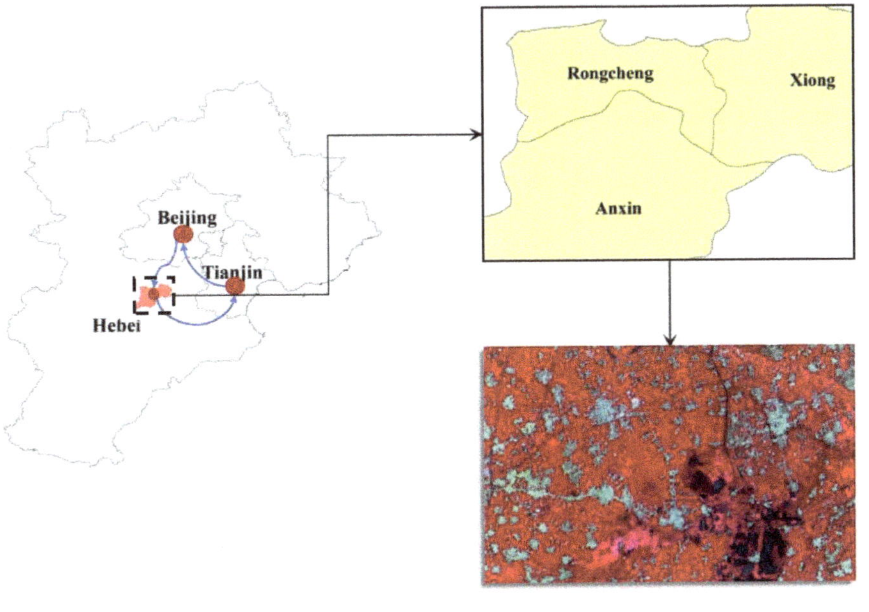

Figure 2. The study area.

The study is based on bi-temporal data obtained by GF-6WFV on 6 September 2018 and 6 September 2019 shown in Figure 3, and data obtained by GF-1WFV on 24 September 2018 and 24 September 2019, and their spatial resolution is 16 m shown in Figure 4. A comparative test was conducted on the two sets of data to study the usefulness of the newly launched GF-6WFVdata in the field of change detection.

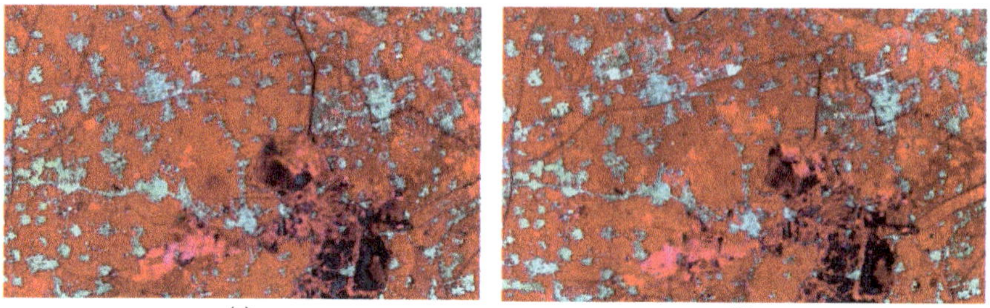

(a) (b)

Figure 3. Two GF-6 Wide Field of View (WFV) images for the study area ((**a**) is for 6 September 2018, (**b**) is for 6 September 2019).

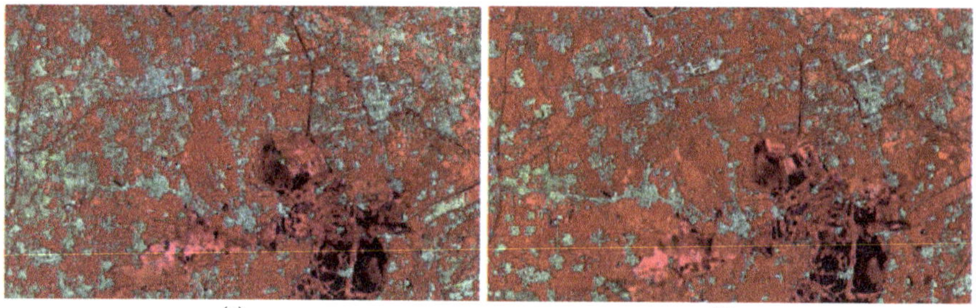

(a) (b)

Figure 4. Two GF-1WFV images for study area ((**a**) is for 24 September 2018, (**b**) is for 24 September 2019).

GF-6 has the characteristics of wide coverage, high-quality imaging, high localization rate, etc., and has a breakthrough in the super-large field-of-view imaging technology in a single projection center, which improves the accuracy of wide-field camera images [30]. Its specific parameters of sensors are shown in Table 2. The additional red edge bands of GF-6 WFV comparing with GF1 WFV can effectively reflect the unique spectral characteristics of crops [23]. In theory, it can provide a greater amount of information for monitoring crop growth, and it has good application potential for the detection of changes in vegetation-covered areas.

Table 2. Technical indicators of sensors payload for GaoFen (GF)-6.

Parameters		**Cameras**			
Spectrum (μm)	panchromatic	0.45–0.90			
	multi-spectral	B1 (blue)	0.45–0.52	B7 (purple)	0.40–0.45
				B1 (blue)	0.45–0.52
		B2 (green)	0.52–0.59	B2 (green)	0.52–0.59
				B8 (yellow)	0.59–0.63
		B3 (red)	0.63–0.69	B3 (red)	0.63–0.69
				B5 (red-edge1)	0.69–0.73
		B4 (near-infrared)	0.77–0.89	B6 (red-edge2)	0.73–0.77
				B4 (near-infrared)	0.77–0.79
Spatial resolution (m)	panchromatic	2		16	
	multi-spectral	8			
Width (km)		90		800	

4. Results

4.1. Feature Analysis and Optimization

Before change detection, the GF-6WFV images in two different years were preprocessed by radiation correction and geometric correction, and then the images were segmented by object-oriented multi-scale [31–33]. After multiple experimental comparisons, a set of optimization segmentation parameters was selected, including segmentation scale of 40, shape factor of 0.1, and compactness of 0.9. Through these processes, a more homogeneous patch can be obtained, that is, the segmentation result is not too broken and reflects the difference information between the patches, making the interior of the object more homogenous and is consistent with the actual boundaries of the features. Figure 5 shows two sets of tiles taken from the study area after multi-scale segmentation.

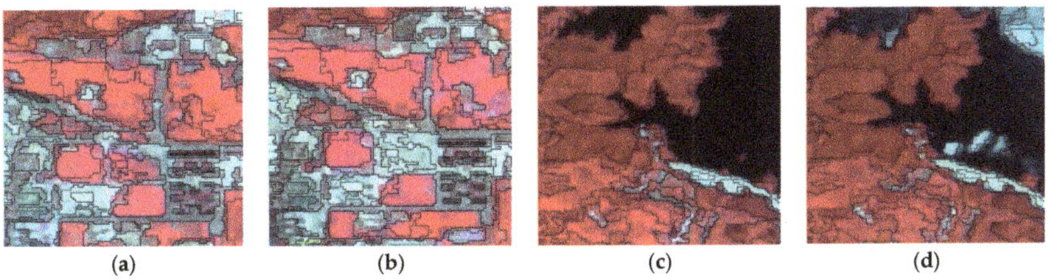

Figure 5. Segmentation tile diagram for GF-6WFV (The segmentation scale is 40, and the shape factor and compactness are 0.1 and 0.9, respectively). (**a**,**b**) represent tile in the same area in 2018 and 2019 of GF-6, respectively; (**c**,**d**) represent another set of tiles in 2018 and 2019 of GF-6, respectively.

Spectrum statistical characteristic values of each band for GF-6WFV are shown in Table 3. Among them, the near-infrared band (B4) has the largest standard deviation, followed by the red band (B3), the red-edge band 2 (B6), and the green band (B2), respectively. The blue band (B1) and the red-edge band 1 (B5) have a smaller standard deviation than the first three bands and the standard deviations of these two bands are similar, while the standard deviation of the purple band (B7) is the smallest. The standard deviation reflects the degree of discrete distribution of DN of each band. The higher the degree of dispersion, the greater the contrast of the image in this band, and the more abundant the information is. It is found that the near-infrared band contains the most abundant information, the spectral value distribution of pixels in this band is more dispersed, and the purple band has the smallest amount of information. The mean reflects the average size of the spectral value of each band. It shows that the near-infrared band (B4) is the highest, reaching 2778.32, and band 7 is the smallest.

Table 3. Statistical characteristic values of all bands of GF-6WFV.

	Min	Max	Mean	StdDev
Blue (1)	729	3722	944.12	174.37
Green (2)	700	4094	1042.05	243.08
Red (3)	471	4094	820.12	303.88
NIR (4)	665	4095	2778.33	536.60
RE1 (5)	330	4088	701.96	180.05
RE2 (6)	381	4081	1448.52	264.46
Purple (7)	612	3064	713.01	77.99
Yellow (8)	437	4094	698.25	207.76

In summary, the new additional bands for GF-6WFV have the characteristics that the near-infrared band contains the largest amount of information, the red-edge band 1 contains less information than the red-edge band 2, and the purple band has the smallest amount of information.

After above analysis, we used GF-6WFV data to analyze the spectral values of different wavebands in each wave band and selected the features that can best express the change information. The features selected in this study are: contrast, dissimilarity, correlation, entropy, and homogeneity. The extracted features of images are shown in Figure 6.

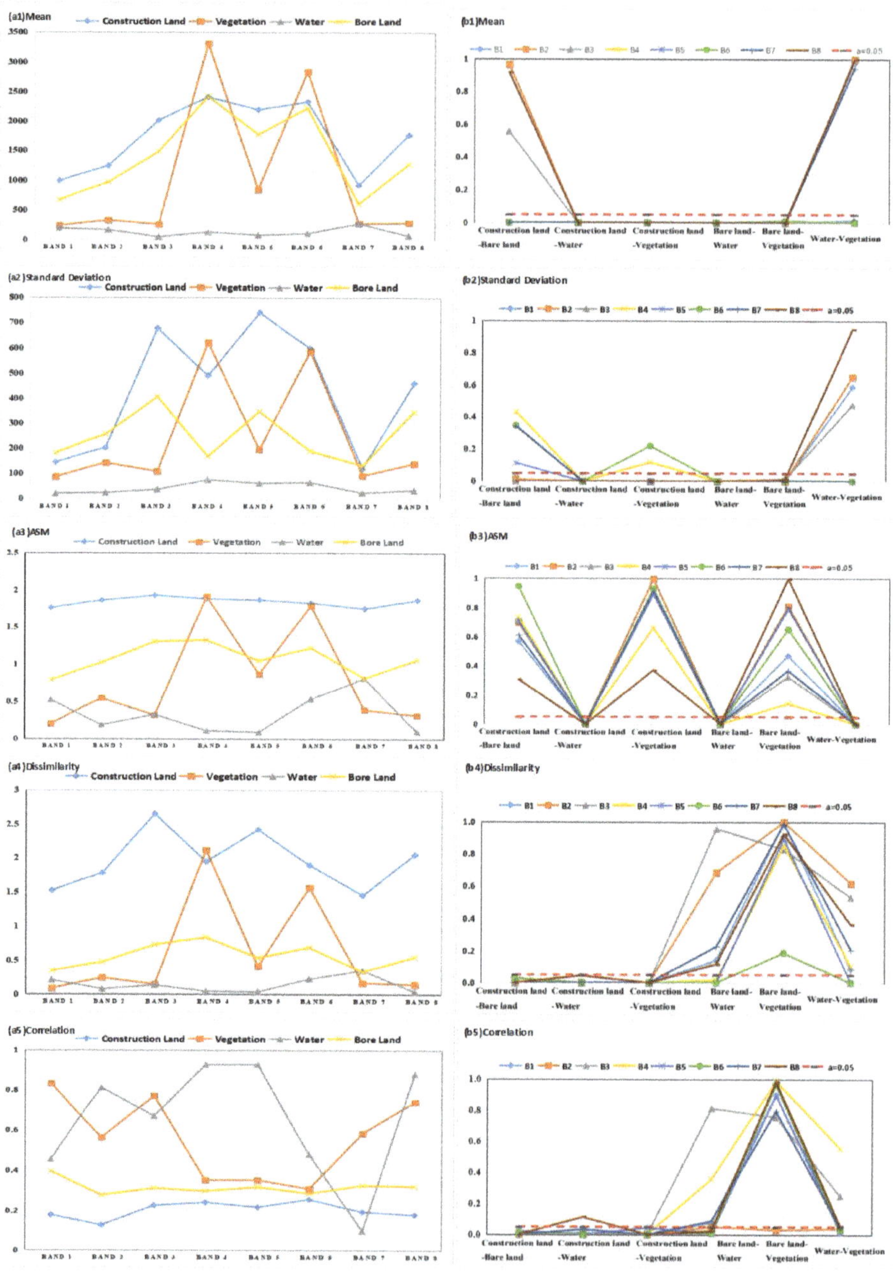

Figure 6. Comprehensive feature analysis for GF-6 WFV data ((**a1**~**a6**): spectral curves for each feature, (**b1**~**b6**): the ANOVA results for each feature, the *p*-value here is 0.05 (95% CI), which is represented as red dotted line in (**b1**~**b6**)).

The F value of each feature for the ANOVA (A 95% confidence interval, $F0.05 = 5.117$) is shown in Table 4.

Table 4. Statistical results of ANOVA.

	B1	B2	B3	B4	B5	B6	B7	B8
Mean								
F	83.7	65.57	92.28	174.34	67.43	150.93	53.98	83.26
p	6.64×10^{-20}	1.11×10^{-17}	8.04×10^{-21}	4.08×10^{-27}	3.18×10^{-22}	1.20×10^{-25}	6.22×10^{-22}	7.44×10^{-20}
Standard Deviation								
F	46.95	47.45	31.25	27.21	37.15	27.12	20.81	25.44
p	1.48×10^{-16}	1.16×10^{-16}	8.20×10^{-13}	1.13×10^{-11}	3.14×10^{-9}	1.20×10^{-11}	5.99×10^{-13}	3.78×10^{-11}
Angular Second Moment (ASM)								
F	10.25	17.58	15.87	17.69	15.33	16.62	11.64	10.17
p	1.76×10^{-5}	3.64×10^{-8}	1.38×10^{-7}	3.35×10^{-8}	5.83×10^{-8}	7.62×10^{-8}	2.19×10^{-6}	1.91×10^{-5}
Dissimilarity								
F	16.81	14.83	16.38	15.19	21.59	24.74	9.76	9.81
p	6.58×10^{-8}	3.19×10^{-7}	9.24×10^{-8}	2.39×10^{-7}	2.11×10^{-9}	2.54×10^{-10}	4.29×10^{-8}	2.65×10^{-5}
Correlation								
F	16.68	13.24	15.69	11.83	15.12	14.29	10.68	12.00
p	7.27×10^{-8}	1.21×10^{-6}	1.59×10^{-7}	4.15×10^{-6}	7.38×10^{-8}	5.00×10^{-7}	3.13×10^{-8}	3.56×10^{-6}

B1–B8 represent the eight bands of GF-6WFV, respectively.

As it can be seen from Table 4, the F value of each feature significance analysis is greater than $F0.05$, which shows that the difference between the five types of features selected in this paper between different types of features is significant. However, it is still uncertain whether the difference between any two feature types is significant. Therefore, the multiple comparisons were carried out for further analysis.

The results of spectral analysis and multiple comparisons for each band are summarized in Figure 6. In the spectral feature space, the difference in the Euclidean distance of the mean of various classes on B5 is the largest. In its corresponding significance analysis, the mean of B1, B4, B5, and B6 are significantly different between any two classes. The difference in standard deviations of Euclidean distance of all the classes on band B3, B4, and B5 is large, but in ANOVA, the difference between the construction land and bare land on these three bands is not significant, neither between the vegetation and water. The difference in the Euclidean distance of ASM of classes on B1, B2, and B8 is large, but any two classes in all bands have poor significance; The difference in the Euclidean distance of the dissimilarity of classes on B2 and B6 is significant. However, in the significance analysis, three pairs of classes are less significant on B2, while on B6, the significance between vegetation and bare land is poor. The difference in Euclidean distance of correlation of these classes on B2, B7, and B8 is large, but in the corresponding significance analysis, only B2 can be selected for the following change detection.

Among the four new bands in GF-6 WFV, B7 cannot be used as an effective feature in land change detection because of the little information contained. Although the amount of information in the B8 band is moderate, it means that it cannot provide sufficient effective information for change detection after significance analysis. The mean value of the two new red bands is significantly different between classes, but after the spectrum statistics, it can be found that the red-edge band 2 (B6) has a higher standard deviation than that of the red-edge band 1 (B5), indicates B6 has a larger image contrast.

4.2. The Accuracy of LULC Change Detection Results

After multi-scale segmentation, the feature vector of the object needs to be constructed through feature optimization. The analysis of variance in the previous section has selected the mean of bands 1, 4, 5, 6 of GF-6 WFV, respectively, the auto-correlation of band 2, and NDVI and NDWI to generate feature vectors. Then, through CVA and correlation analysis to obtain the difference intensity information between two images. Finally, determine the areas of change and non-change according to the threshold. The change detection results are shown in Figures 7 and 8.

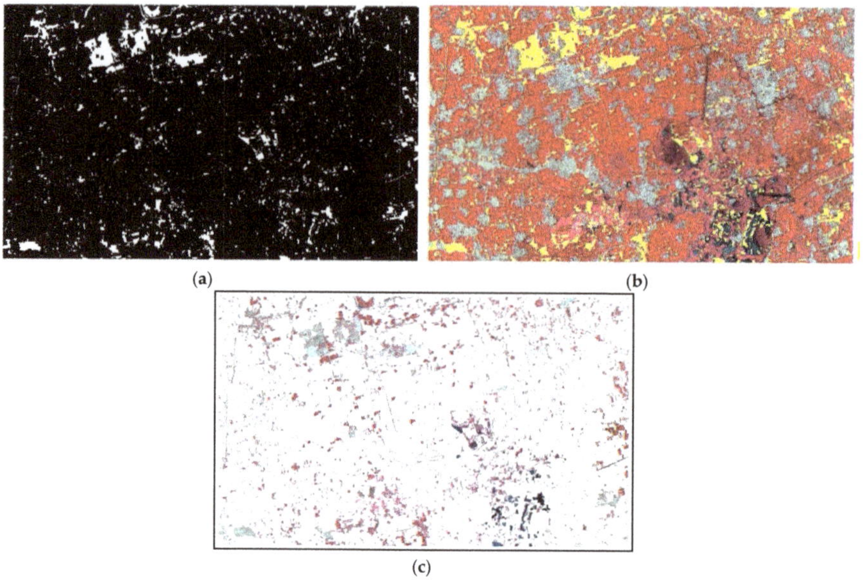

Figure 7. Change detection results by GF-6 WFV images of September of 2018 and 2019 (**a**) is binary (white represents changed areas); (**b**) is the superposition of the change detection results and the image (the yellow are the detected change areas; (**c**) is the base map of changed areas (white represents unchanged areas)).

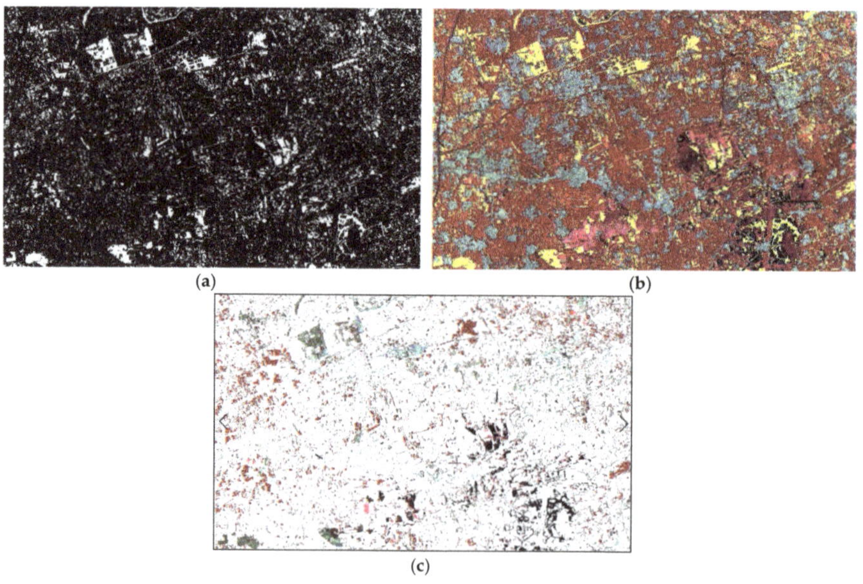

Figure 8. Change detection results by GF-1 WFV images of September of 2018 and 2019 (**a**) is binary (white represents changed areas); (**b**) is the superposition of the change detection results and the image (the yellow are the detected change areas; (**c**) is the base map of changed areas (white represents unchanged areas)).

In order to verify the accuracy of the model for GF-6WFV data, 300 samples were selected, including 133 samples without change and 167 samples with change. The selection of samples is carried out by the stratified sampling, randomly selecting the corresponding number of objects with change or without change individually, and then locate these objects through the generated ID numbers. The results are shown in Table 5.

Table 5. Confusion matrix of change detection using GF-6WFV images.

Detection Results \ Evaluation Index	Unchanged	Changed	Total
Unchanged	119	19	138
Changed	14	148	162
Total	133	167	300

According to the calculations via the formulas, the commission error, omission error, and Kappa are 8.64%, 11.38%, and 0.7780, overall accuracy of 89%, respectively. For GF-6 WFV data with the resolution of 16 m, this accuracy can fully meet the application requirements of LULC change detection. Among the sample of the wrong and the missing, the proportion of each category is shown in Table 6. Among them, the change between bare land and vegetation has the highest probability of being incorrectly detected, reaching 42.4%, most of them are the farmland after harvest. In fact, such changes are often not regarded as real changes between classes. Due to the large proportion of farmland in the study area, the sample of farmland included in a random sampling sample will be higher than that of other land cover types, which leads to the largest proportion of farmland in the samples that were falsely detected and missed. Theoretically, the additional red-edge bands can provide more information for identifying changes in vegetation. For construction land, it has achieved good results for GF-6WFV data.

Table 6. Samples analysis for false/missing detection by GF-6WFV images.

Class	Vegetation-Bare Land	Vegetation-Construction Land	Vegetation-Water	Bare Land-Water	Bare Land-Construction-on Land	Water-Construction Land
Number of samples	14	4	2	1	11	1
percentage	42.4%	12.1%	6.1%	3.0%	33.3%	3.0%

It can be seen from Figure 7 that the Xiong'an New Area has undergone many changes in LULC from 2018 to 2019, and the types of changes mainly include construction land, vegetation, water bodies, and bare land. Moreover, the construction land changes account for a relatively large amount, including demolition and engineering facilities under construction. These changes are evenly distributed roughly throughout the whole study area.

4.3. Comparison Results of GF-1WFV Images

To verify whether the GF-6WFV data has an advantage in change detection, a comparison test was performed on GF-1WFV images in the same area and the same phase. The results are shown in Figure 8 and Table 7.

Table 7. The comparison of the change detection results for LULC by GF-6WFV and GF-1WFV data.

Data Type \ Evaluation Index	Overall Accuracy	Kappa	Commission Errors	Omission Errors
GF-1 WFV	87%	0.7351	13.21%	9.58%
GF-6 WFV	89%	0.7760	8.64%	11.38%

The experiment found that the double-constrained change detection method can detect most of the change areas in the GF-1WFV images. Compared with the GF-6 data obtained on September 6, the GF-1 data obtained on September 24 obviously have many smaller blocky objects, most of which are harvested arable land, cause the late September coincides with the harvesting season of corn and sorghum. Due to the large amount of cloud interference in the data of GF-1 in the first ten days of September, we can only select the data in late September as a substitute in the experiment. As shown in Figure 8, there are many regular rectangular objects evenly distributed in the study area. Most of these are pseudo-changes caused by the harvest of cultivated land. In fact, the land cover type of the features themselves has not changed. To further verify the accuracy of the results, 300 samples were selected for comparative analysis, including 133 unchanged samples and 167 changed samples. Through comparison and calculation, the commission error is 13.21%, the omission error is 9.58%, Kappa is 0.7351, and the overall accuracy is 87%.

The result shows that the commission error of change detection under GF-1WFV data is significantly higher than that of GF-6WFV data using the same method, and the overall accuracy is reduced by 2%. Therefore, the accuracy of change detection that used the GF-6WFV data with the contained information of new additional bands is significantly improved. This indicates that these new red-edge bands contribute to the improvement of change detection accuracy and possess good potentials in the application of change detection.

5. Discussion

Through this research, it is necessary to discuss the following related issues.

In this paper, we only used the 16 m-resolution imageries of GF-6 WFV, not include other sensors' data of GF-6, such as panchromatic camera data and 8 m multi-spectral data for change detection of LULC. Although an accuracy of 89% is achieved, if the spatial resolution of the image can be improved, the accuracy of change detection should be higher. In order to fully test the performance of GF-6, the subsequent experiments can be considered, respectively. And the DCDM method used in the research is not an exclusive method proposed for GF series satellites, it is a general method that can be used for the detection of changes in various commonly used satellite data. In theory, if the appropriate features of objects are selected, the DCDM method can be suitable for any common used multi-spectral satellite imageries.

In terms of the change detection accuracy, the errors of registration preprocessing for different time-phase images, and vegetation phenology have some impacts on the final accuracy. The registration error is mainly caused by the change of the satellite parameters, the observation angle, topography and etc., which leads to the position of the same ground object in the images of different time phases to change. Although image registration is performed, this type of error cannot be completely eliminated, and we can try other methods to test whether such errors can be reduced. Moreover, taking the impact caused by phenology into consideration will help to detect the vegetation more accurately.

The experiments in this study are only for the change detection of bi-temporal images. For further validation, the more time phases for multi-temporal change detection should be tested and fully demonstrate the response of the new red-edge bands to change. This will be able to provide the advantages for resource monitoring or disaster assessment. And the automatic identification of the classes of LULC changes has not been done in this paper, it can be completed in the near future by learning the different types of features through machine learning methods.

6. Conclusions

The main conclusions of this study are as follows:

Firstly, through the spectral analysis and significance analysis of images of GF-6WFV, we found that its two new red-edge bands can provide effective information for change detection of LULC, while the purple and the yellow band cannot, and the addition of the new added two red-edge bands of

GF-6WFV increased the overall accuracy of 89% for change detection and improved by 2% compared with using the same temporal GF-1WFV data.

Secondly, the DCDM method based on change vector intensity and correlation coefficient can effectively detect the changes between different time-phase images for the same area. The object-oriented multi-scale segmentation method can more clearly describe the boundary of the changing area. It enables a more complete detection of the large change objects, as well as less omission of small change objects, making better use of the spectral and spatial information of images to obtain the changes. In addition, it can effectively reduce the generation of "salt and pepper" noise and increase the accuracy of change detection.

However, according to the validation, some narrow and long roads are incorrectly detected due to the errors caused by the registration, which affects the overall accuracy.

In summary, it can be concluded from the study that GF-6, as the first high-resolution satellite for precision agricultural observation in China, can meet the accuracy requirements of change detection of LULC with the combination of the new red-edge bands, and there is a good potential in monitoring resource and environment, for example, agriculture and forestry, and disaster mitigation, etc. This study also provides a theoretical support for the in-depth applications in related fields.

Author Contributions: All authors contributed in a substantial manner to the manuscript. Conceived, designed, and performed the research and wrote the manuscript, J.Y.; made contributions to conceive, and design the research, data analysis, and wrote some parts of the manuscript, Y.L.; contributed the model and results analysis and the conclusion, Y.R.; contributed to the programming, H.M.; contributed to the data preprocessing, Y.J., D.W. and L.Y. All authors discussed the basic structure of the manuscript. All authors have read and agreed to the published version of the manuscript.

Funding: This study was funded by the major projects of High resolution Earth Observation System of China (30-Y20A07-9003-17/18).

Acknowledgments: We would like to thank the staffs who provided the data and thank the reviewers for their comments and suggestions.

Conflicts of Interest: The authors declare no conflict of interest.

References

1. Liu, L.W.; Lin, J.; Chen, M.; Zhang, Y.; Feng, H.; Ye, Z.; Chu, J. Progress Review on Land Science Research in 2018 and Prospects for 2019: The Sub-report of Land Resources. *Land Use Plan.* **2019**, *33*, 106–114.
2. Ze, Y.W.; Hua, X.J.; Hua, X.L. An analysis on eco-environmental effect of urban land use based on remote sensing images: A case study of urban thermal environment and NDVI. *Acta Ecol. Sin.* **2006**, *26*, 1450–1460.
3. Teng, S.P.; Chen, Y.K.; Cheng, K.S.; Lo, H.C. Hypothesis-test-based landuse/cover change detection using multi-temporal satellite images—A comparative study. *Adv. Space Res.* **2008**, *41*, 1744–1754. [CrossRef]
4. Zhou, Q.M. Review on Change Detection Using Multi-temporal Remotely Sensed Imagery. *Geomat. World* **2011**, *9*, 28–33.
5. He, X. The development of direct change detection methods for remote sensing images. *Sci. Technol. Inf.* **2010**, *13*, 29.
6. Yu, X.F.; Luo, Y.Y.; Zhuang, D.F.; Wang, S.K.; Wang, Y. Comparative analysis of land cover change detection in an Inner Mongolia grassland area. *Acta Ecol. Sin.* **2014**, *34*, 7192–7201.
7. Bruzzone, L.; Prieto, D.F. An adaptive semiparametric and context-based approach to unsupervised change detection in multitemporal remote-sensing images. *IEEE Trans. Image Process. A Publ. IEEE Signal Process. Soc.* **2002**, *11*, 452–466. [CrossRef] [PubMed]
8. Bruzzone, L.; Marchesi, S.; Bovolo, F. Change Detection in VHR Multispectral Images: Estimation and Reduction of Registration Noise Effects. In *Optical Remote Sensing*; Springer: Berlin/Heidelberg, Germany, 2011.
9. Singh, A. Digital change detection techniques using remotely sensed data. *Int. J. Remote Sens.* **1988**, *10*, 989–1003. [CrossRef]

10. Lorena, R.; dos Santos, J.R.; Shimabukuro, Y.E.; Brown, F. A change vector analysis technique to monitor land use/land use/cover in sw Brazilian amazon: Acre state. In Proceedings of the Mid-Term Symposium in Conjunction with Pecora 15/Land Satellite Information IV Conference, Denver, CO, USA, 10–15 November 2002.
11. Bovolo, F.; Marin, C.; Bruzzone, L. A multilevel approach to change detection for port surveillance with very high resolution SAR images. In Proceedings of the 6th International Workshop on the Analysis of Multi-temporal Remote Sensing Images (Multi-Temp), IEEE, Trento, Italy, 12–14 July 2011.
12. Zhao, M.; Zhao, Y. Object-oriented and multi-feature hierarchical change detection based on CVA for high-resolution remote sensing imagery. *J. Remote Sens.* **2018**, *22*, 123–135.
13. Yang, C.; Hu, J. Test and Analysis on Application of Change Detection Method for Multitemporal Remote Sensing Images. *Geomat. Spat. Inf. Technol.* **2015**, *3*, 54–56.
14. Wei, H.; Hu, J.L.; Wang, L.H.; Hu, Y.X.; Han, P. Remote sensing image change detection based on change vector analysis of PCA component. *Remote Sens. Land Resour.* **2016**, *28*, 22–27.
15. Wang, L.; Yan, W.; Yu, Q. Research on Land Use Change Detection Based on an Object-oriented Change Vector Analysis Method. *J. Geo-Inf. Sci.* **2014**, *27*, 74–80.
16. Han, S.; Li, H.; Gu, H. The Study on Land Use Change Detection Based on Object-oriented Analysis. *Remote Sens. Inf.* **2009**, *3*, 23–29.
17. Wang, Z.; Liu, Y.; Ren, Y.; Ma, H. Object-Level Double Constrained Method for Land use/cover Change Detection. *Sensors* **2019**, *19*, 79. [CrossRef] [PubMed]
18. Li, X.; Shu, N.; Yang, J.; Li, L. The land-use change detection method using object-based feature consistency analysis. In Proceedings of the 19th International Conference on Geoinformatics, Shanghai, China, 24–26 June 2011.
19. Zhang, Z.; Zhang, X.; Xin, Q.; Yang, X. Combining the Pixel-based and Object-based Methods for Building Change Detection Using High-resolution Remote Sensing Images. *Acta Geod. Cartogr. Sin.* **2018**, *47*, 102.
20. Li, C.; Liang, W. Forest change detection using remote sensing image based on object-oriented change vector analysis. *Remote Sens. Land Resour.* **2017**, *29*, 77–84.
21. Ma, J.; Pei, L. Study of Urban Expansion Change Monitoring Based on GF-1. *Geomat. Spat. Inf. Technol.* **2017**, *40*, 61–64.
22. Zhang, C.; Yang, C.; Jia, M.; Jia, T. Change Detection of Land Use in Wuhan City Based on Multi Temporal Band Combination of GF-1. *Geomat. Spat. Inf. Technol.* **2016**, *39*, 33–35.
23. Liu, J.Y.; Xin, C.; Wu, H.; Zeng, Q.; Shi, J. Potential Application of GF-6 WFV Data in Forest Types Monitoring. *Spacecr. Recovery Remote Sens.* **2019**, *40*, 111–120.
24. Zhao, Y. *Principle and Method of Remote Sensing Application Analysis*; The Science Publishing Company: Beijing, China, 2003.
25. Li, L.; Shu, N.W.; Kai, G.; Yan, C.E. Detection Method for Remote Sensing Images Based on Multi-features Fusion. *Acta Geod. Cartogr. Sin.* **2014**, *43*, 945–953.
26. Wang, S.; Yang, X.; Dong, Z.; Shi, C. Remote Sensing Image Change Detection Based on Relief-PCA Feature Selection. *J. Graphics* **2019**, *40*, 119–125.
27. Liu, J.F.; Zhao, W.; Zhu, H.; Jin, Y. *Pattern Recognition*; Harbin Institute of Technology Press: Harbin, China, 2014.
28. Zhang, Y. The sample size is not "the more the merrier"-on the scientific determination of sample size in sampling survey. *Chin. Stat.* **2008**, *5*, 46–48.
29. Xi, J. Secure a Decisive Victory in Building a Moderately Prosperous Society in All Respects and Strive for the Great Success of Socialism with Chinese Characteristics for a New Era. In Proceedings of the 19th National Congress of the Communist Party of China, Beijing, China, 18 October 2017.
30. Qiu, C.H. Gaofen 6 lifts off: The national "Sky Eye" project data system is basically formed. *Public Commun. Sci. Technol.* **2018**, *10*. Available online: https://xueshu.baidu.com/usercenter/paper/show?paperid=7878de1d46af59d15d227fcbb4ddf8cb&site=xueshu_se (accessed on 28 July 2020).
31. Tan, Q.; Liu, Z.; Shen, W. An Algorithm for Object-Oriented Multi-Scale Remote Sensing Image Segmentation. *J. Beijing Jiaotong Univ.* **2007**, *31*, 111–114.

32. Schiewe, J. Segmentation of high-resolution remotely sensed data-concepts, applications and problems, Symposium on Geospatial Theory. In Proceedings of the Processing and Applications, Ottawa, IL, Canada, 9–12 July 2002.
33. Dian, Y.Y.; Fang, S.H.; Yao, C.H. Change detection for high-resolution images using multilevel segment method. *J. Remote Sens.* **2016**, *20*, 129–137.

remote sensing

MDPI

Article

Coastline Vulnerability Assessment through Landsat and Cubesats in a Coastal Mega City

Majid Nazeer [1], Muhammad Waqas [2,3], Muhammad Imran Shahzad [2,*], Ibrahim Zia [4,5] and Weicheng Wu [1]

1 Key Laboratory of Digital Land and Resources, East China University of Technology, Nanchang 330013, China; majidnazeer@ecit.cn (M.N.); wuwch@ecit.cn (W.W.)
2 Earth & Atmospheric Remote Sensing Lab (EARL), Department of Meteorology, COMSATS University Islamabad, Islamabad 45550, Pakistan; waqas.butt07.vb@gmail.com
3 Department of Land Surveying and Geo-Informatics, The Hong Kong Polytechnic University, Kowloon, Hong Kong
4 Department of Physical Oceanography, National Institute of Oceanography, Karachi 75600, Pakistan; ibrahimzia@hotmail.com
5 Department of Geography, University of Karachi, Karachi 75270, Pakistan
* Correspondence: imran.shahzad@comsats.edu.pk

Received: 22 January 2020; Accepted: 23 February 2020; Published: 25 February 2020

Abstract: According to the Intergovernmental Panel on Climate Change (IPCC), global mean sea levels may rise from 0.43 m to 0.84 m by the end of the 21st century. This poses a significant threat to coastal cities around the world. The shoreline of Karachi (a coastal mega city located in Southern Pakistan) is vulnerable mainly due to anthropogenic activities near the coast. Therefore, the present study investigates rates and susceptibility to shoreline change using a 76-year multi-temporal dataset (1942 to 2018) through the Digital Shoreline Analysis System (DSAS). Historical shoreline positions were extracted from the topographic sheets (1:250,000) of 1942 and 1966, the medium spatial resolution (30 m) multi-sensor Landsat images of 1976, 1990, 2002, 2011, and a high spatial resolution (3 m) Planet Scope image from 2018, along the 100 km coast of Karachi. The shoreline was divided into two zones, namely eastern (25 km) and western (29 km) zones, to track changes in development, movement, and dynamics of the shoreline position. The analysis revealed that 95% of transects drawn for the eastern zone underwent accretion (i.e., land reclamation) with a mean rate of 14 m/year indicating that the eastern zone faced rapid shoreline progression, with the highest rates due to the development of coastal areas for urban settlement. Similarly, 74% of transects drawn for the western zone experienced erosion (i.e., land loss) with a mean rate of −1.15 m/year indicating the weathering and erosion of rocky and sandy beaches by marine erosion. Among the 25 km length of the eastern zone, 94% (23.5 km) of the shoreline was found to be highly vulnerable, while the western zone showed much more stable conditions due to anthropogenic inactivity. Seasonal hydrodynamic analysis revealed approximately a 3% increase in the average wave height during the summer monsoon season and a 1% increase for the winter monsoon season during the post-land reclamation era. Coastal protection and management along the Sindh coastal zone should be adopted to defend against natural wave erosion and the government must take measures to stop illegal sea encroachments.

Keywords: shoreline change; landsat; planet scope; coastline; morphological changes

1. Introduction

A shoreline is defined as the boundary between land and water. The position of the shoreline is dynamic both spatially and temporally, due to hydrological, geological, climatic, and economic developments in coastal areas [1,2]. Therefore, shoreline indicators are used for a consistent representation of a shoreline. The most common shoreline indicators are the tidal datum (e.g., a specific elevation at the land–ocean boundary) and visually discernible features (e.g., a revetment structure, an erosion scarp, a previous high-tide high-water level, low-tide low-water level, or dune vegetation line, etc.) [3–8]. The shoreline is an indicator of the ecological health of coastal areas [3,9–11]. In past decades, the development of mega-infrastructure projects in coastal areas for urbanization/industrialization has placed more stress on the coastal ecosystem and changed ocean hydrodynamics [12–16]. Unplanned coastal development coupled with the natural action of ocean processes has led to the increasing vulnerability of low lying coastal environments [16–18]. This also induces spatiotemporal changes in the shoreline. Short-term geomorphological changes in the shoreline are caused by extreme geological, climatic, and oceanic events (i.e., earthquakes, tsunami, seasonal variation in waves, tides, and storms conditions), hence, such changes are less predictable, while long-term changes to the shoreline are caused by relative changes in astronomical, meteorological, and regional climatic variations (i.e., tides, waves, sea-level rise, and storm surges), and are somewhat predictable [6,19]. Both types of shoreline change are important for understanding trends in coastal sustainability for different times and spaces [3].

Anthropogenic activities coupled with global warming intensify ocean controlling factors that modify the coastal environment [20,21]. Morphological shoreline changes are directly associated with wave height, tide level, wind speed, periodic storms, and sea level [15,22]. Specifically, an accelerated rise in sea level and cyclones pose a considerable threat to populations living within a 100 km vicinity of the shoreline, as well as their economic assets. It is considered that more than 40% of the world's population is living in coastal cities [18]. For the monitoring of coastal zones, a common method for the assessment of spatiotemporal variability of the shoreline consists of the extraction of the shoreline position using multi-temporal data sources [5,6,22,23]. The primary data sources used for shoreline extraction and evaluation are historical photographs, coastal maps/charts, aerial and beach surveys, global positioning system based shoreline surveys, and images acquired through satellite sensors [15,24,25]. Each source has its own measurement and positional uncertainty [1,4,7,26]. Historical archives of satellite remote sensing open source data with a high spatial resolution of 3 m such as from Planet Scope to a low spatial resolution of 1000 m from the Moderate Resolution Imaging Spectroradiometer (MODIS) have revolutionized science. These datasets have proved their potential for the detection of historically changing trends in extensive coastal land masses, with an insignificant uncertainty in the case of high spatial resolution, and a relatively high uncertainty in cases of moderate to low spatial resolution [4,12,15,16,27–30].

Different methods, including image classification [4,31], band ratio [15],principal component analysis [32], overlay operation [30], and density slicing [23] have been used for the delineation, mapping, and estimation of the variability of proxy-based shoreline positions through satellite imagery, and have differing levels of measurement accuracy [4,5]. The most commonly used indices for the delineation of the shoreline include the Normalized Difference Water Index (NDWI) [33], the Automated Water Extraction Index (AWEI) [34], and the Modified Normalized Difference Water Index (MNDWI) [35]. Several studies have been carried out globally, of both short-term and long-term shoreline changes using different image analysis techniques. For example, a recent study by Nassar et al. (2018) [36] used the archive of Landsat multi-sensor images from 1989 to 2016 to assess the rates of change by the accretion and erosion of the coastal area of North Sinai that were caused by changes in hydrodynamics and alongshore currents. Cenci et al. (2018) [37] performed a multi-proxy analysis of two different shorelines (i.e., in high and low energy areas) and assessed the rate of shoreline change to highlight the uncertainty in coastal risk management. Bheeroo et al. (2016) [38] carried out a risk assessment of the north-western side of the Mauritius coastline and addressed changes in the coastal environment with reference to high energy wave action and storms. Kermani et al. (2016) [39] used aerial photographs and high resolution satellite images of the Bay of Jijel in Eastern Algeria to investigate changes in morpho-dynamics of the shoreline. Ozturk and Sesli (2015) [15] used historical multi-temporal satellite data to assess changes in the coastal environment in the lagoons of Kizilirmak

Delta due to ocean controlling factors (e.g., wind and wave erosion). They also addressed the rate of shoreline change and shrinkage of the lagoon series. Kuleli et al. (2011) [40] assessed the rates of shoreline change on the Ramsar beaches of Turkey using Landsat images from 1975 to 2009 and found that most parts of the wetland areas were under stress with a significant withdrawal rate.

Pakistan is considered the seventh most vulnerable country to climate change induced events. These include four major floods and six historical cyclones, which have damaged the socio-economy of low lying areas within the Sindh coastal region [41]. The coastal belt of the Indus Delta Region is continuously changing due to physical processes such as tidal actions, waves, wind speed, sea level rise, and anthropogenic factors, for example land reclamation and modification [12,16,42,43]. Rising sea levels are contributing to the erosion of less resistant soil in the region. It has been claimed that the mean sea level rose from 1.7 mm/year between 1900 and 2010 and 3.2 mm/year globally over the past decade [44]. Similarly, the Intergovernmental Panel on Climate Change (IPCC) predicts that the global mean sea-level will rise from 0.43 m to 0.84 m in the current scenario and 0.61 m to 1.10 m in the worst case scenario by 2100 [45]. It has been reported that the sea level has already risen from 1.1 mm/year to 1.8 mm/year regionally over the past decade [42]. If sea levels rise above the predicted rate, shorelines in the region will experience erosion in both the long- and short-term. Unfortunately, there has been no study to quantitatively estimate future shoreline changes along the coastline of Karachi, Pakistan. Therefore, this study aims to (i) merge multi-temporal datasets i.e., topographic maps and medium (30 m) to fine spatial resolution (3 m) satellite imagery, (ii) investigate a change in the position of the Karachi shoreline from 1942 to 2018, and (ii) assess the rate of systematic land loss and/or gain due to oceanic processes and anthropogenic activities along the coastline.

2. Study Area

Karachi is considered Pakistan's largest city, hosting about 7% of the country's population, making it the fifth largest coastal city in the world [46]. The coast of the Karachi city is about 100 km between the Gharo creek on the east and the river hub on the west, and constitutes a number of tourist beaches [43,47]. The coastline of Karachi is considered an economic hub for Pakistan as 90% of seaborne trade is carried out through two of its international ports i.e., the Port of Karachi and the Port Qasim [43] (Figure 1). The elite class of Pakistan's population enjoy life along the Karachi coast, and a number of tourist spots have been developed, which has increased its residential property values [16,47]. The study area, the coast of Karachi, is divided into two zones i.e., eastern and western zones. The eastern zone starts from the Defense Housing Authority (DHA) phase 8 to the South Asian terminal on the west and is 25 km long (Figure 1), while the western zone from Manora beach on the east to the Engro beach huts on the west (Figure 1) is 29 km long. The western zone is mostly sandy but toward its western part the nearshore region becomes steeper, with irregular rocky outcrops associated with the underlying geology. During the last decade, several housing societies including the DHA Phase 8 and EMMAR project have encroached on the coastline in the Karachi metropolitan and Crescent Bay areas. The coastline of the eastern zone constitutes a mostly loose stone structure and sandy beaches.

The Sindh coastal zone is mixed wave dominated, with continuous accretion and erosion due to waves and tides of up to 3 m in height, modifying the coastal environment [16]. Strong westerly winds prevail during the summer monsoon and influence ocean circulation with the dominant direction of surface water flow [42]. The pre- and post-monsoon periods have long been associated with cyclone development in the Arabian Sea, which increase erosion and move the shoreline landward. The alongshore sediment transportation is from west to east due to a longshore drift in a clockwise direction. Water depth within 1 km offshore in the study area is shallow (3 m–10 m) with a maximum depth of about 10 m at Hawk's Bay. Seawater surface temperatures in the near shore water off the coast of Karachi range from 24 °C to 28 °C in summer and 20 °C to 24 °C in winter [43].

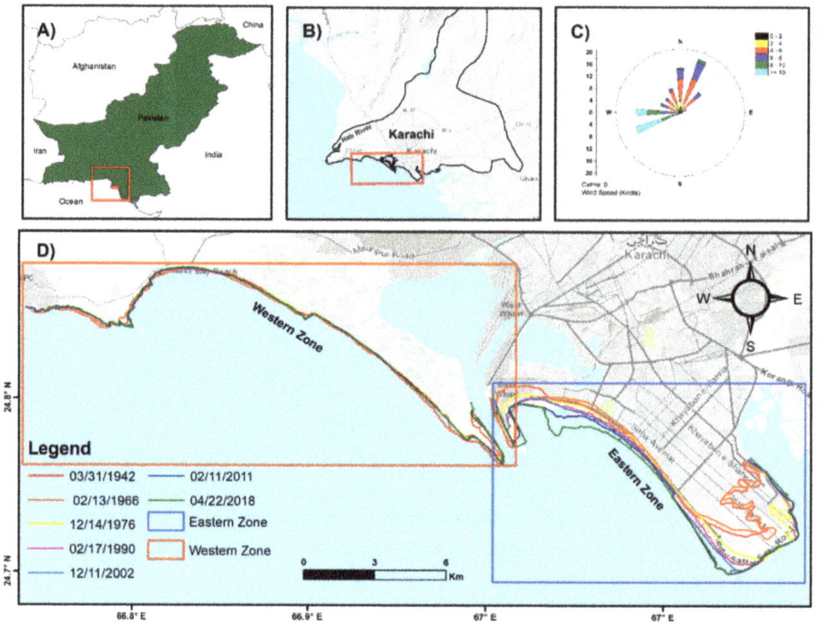

Figure 1. (**A**) Map of Pakistan with a demarcation of the study area (filled red rectangle). (**B**) Geographical location of the coastal city, Karachi, where the red rectangle shows the study area. (**C**) Wind climatology of the study area and (**D**) historical shoreline positions of the Karachi coast in the eastern and western zones.

3. Data Used

3.1. Historical Topographic Maps

In order to use a long-term data record for the monitoring and evaluation of the shoreline, two topographic maps by the Army Map Service (GDPE), Corps of Engineers, at a 1:250,000 scale for the years 1942 and 1966 (AMS U502 NG 42-13, Edition 1), were used.

3.2. Satellite Data

Two types of satellite datasets were used to assess morphological changes in the shoreline positions i.e., data from Landsat at medium spatial resolution (and Planet Scope at a finer spatial resolution). Landsat collection 1 Level-1 images of Multispectral Scanner (MSS) sensor and Level-2 images of Thematic Mapper (TM), Enhanced Thematic Mapper Plus (ETM+), and Operational Land Imager (OLI) sensors were acquired on different dates (Table A1) from the United States Geological Survey (USGS) Earth Explorer (http://earthexplorer.usgs.gov/) and USGS Earth Resources Observation and Science (EROS), as well as the Center Science Processing Architecture (ESPA) On Demand Interface (https://espa.cr.usgs.gov/), respectively. The finer spatial resolution data from Planet Scope (PS) was acquired for the year 2018. Four continuous strips of PS (Table A1) scenes covering the study area and Level 3B product (orthorectified and surface reflectance image product) of PS were acquired.

3.3. Wind-Wave Data

Winds are the driving source of waves and the transformation of wind energy results in strong ocean currents [42,48,49]. Wave action is considered the most important factor for a change in a coastal environment during the summer and winter monsoon seasons, as prevailing winds from the south-west and northeast intensify attacks on the coastline. Long-term offshore daily-hourly wind-wave data collected from National Oceanic and Atmospheric Administration (NOAA) Wave-Watch III Model [50] were used to evaluate the influence of wind-wave climatology on the coastline of Karachi before and after the mega land reclamation project of 2008 to 2017.

3.4. Tidal Data

Tides significantly affect wave dominant shorelines [49,51] as they induce longshore drift, which can nourish or erode the coast. The variations of tides at Karachi can be used to analyze the impact of tidal currents along the coastline and identify the coastal waterlines. In this study, tide height was used as a tool in the image selection criteria to extract the true shoreline position. Port Muhammad Bin Qasim is located about 5 km east of the coast of Sindh where the tide height has been monitored by the National Institute of Oceanography (NIO) Karachi, Pakistan from 1976 to 2018. Historical tide height records show that maximum and minimum tide heights recorded between 1976 and 2018 at the station were 3.8 m and a little less than 0.1 m.

4. Methodology

4.1. Pre-Processing of Data

4.1.1. Rectification of Topographic Maps

The geo-referencing of scanned topographic maps was made in-house using the Universal Transverse Mercator (UTM) coordinate system (WGS1984-UTM-Zone-42N; EPSG-32642 OTF) to project over the study area so as to estimate the shoreline position. The topographic sheets were then rectified using 35 Ground Control Points (GCP) and the NIO shoreline profile along with the 2016 image to reduce measurement uncertainty in the shoreline position [4,26,52].

4.1.2. Atmospheric Correction of Satellite Imagery

The pre-processing of images from different sensors in order to make it consistent with other images by normalizing them is very important in shoreline change detection studies [4,7,31,53,54]. The image based Dark Object Subtraction (DOS) model was used to convert the digital number recorded by the MSS sensor to generate a surface reflectance product [55–57]. Then, the surface reflectance products of L2 MSS images with a 60 m spatial resolution were resampled at 30 m to conform to the spatial resolution of other Landsat sensors (TM, ETM+, and OLI) [16,56].

The Scan Line Corrector (SLC) error encountered on all images acquired after 31 May 2003 from ETM+ caused about a 22% loss of data [57]. Therefore, a correction of the SLC error was carried out using the "Fill no data tool" under the raster tools available in Quantum Geographical Information System (QGIS 2.8.8) software, though some degree of error remains [4,26,52].

4.2. Satellite Data Selection Criteria

Images for the same months of different years for several decades enabled us to investigate the variability of shoreline morphology with nominal inherited measurement and positional uncertainty [15,43,58,59]. Therefore, we only selected Landsat (L2, L5, L7, and L8) and PS images that met the following conditions: (1) 0% cloud cover over the study area, (2) non-flooding months (December to April), and (3) tide height of <1 m within a time window of ±1 h of the image acquisition time. Non-flooding months provided information about the stability of the ocean condition because tropical storms mainly prevail during the months of May to June and September to November [16]. Furthermore, low tide height provided stable outer boundaries of the coastline [4,5,26,59]. This image selection criteria (Figure 2) resulted in a limited number of images available, including four images from Landsat sensors and one from PS which was prioritized over the OLI images due to a higher resolution (Table A1). The image selection criteria in this study was helpful in reducing or removing the inherited seasonal uncertainty in shoreline change analysis.

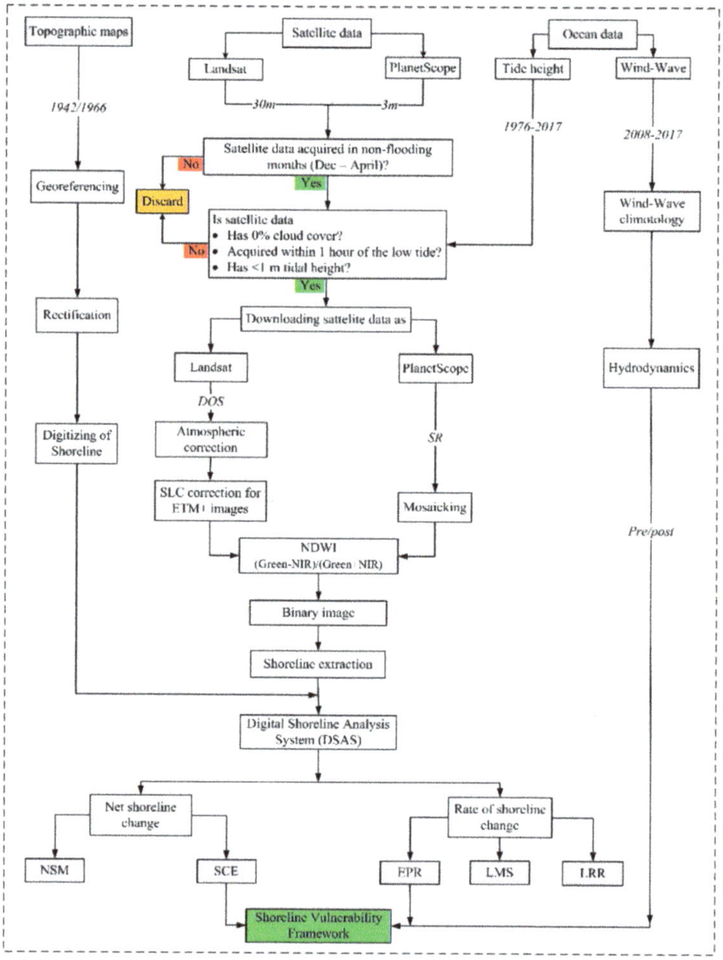

Figure 2. Flowchart of the methodology for the shoreline change detection along the coast of Karachi from 1942 to 2018. (SLC = Scan Line Corrector, SR = Surface Reflectance, NSM = Net Shoreline Movement, SCE = Shoreline Change Envelope, EPR = End Point Rate, LMS = Least Median of Square, and LRR = Linear Regression Rate).

4.3. Shoreline Identification and Extraction

The basic requirement for detection of the true shoreline position is to determine the shoreline indicator i.e., the high tide line, the high-water line, dune line, or vegetation line [3–5,7,26,60]. These indicators show a variability in the shoreline position for the same time and space [3–5]. For the present study, detection of a high water line (HWL) under stable oceanic conditions (low wave-tide condition) from satellite images was calculated by band ratioing. The HWL shoreline indicator for images acquired during non-flooding months at a low tide condition reduced the overall positional uncertainty to a minimum as HWL migrates to contemporary low water line (LWL) or mean water line (MWL) to draw satisfactory continuity of the shoreline positions during different time periods [26,27]. The best results for shoreline identification were obtained from NDWI (Equation (1), which is used to monitor changes in water content) by evaluating the results of all indices (NDWI, MNDWI, and

AWEI) [33–35] visually with true color satellite images. Mud-flats were also considered to represent the shoreline [48].

$$NDWI = \frac{Green - NIR}{Green + NIR} \quad (1)$$

The land and water mask were then developed by image thresholding [61]. Raster binary images were then converted to linear vectors by using the raster to vector conversion tool available in QGIS 2.8.8 software. Minor edits were made to the shoreline position by visual interpretation at a scale of 1:20,000 for mapping the shoreline position with more accuracy than the pixel size [62,63]. The maximum measurement uncertainty for the MSS sensor was presumed as ±37 m (about one pixel of resampled MSS image) while for TM and ETM+ sensors it was ±34 m (about one pixel of TM/ETM+) and for PS ±3.4 m (about one pixel of PS). Historical shoreline positions for 1942 and 1966 prior to the available satellite images were acquired from rectified GeoRef topographic sheets through digitization using GIS state of the art tools at a scale of 1/8000, having a maximum measurement uncertainty of about ±37 m [26,52,62]. The MWL for topographic maps could be located to within 0.150 mm on a map scale or to within 37 m on the ground for a map of scale 1:250,000. The comparison of MWL between topographic sheets and HWL for satellite images was possible due to their near equivalence [52]. The best estimate of maximum annualized average shoreline uncertainty for the study was considered ±0.43 m. Here it should be noted that the uncertainty 0.43 m/year is an average value for change in rate i.e., not constant for all transects of the local area. For example, in the case of Hawks Bay (in the western zone), only 105 transects out of the total 268 transects encounter accretion and erosion less than 0.07 m/year, which is less than the annualized uncertainty rate of 0.43. Furthermore, it needs to be clarified that the value of 0.07 m/year (by the EPR method) reflect the averaged value of shoreline change (for all transects drawn including erosional and accretional) over the period of 76 years for the Hawks Bay area (Table 1), and do not reflect a change within a pixel size. These low rates in change represent very low or moderate anthropogenic activity compared to the eastern zone where there are more anthropogenic activities (Section 5).

Table 1. Average shoreline change rate of the eastern and western zone.

					Eastern Zone				
	DHA ZB	**DHA GC**	**DHA ZC**	**DHA ZD**	**DHA EXT**	**S V**	**SJC**	**SAT**	
TT	86	104	65	236	149	257	162	135	
T ID	1–86	87–190	19–255	256–491	492–640	641–897	898–1059	1060–1194	
ShoreT	RS	RS	RS	RS	SGB	SB	SedB	RS	
EPR	14.3	22	22.4	21	8.7	5.9	12	8.38	
LMS	2.51	5.05	6.98	16.98	11.54	7.2	4.78	7.01	
LRR	16.29	24.96	23.95	25.8	11.26	5.8	9.95	5.57	
SCE	1.73	1.73	1.77	1.89	0.90	0.57	1.01	0.82	
NSM	1.63	1.68	1.71	1.59	0.67	0.45	0.91	0.64	
					Western Zone				
	MB	**SS**	**KP**	**HB**	**JG**	**SG**	**ARG**	**HAG**	**EBH**
TT	342	291	62	268	77	46	40	96	100
T ID	1–342	343–633	634–695	696–963	964–1040	1041–1086	1087–1126	1127–1222	1223–1322
ShoreT	SGB	SB	RB	SB	SB	SB	SB	SB	RB
EPR	−0.61	−0.77	0.18	0.07	−1.55	−0.95	−0.73	−0.75	−0.87
LMS	−0.07	−0.19	0.12	−0.01	−1.7	−0.54	−0.82	−0.28	−1.34
LRR	−0.32	−0.72	0.07	−0.22	−1.79	−0.65	−0.12	−0.74	−1.2
SCE	0.23	0.10	0.10	0.11	0.15	0.11	0.14	0.17	0.13
NSM	−0.05	−0.06	0.01	0.01	−0.12	−0.07	−0.06	0.17	−0.07

Notes: (1) TT = Total Transects drawn, T ID = Transect Identity, ShoreT = Shoreline Type; (2) RS = Riprap Structer, SB = Sand Beach, SGB = Sand and Gravel Beach, SedB = Sediment Bank, RB = Rocky Beach; (3) Eastern Zone: DHA = Defense Housing Authority, ZB = Zone B, GC = Golf Club, ZC = Zone C, ZD = Zone D, Ext = Extension, SV = Sea-view, SJC = Shireen Jinnah Colony, SAT = South Asian Terminal; Western Zone: MB = Manora Beach, SS = Sandspit, KP = Kaka Pir, HB = Hawks Bay, JG = Jamali Goath, SG = Somar Goath, ARG = Abdul Rehman Goath, EBH = Engro Beach Huts; (4) EPR, LMS, LRR = Rate of shoreline change (m/year) and NSM, SCE = Shoreline movement (Km).

4.4. Digital Shoreline Analysis System (DSAS)

The Digital Shoreline Analysis System (DSAS) version 4.4 is now provided by the Environmental Systems Research Institute (ESRI) as an add-in utility for the ArcGIS version 10.5 service pack, which was developed by the USGS under the Coastal and Marine Geology Program [64]. The statistical algorithm in DSAS allows the measurement of shoreline change statistics for multiple historical shorelines at each transect drawn at a user specified interval and length, relative to the baseline. The baseline is drawn at a user specified distance from the source shoreline, which remains parallel and offset to the reference shoreline orientation. Then transects were cast, which must intersect each shoreline position to calculate the change in shoreline positions. The DSAS executes statistical operations based on measured differences between historic shoreline positions to measure the rate of change in the following terms:

- SCE: Shoreline change envelopes (distance measurement);
- NSM: Net shoreline movement (distance measurement);
- EPR: End point rate (point rate);
- LRR: Linear regression rate (regression statistics);
- WLR: Weighted linear regression (regression statistics);
- LMS: Least median of squares (advanced statistics).

The extracted shorelines (i.e., from the historical topographic sheets and multi-temporal images) and the relative discrete baselines were placed in a personal geo-database prepared using Arc Catalog and prerequisite attribute fields (i.e., date, shape length, cast direction, etc.) were added to them. The movement of shoreline recorded at each transect drawn relative to the orientation of baseline, was considered as a landward movement if there was a negative (–) value recorded at the transect drawn, while positive (+) values indicated seaward movement (accretion) at that specific transect [64]. In this study, NSM, SCE, EPR, LMS, and LRR metrics were used to determine changes in the shoreline during the study period. Here SCE and NSM measured the net change in historical shoreline positions (distance) while EPR was used to measure the net rate of change in the shoreline [15]. The LMS and LRR were used to calculate the median and mean susceptibility rate of the shoreline change by fitting a regression line for each transect [64]. The SCE measured the distance between the two farthest shoreline positions and the distance value from the SCE method was always positive, as it does not record the accretion and erosion in shoreline positions. NSM calculated the distance of the youngest from oldest shoreline position, which represents the change in positive and negative values [40,64]. SCE was independent of the date factor of shoreline position while NSM was associated with the date factor for these two shoreline positions, and both estimated the distance, but not the rate. The EPR was used to estimate the rate of erosion and accretion per year at each transect [52]. In LMS, the median value of the squared residual was used for fitting a regression line shoreline. Its value was obtained by dividing the total coastal change distance, or NSM, by the total number of years at each transect [52]. In LMS, the median value of the squared residual was used for fitting a regression line relative to different positions of shoreline at each transect, which exhibited the rate of shoreline change. Similarly, the LRR method used the mean value of residuals to fit a line for each transect in which the slope of the line predicted the rate of shoreline change. A great variation between these two methods was observed because in the LRR method, each input offset data point had an equal influence on the estimation of the slope's change rate, while in the LMS method each offset data point had less influence on the slope as the median value was considered the rate. The LRR rate of shoreline change was susceptible to effects of outlier values because it could underestimate or overestimate shoreline change rate values observed by the EPR method [58].

4.5. Shoreline Change Analysis

Different parts of the study area showed different characteristics of coastal change [7,65]. Therefore, we divided the study area into two major zones as eastern and western zones of a length of 25 km and 29 km, respectively. These major zones were further divided by their local area name to better understand shoreline change at a micro level. The eastern zone had local areas named DHA Zone B, DHA Golf Club, DHA Zone C, DHA Zone D, DHA Extension, Sea-view, Shireen Jinnah Colony, and

South Asian Terminal (Figure 3). The shorelines of DHA Zone B, DHA Golf Club, DHA Zone C, DHA Zone D, and South Asian Terminal have loose stone structures while DHA Extension has a sand and gravel beach (Table 1). The shoreline of Sea-view has a tidal flat sand beach and Shireen Jinnah Colony has a sediment bank (Table 1). Similarly, the western zone has local names Manora Beach, Sandspit, Kaka Pir, Hawks Bay, Jamali Goth, Somar Goth, Abdul Rehman Goth, Haji Ali Goth, and Engro Beach Huts (Figure 4). The shoreline of Sandspit, Hawks Bay, Jamali Goth, Somar Goth, Abdul Rehman Goth, and Haji Ali Goth have sandy beaches while Kaka Pir and Engro Beach Huts have rocky beaches (Table 1). The shoreline of Manora Beach constitutes sand and gravel beaches (Table 1).

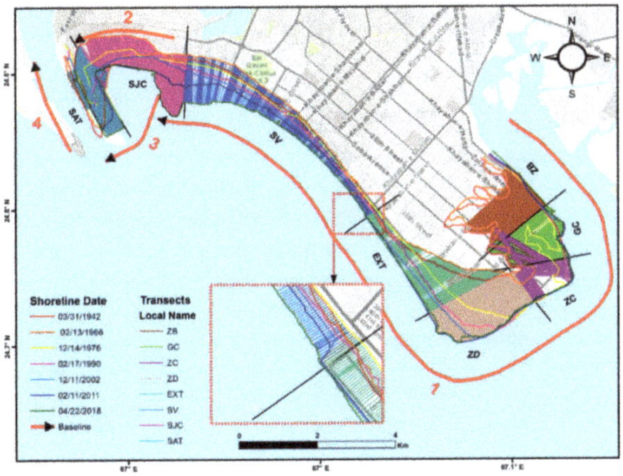

Figure 3. Digital shoreline analysis on the eastern zone (inset red doted rectangle shows a part of the study area) (ZB = Zone B; GC = Golf Club; ZC = Zone C; Ext = Extension, SV = Sea-view, SJC = Shireen Jinnah Colony, and SAT = South Asian Terminal) (Table 1).

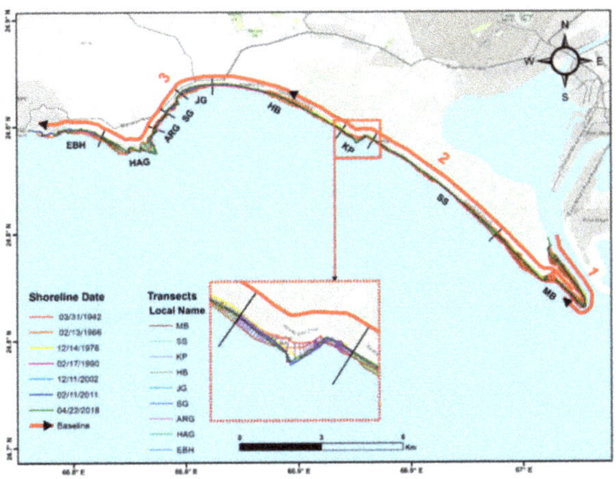

Figure 4. Digital shoreline analysis on the western zone (inset red doted rectangle shows a part of the study area) (MB = Manora Beach, SS = Sandspit, KK = Kaka Pir, HB = Hawks Bay, JG = Jamali Goth, SG = Somar Goth, ARG = Abdul Rehman Goth, HAJ = Haji Ali Goth, and EBH = Engro Beach Huts) (Table 1).

A baseline was created for the two zones offshore and onshore of the coast, which remained parallel to the shorelines, in order to draw transects for determining shoreline changes. The baseline of the eastern zone (Figure 3) constitutes segments of baseline and a baseline of the western zone is divided into three segments because of the discrete shoreline orientation to draw transects of length 3500 m and 1000 m respectively from the baseline as it intersects all the shorelines. A total of 2516 transects were drawn for the study area at 20 m intervals along the baseline (1194 for the eastern zone, and 1322 for the western Zone) (Figures 3 and 4). This transect interval was small enough comparatively to the satellite image resolution, but transect intervals below than this would not provide a better understanding of shoreline change.

5. Results

An analysis of the shoreline change rate revealed that both zones exhibited both types of change, with erosional trends as negative values and accretional trends as positive values. The eastern zone exhibited accretional action in the shoreline position throughout the study period while the western zone mostly showed both accretion and erosion. Therefore, the results of the present study identified areas with accretion and erosion in the shoreline position at each transect drawn (Tables 1 and 2). Table 1 represents the types of shoreline, transect-based change rates from selected metrics (EPR, LMS, LRR, SCE, and NSM) for both zones with respect to their local areas, and a total number of transects drawn for each local area. Table 2 shows maximum accretion and erosion rates recorded for transects in both zones, and the percentage of transects that recorded erosion or accretion.

Table 2. Maximum and mean actions (accretion and erosion) at eastern and western zones.

	Rate	Eastern Zone			Western Zone		
Accretion		**Max**	**Location**		**Max**	**Location**	
	EPR	32	South Asian Terminal		3.88	Haji Ali Goth	
	LMS	27	DHA zone D		4.37	Haji Ali Goth	
	LRR	35.49	DHA zone D		5.66	Haji Ali Goth	
	NSM	2.43	South Asian Terminal		0.30	Haji Ali Goth	
Erosion		**Max**	**Location**		**Max**	**Location**	
	EPR	−1.85	South Asian Terminal		−3.96	Manora Beach	
	LMS	−14	Shireen Jinnah colony		−3.98	Manora Beach	
	LRR	−2.21	South Asian Terminal		−3.27	Manora Beach	
	NSM	−0.14	South Asian Terminal		−0.30	Manora Beach	
Accretional Transects		**Mean**	**Count**	**Percent %**	**Mean**	**Count**	**%**
	EPR	14.05	1130/1194	94.6	1.37	339/1322	25.6
	LMS	9.99	1084/1194	90.8	0.58	523/1322	40
	LRR	15.31	1130/1194	94.6	0.98	296/1322	22.4
	NSM	1.06	1130/1194	94.6	0.08	338/1322	25.6
Erosional Transects		**Mean**	**Count**	**Percent %**	**Mean**	**Count**	**%**
	EPR	−1.48	64/1194	5.4	−1.15	983/1322	74.4
	LMS	−3.23	111/1194	9.3	−0.85	799/1322	60.4
	LRR	−1.89	64/1194	5.4	−0.99	1026/1322	77.6
	NSM	−0.11	64/1194	5.4	−0.08	984/1322	74.4

Note: EPR, LMS, LRR = Rate of shoreline change (m/year) and NSM = Shoreline movement (Km).

5.1. *Rate of Shoreline Change for the Eastern Zone*

The maximum average accretion rate in this zone was observed at DHA Phase 8 Zone C with a rate of 22 m/year at a transect identity (TID) from 191–255 (Table 1). The lowest average rate of accretion was observed at Sea-view Clifton beach with an average rate of 6 m/year (TID 641–897) (Table 1). The lowest to highest EPR rate in the eastern zone, from −1.85 to 32 m/year was observed at

South Asian terminal (Table 2). Using the EPR method in the eastern zone, 1130 out of 1194 transects (94.6%) experienced accretion, while 64 (5.4%) exhibited erosion. The mean values of accretion and erosion from EPR were 14 m/year and –1.48 m/year for the total accretional and erosional transects in the eastern zone (Table 2).

The LMS was the second metric used for estimating shoreline change rates. The eastern zone showed a high variation of accretional shoreline change rates with a mean maximum of 17 m/year at DHA Phase8 zone D and a mean lowest change of 2.5 m/year at DHA Phase8 zone B (Table 1). The transects with the greatest erosional LMS rate of –14 m/year and greatest accretional rate of 27 m/year were estimated at Shireen Jinnah Colony and DHA phase8 zone D respectively (Figure 5, Table 2). The LMS rate showed that 90.7% of transects were accretional with an average rate of 10 m/year while 9.3% of them were erosional with a mean rate of –3.23 m/year. Similarly, observed values of the LRR for eastern zone showed that the DHA golf club exhibited a maximum mean accretional rate of 25 m/year and South Asian terminal had the lowest mean rate of 5.6 m/year (Table 1). The greatest significant LRR erosional rate of -2.21m/year and accretional rate of 35 m/year (Table 2) were estimated for the transects of DHA Phase8 zone D and South Asian terminal respectively (Figure 5). Furthermore, the overall LRR records show that 94.6% of transects drawn in eastern zone experienced shoreline accretion, with mean rate of 15.3 m/year while only 5% of transects recorded erosion, with mean rate of -1.89 m/year (Table 2).

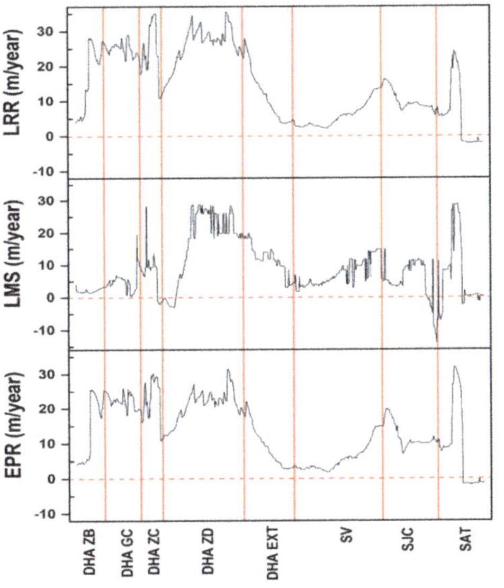

Figure 5. Shoreline change rate (EPR, LMS, LRR) of eastern zone (DHA = Defense housing Authority; ZB = Zone B; GC = Golf Club; ZC = Zone C; Ext = Extension, SV = Sea-view, SJC = Shireen Jinnah Colony, SAT = South Asian Terminal).

The estimation of the shoreline changes in the study area for the eastern zone constitutes on the predicted values of shoreline change envelope (SCE) and net shoreline movement (NSM). The results of SCE reveal that a mean highest accretional SCE value of 1.9 km was obtained for the DHA phase 8 zone D (TID from 256-491) (Table 1). However, it cannot be concluded that the shoreline moved seaward. Similarly, the lowest mean accretional SCE value of 0.57 km was observed for the Sea View Clifton beach (Table 1). The highest and lowest SCE values of 2.4 km and 0.15 km were obtained for the South Asian terminal (TID from 1060-1194) (Table 2). Furthermore, the NSM values recorded in the analysis indicate that the maximum mean accretion distance between the oldest shoreline position of 1942 and the recent shoreline position of 2018 was 1.7 km, obtained for DHA phase 8 zone C (TID from 191–255) (Table 1). Similarly, the lowest mean NSM value of 0.4 km was obtained for Sea View Clifton Beach (TID from 641–897) (Table 1). The highest accretional and erosional values of NSM were obtained for South Asian terminal ranging from 2.43 km to -0.14 km respectively (Table 2, Figure 6). Overall results for NSM shows that 95% of the eastern zone transects experienced accretion with a mean seaward movement of the shoreline of 1.06 km, while 5.4% of transects in the eastern zone show erosion, with a mean retreat of -0.11 km (Table 2) (Figure 6).

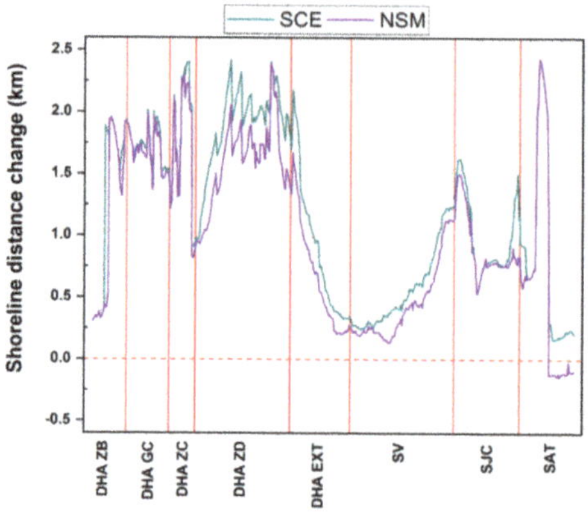

Figure 6. Shoreline change (NSM, SCE) of eastern change (DHA = Defense housing Authority; ZB = Zone B; GC = Golf Club; ZC = Zone C; Ext = Extension, SV = Sea-view, SJC = Shireen Jinnah Colony, SAT = South Asian Terminal).

5.2. Rate of Shoreline Change for Western Zone

The western Zone (Figure 4) constitutes rocky and sandy beaches, and exhibited both erosional and accretional trends in the shoreline (Figures 7 and 8). Hawks Bay, Haji Ali Goth (French Beach) and Manora Beach are recreational beaches with beach huts and water sports parks with thousands of visitors daily. The maximum mean accretion rate of 0.18 m/year by EPR was observed for the Kaka Pir area, having TID ranges from 634–695 (Table 1, Figure 7). The lowest mean erosional rate of −1.5 m/year by EPR (Table 1) was observed at Jamali Goth area (TID from 964–1040) (Table 1). The highest rate of accretion and erosion shoreline change was recorded at Manora Beach with a value of 3.86 m/year and −3.96 m/year (Table 2). The EPR method shows that 26% of the 1322 transects 1322 experienced accretion, with a mean linear rate of 1.37 m/year (Table 2). Similarly, erosive action was dominant in the western zone, with 1322, or 74% of transects drawn showing erosion, at a mean rate of −1.15 m/year.

LMS values for the western zone showed that the highest mean accretional rate occurred at Kaka Pir area at a rate of of 0.12 m/year (Table 1) and highest mean erosional rate of −1.7 m/year for Jamali Goth (Table 1). The highest accretional rate of shoreline change from the LMS method was observed over Haji Ali Goth, with a rate of 4.37 m/year (Table 2). Similarly, the highest erosional rate of change by LMS was obtained for Manora beach with a value of −3.94 m/year (Table 2) (Figure 7). The overall results for the LMS method indicated that 40% of transects drawn encountered accretion, with mean accretional rate of 0.5 m/year and 60% of them recorded erosion with the mean erosional rate of −0.85 m/year.

Moreover, the values of LRR showed that the highest mean accretional rate of 0.07 m/year was observed for Kaka Pir and the highest mean erosional rate of −1.79 m/year for Jamali Goth in the zone (Table 1). Similarly, the highest accretional and erosional rate by LRR method for each transect drawn shows that Haji Ali Goth had a highest positive rate of 5.66 m/year while Manora beach with a highest erosive rate of −3.27 m/year (Table 2). The LRR showed accretion in 22% of transects in the western zone with mean accretion rate of 0.98 m/year, and erosion in 78% at a rate of −1 m/year (Table 2).

Movement of the shoreline position was estimated by the SCE and NSM methods (Figure 8). The highest mean distance change by SCE shows value of 0.23 km obtained for the Manora beach (TID from 1-342) which shows the highest variability in shoreline position during the study period (Table 1). Similarly, the lowest mean distance change by SCE method shows 0.10 km value for Sandspit beach area (TID from 343-633) (Table 1). The highest variability in shoreline position at any transect drawn in western zone by SCE method was obtained for the Manora beach with a value of 0.769 km and minimum distance variability in shoreline position for Haji Ali Goth observed with a value of 0.016 km (Figure 8) (Table 2). The estimated NSM values show displacement of the shoreline on both landward and seaward sides. The NSM result indicate that highest mean accretion occurred at Haji Ali Goth (TID from 1127–1222) with a value of 0.17 km (Table 1). Similarly, the highest mean landward shifting (regression) of the shoreline encountered at Somar Goth (TID from 1041–1086) with a value of −0.07 km (Table 1). The highest net shoreline shift towards sea (accretion) encountered at Haji Ali Goth with a value of 0.29 km and highest landward shoreline shifting (erosion) was observed at Manora beach with a value of 0.3 km (Figure 8) (Table 2). The NSM results show that 26% of transects drawn in the western zone moved sea, with mean net shoreline change of 0.08 km, while 74% of transects showing landward movement, with a mean retreat of −0.08 km (Table 2).

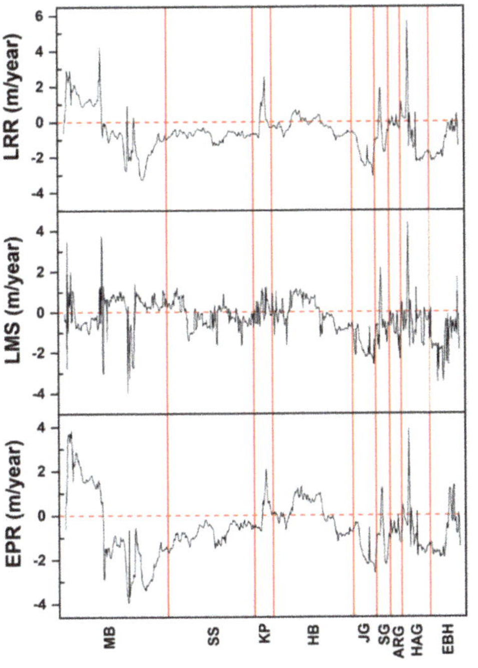

Figure 7. Shoreline change rate (EPR, LMS, LRR) of the western zone (MB = Manora Beach, SS = Sandspit, KP = Kaka Pir, HB = Hawks Bay, JG = Jamali Goath, SG = Somar Goath, ARG = Abdul Rehman Goath, HAJ = Haji Ali Goath, EBH = Engro Beach Huts).

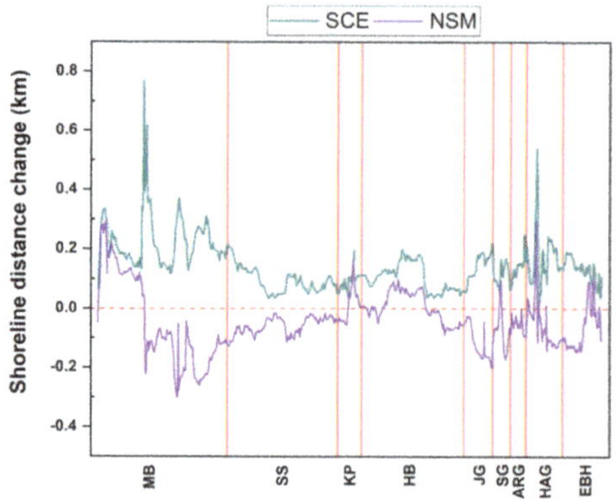

Figure 8. Shoreline change (NSM, SCE) of the western zone (MB = Manora Beach, SS = Sandspit, KK = Kaka Pir, HB = Hawks Bay, JG = Jamali Goath, SG = Somar Goath, ARG = Abdul Rehman Goath, HAJ = Haji Ali Goath, EBH = Engro Beach Huts).

6. Discussion

This study examines variability in historical shoreline position along the coastal city of Karachi, using historical topographic maps (1942 & 1966), Landsat satellite imagery (1976, 1990, 2002, and 2011) and Planet Scope satellite imagery (2018). Historical shoreline positions were extracted using NDWI and shoreline changes were observed using the DSAS approach. The study provides in-depth fine spatio-temporal shoreline change analysis and contributes to exploring the causes of such changes. Results show significant accretionary change in shoreline position in the eastern zone (Figure 3) (Table 2), where large residential projects have encroached on the sea in the past decade. On the other hand, the western zone (Figure 4) (Table 2) has the highly stable shoreline. The results also demonstrate that the shoreline changes in the western zone were related to the changes in hydrodynamics due to land reclamation projects in the eastern zone, associated with the local longshore current. Similar findings have been reported by recent studies conducted in other parts of the Indus delta region where large land reclamation projects have increased coastal vulnerability by altering the ocean hydrodynamics. For instance, Waqas et al. (2019) [16] found that 38% of the barrier islands in the Indus delta were vulnerable to oceanic factors. Similarly, Salik et al. (2015) [41] reported the vulnerability of the communities living in the lower Indus deltaic plain due to climatic and anthropogenic impacts on the coast.

6.1. Qualitative Rate of Shoreline Change

The shoreline positions in both zones have experienced accretion and erosion in different ways. The length of shorelines in the eastern and western zones have increased by 2.6% and 16.6% respectively, over a period of 76 years. The maximum accretional and erosional rates of shoreline recorded at each transects by EPR and NSM methods help to quantify the length of shoreline vulnerable to extreme ocean events (Table 3) while SCE can help to track the overall movement in the shoreline position. This study revealed that 5.4% of transects drawn in the eastern zone (25 km), representing 1.28 km of the whole shoreline length, experienced erosion (Table 2) and of this, 1.22 km at South Asian terminal underwent high erosion rates (Table 3). Similarly, 94.6% of transects drawn in the eastern zone, corresponding to a 22.6 km shoreline length, showed accretion (Table 2), of which 22.4 km exhibited very high accretion rates (Table 3) at DHA Zone B, DHA Golf Club, DHA Zone C, DHA Extension, Sea-view, Shireen Jinnah Colony, and South Asian Terminal. The DHA Zone B, DHA Golf Club, DHA Zone C, and DHA Extension corresponding to a combined shoreline length of 12.8 km showed high accretional rates (Table 3) of shoreline change because of massive land reclamation projects for urbanization and industrialization (Figure 9). Similarly, Sea-view is a recreational beach with a low substrate slope and had high accretion rates (Table 3). Table 3 indicates that 4.94 km of the beach length was nourished by trapping, dredging, and constructional materials from offshore drift. This beach regularly receives large numbers of visitors from all over Pakistan (Figure 9). Furthermore, Shireen Jinnah Colony had high accretion rates (Table 3) due to soft sediments deposited by longshore drift after construction of breakwaters preventing seawater and high energy waves from entering the residential areas (Figure 9) (Section 6.2).

Table 3. Shoreline classification based on EPR [39].

Rate of Shoreline Change (x m/y)	Shoreline Classification	Affected Shoreline of Easter Zone (%)	Affected Shoreline of Western zone (%)
$x \leq -2$	Very High Erosion	0.0	9.9
$-1 < x \leq -2$	High Erosion	5.1	27.1
$0 < x < -1$	Moderate Erosion	0.3	37.4
$x = 0$	Stable Shoreline	0	0
$0 \leq x < 1$	Moderate Accretion	0.0	13.7
$1 \leq x < 2$	High Accretion	0.8	8.4
$x \geq 2$	Very High Accretion	93.8	3.6

Note: Affected shoreline length reflects the percentage (%) of total shoreline length of each zone.

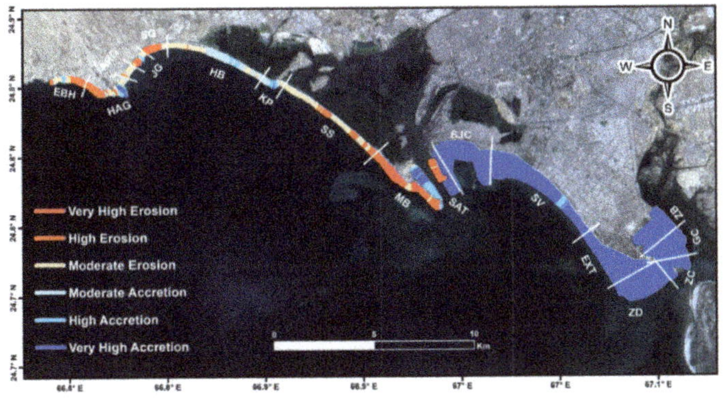

Figure 9. Shoreline vulnerability map of the study area (the width indicating the spatial extent of the overall shoreline development) (DHA = Defense housing Authority; ZB = Zone B; GC = Golf Club; ZC = Zone C; Ext = Extension, SV = Sea-view, SJC = Shireen Jinnah Colony, SAT = South Asian Terminal, MB = Manora Beach, SS = Sandspit, KK = Kaka Pir, HB = Hawks Bay, JG = Jamali Goath, SG = Somar Goath, ARG = Abdul Rehman Goath, HAJ = Haji Ali Goath, and EBH = Engro Beach Huts).

The study also revealed that during this period, 74.4% of transects drawn in the western zone (29 km) corresponding to 19.7 km of the total coastline experienced erosion (Table 2), and of this, 9.9 km experienced moderate erosion (Table 3) at Manora Beach, Sandspit, Kaka Pir, Hawks Bay, Jamali Goth, Somar Goth, Abdul Rehman Goth, Haji Ali Goth, and Engro Beach Huts (Figure 9). A total of 2.5 km out of the 5.3 km shoreline length of Hawks Bay and 4 km of the 5.82 km of Sandspit shows moderate erosion action (Tables 1 and 3) because they have almost flat substrate slopes that are being eroded by waves and current action during the south-west monsoon season (Section 6.2). Similarly, 1.82 km out of the shoreline length of the 6.64 km western part of Manora beach and 0.7 km out of the 1.54 km of Jamali Goth showed severe erosion from strong wind-driven longshore currents in the western part of study area (Figure 9). Furthermore, 25.6% of transects drawn in the western zone corresponding to 6.8 km of the shoreline experienced accretion, of which 3.62 km showed moderate accretion (Table 3). These involved some parts of Manora beach, Hawks Bay, Kaka Pir, and Haji Ali

Goth. High to very high accretion was encountered at the northeastern part of Manora beach due to soft sediments deposited by longshore drift from the construction of the deep-water container port, corresponding to 1.38 km to 0.88 km of shoreline (Figure 9).

6.2. Hydrodynamics of the Sindh Coastal Zone

Pakistan's coastal hydrodynamics are controlled by the reversal of the monsoonal system. Significant oceanic factors such as wind and wave climatology influence the coastal environment of Karachi because these changes are associated with the local wind driven longshore current [16,51,66,67]. To analyze the wind and wave climate influencing the study area seasonally before and after the major encroachment by urban development (Figures 10 and 11), the long-term offshore daily-hourly data from the National Oceanic and Atmospheric Administration (NOAA) was divided into two periods: Pre-encroachment (2008–2010) and post-encroachment (2011–2017). During the pre-summer monsoon season (March–May) of the pre-encroachment period, the wave height (WH) results showed an absence of swell for 35% of the time, while waves exceeding the 2 m height were recorded for only about 2% of the whole data (Figure 13). During this season, winds were northwesterly, with an average wind speed (WS) of 4.4 knots having an average wave period (WP) of 13 s, and only 0.5% of the time did WS exceed 9 knots, generating large wavelets (Figure 12). However, during the summer (Jun–Sep) i.e., the southwest (SW), monsoon season winds were southwesterly (Figure 10) with an average WS of 9 knots, generating strong waves with a mean significant WH of 3 m, and an average WP of 11 s.

Surprisingly, 22% of the time WH exceeded 4 m, and occasionally reached 6 m due to a rising tide, generated by heavy SW winds of WS above 12 knots (Figure 12). During the SW monsoon, high-energy southwesterly waves strikes the coast, accelerating erosive action in the western zone and sediment deposition in the eastern zone from longshore drift. During the retreat of the monsoon season (Oct–Nov), northeasterly winds with an average WS of 5 knots generated waves with an average WH lower than 1 m and a mean WP of 11 s. During this season, 86% of the observed time WS was observed to be between 3–9 knots, which generated small waves with a significant WH lower than 2 m (Figure 13). Furthermore, during the northeastern (NE) monsoon transition season (Dec–Feb), northeasterly winds (Figure 10) with a low mean WS of 5 knots generated small wavelets of an average 1 m wave height, of 1 m, which allowed the settling of suspended sediments. During this season, the NE wind remained lower than 9 knots for 99% of the time (Figure 12) and WH remained less than 2 m for 98% of the time (Figure 13).

During the pre-summer monsoon (NW) season (Mar–May) of the post-encroachment (2011–2017) period (Figure 11), the average WS was more than 4 knots and only 14% of the observed time the WS remained between 6–9 knots (Figure 13). The persistent NW winds generated a small swell of which about 97% of the swell's significant WH remained below 2 m (Figure 13) with an average WP of about 13 s. Similarly, average WS of about 9 knots during the SW monsoon season (Jun–Sep) (Figure 11) generated strong wave swells with a mean wave height of more than 3 m having a mean WP of 11 s. During the SW season, 26.6% of the time WH exceeded more than 4 m generated by WS reached to 22 knots (Figure 12) and a maximum 7.7 m of WH recorded during the post-encroachment period (Figure 13). During the retreating period of the monsoon season (Oct–Nov), wind prevailed from the NE side with an average WS of about 5 knots, which generated a swell wave of an average significant wave height of 1 m. The average WP of 13 s during the season showed the calmness of the sea where a total 98% of the time WH recorded below 2m, which helps in the settlements of the sediments in the bottom (Figure 13). During the winter monsoon season (NE) (Dec–Feb), winds with an average speed of more than 5 knots generated swells with an average significant WH of 1 m. Consistent and normal wind generated 98% of swell wave less than 3 m (Figure 13) with an average WP of about 11 s.

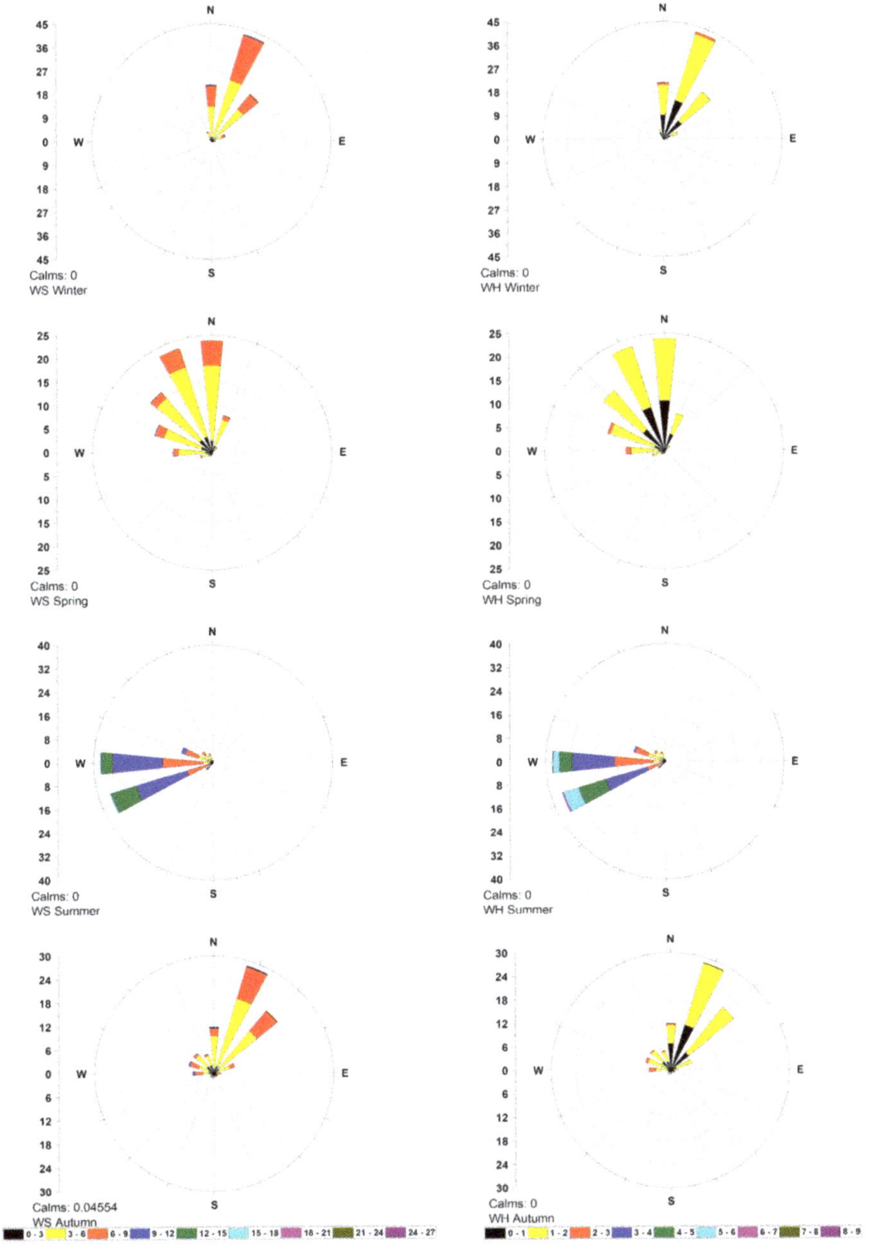

Figure 10. Pre-encroachment (2008–2010) seasonal hydrodynamics of the Karachi (WS = wind speed, WH = wave height).

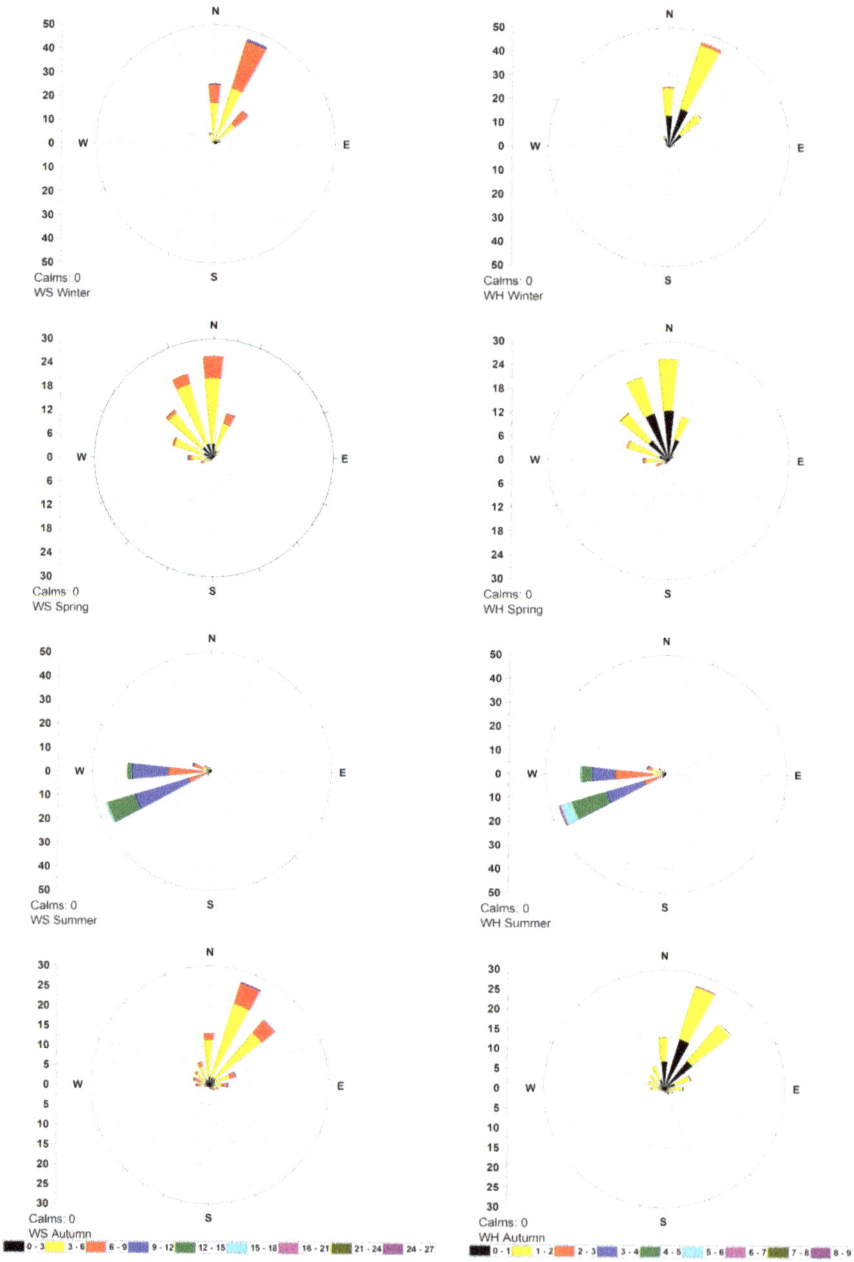

Figure 11. Post-encroachment (2011–2017) seasonal hydrodynamics of the Karachi (WS = wind speed, WH = wave height).

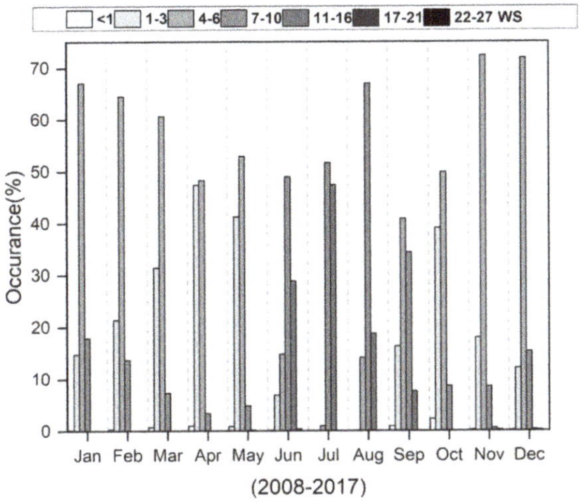

Figure 12. Wind (knots) climatology of the Sindh Coastal zone with respect to the modern Beaufort scale.

Generally, the wave height and its energy in the study area directly or indirectly subject of the wind speed, wave period, surface area over which the wind blows, and bathymetry of the area. The wave packets remain in action until the energy of the waves is absorbed by coastal region. Therefore, shoreline vulnerability of Karachi will probably be accelerated by high wave-height generated by high-speed winds at a high frequency (Figures 12 and 13). The pre- and post-hydrodynamic study revealed that about a 2.95% average WH increased significantly during the SW summer monsoon season and a 0.97% average for the NW winter monsoon season indicating the vulnerability of the shoreline position in the future. The sediments usually deposited during the stable ocean condition when the energy of waves is low and the deposited soft sediments in the shelves are driven away from the coast by the action of longshore drift and high wave energy [43]. The wave and wind are important factors for changing the nourishment and erosion of the coastal environment and development of urban areas along the Sindh coastal zone [15,68,69].

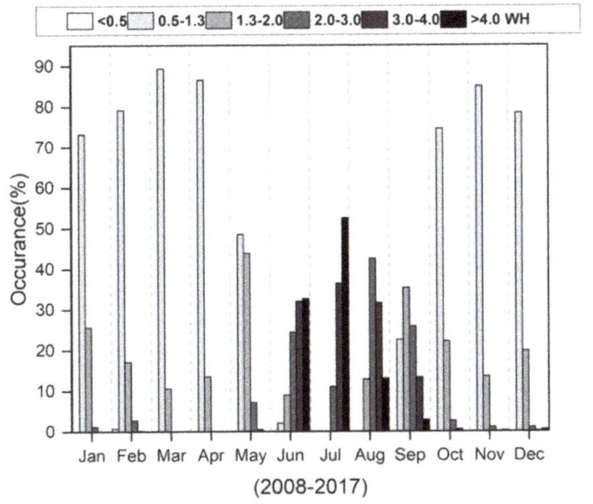

Figure 13. Wave (m) climatology of the Sindh Coastal zone with respect to the modern Beaufort scale [70].

7. Conclusions

This study presented tracks changes along a coastal mega city Karachi, Pakistan by employing historical topographic sheets dated back to 1942 and images from different missions of Landsat and Cubesats, until 2018. The total shoreline length was divided in two eastern zone and western zones. The mean accretional End Point Rate (EPR) for the eastern zone was observed to be 14 m/year, which indicated that 95% of the total transects of the eastern zone encountered accretion. This implies that the eastern zone remained under extensive land reclamation activities. Similarly, the EPR rate of change for the western zone revealed a mean accretional rate of 1.37 m/year, and with that 26% of the total transects underwent positive change. This indicates a small positive change rate for a quarter of the study area, while 74% of the transects in the western zone experienced mean an erosional EPR value of −1.15 m/year, indicating that a major proportion of the study area faced erosional activities. Furthermore, a total of 23.5 km of the shoreline length of eastern zone was designated as very high accretional while 2.9 km of western zone shoreline was found prone to very high erosional activity. The pre- and post-land reclamation hydrodynamics revealed that an average wave height increased by 2.95% during the south-west summer monsoon season and 0.97% during the NW winter monsoon season, indicating potential future threats to the shoreline position. Most of the western part of the shoreline is still undisturbed by human intervention. The shoreline position at a place is mostly determined by sand supply for nourishment, geomorphology, and oceanic forcing factors coupled with sea level change. Coastal land development via encroaching sea is a two-way process and triggers disturbance in the ocean hydrodynamics, which may create future stress on communities living in low-lying coastal areas. There are potential chances of land subsidence on reclaimed coastal areas following extreme events. Our study shows the potential of GIS and remote sensing techniques for comprehensive coastal risk assessment and our method would be very useful for the entire delta or other deltaic areas. Furthermore, it was observed that low to medium spatial resolution data sets (i.e., topographic sheets, Landsat MSS, TM, ETM+, and OLI) have higher uncertainty than the higher spatial resolution (3 m) Planet Scope imagery. The availability of Planet Scope imagery for educational and research purpose has opened new doors for the exploration of new techniques to preserve the environment. It is recommended that proper coastal protection and coastal management along ocean facing coastal areas should be adopted to defend against the erosive action of the ocean. The master plan for Karachi coastal areas should be revised to be friendlier to the coastal environment, with more emphasis on sustainable coastal development and coastal safety. In future, we will employ coastal vulnerability tools to quantify physical and socio-economical vulnerabilities of the Sindh coastal region to examine the effect of seawater intrusion due to sea level rise.

Author Contributions: Conceptualization, M.N. and M.I.S.; Data curation, M.W.; Formal analysis, M.W.; Funding acquisition, M.N., M.I.S. and W.W.; Investigation, M.N. and M.I.S.; Methodology, M.N. and M.I.S.; Project administration, M.I.S.; Resources, M.N., M.I.S. and I.Z.; Supervision, M.N.; Writing – original draft, M.W.; Writing – review & editing, M.W., I.Z. and W.W. All authors have read and agreed to the published version of the manuscript.

Funding: This research is supported by the East China University of Technology, Nanchang China through the Research Startup Funding to the first author (M.N.) through grant number DHBK2019003, the Foresight Lab, Islamabad Pakistan and Interactive Group Islamabad Pakistan under grant numbers FS/CUI/SEP/00 and IAC/CUI/SEP/00, respectively to M.I.S.

Acknowledgments: The authors would like to thank Janet Elizabeth Nichol (University of Sussex, Brighton United Kingdom) for her help in making the manuscript more concise and grammatically correct. The authors would also like to thank USGS for the distribution of Landsat MSS, TM, ETM+, and OLI data products, Planet.com for providing access to, and use of, Planet Scope imagery, the QGIS team for the development and support of the QGIS software, as well as the five anonymous reviewers, and editors for providing constructive feedback on our manuscript.

Conflicts of Interest: The authors declare no conflict of interest. The funders had no role in the design of the study; in the collection, analyses, or interpretation of data; in the writing of the manuscript, and in the decision to publish the results.

Appendix A

Table A1. List of satellite images used in this study.

Sr. #	Satellite	Sensor	Scene Identifier	Path/Row	Acquisition Date
1	PlanetScope	Optical	20180422_053509_1033_3B_ AnalyticMS_SR	53/509	22 Apr 2018
2	PlanetScope	Optical	20180422_053510_1033_3B_AnalyticMS_SR	53/510	22 Apr 2018
3	PlanetScope	Optical	20180422_060440_0f32_3B_AnalyticMS_SR	60/440	22 Apr 2018
4	PlanetScope	Optical	20180422_060441_0f32_3B_AnalyticMS_SR	60/441	22 Apr 2018
4	Landsat-5	TM	LT05_L1TP_152043_20110211_20161010_01_T1	152/43[b]	11 Feb 2011
5	Landsat-7	ETM+	LE07_L1TP_152043_20021211_20170127_01_T1	152/43[b]	11 Dec 2002
6	Landsat-5	TM	LT05_L1TP_152043_19900217_20170131_01_T1	152/43[b]	17 Feb 1990
7	Landsat-2	MSS	LM02_L1TP_163043_19761214_20180425_01_T2	163/43[a]	14 Dec 1976

Note: The superscripts a and b represent the path/row for WRS-1 and WRS-2, respectively.

References

1. Hapke, C.J.; Himmelstoss, E.A.; Kratzmann, M.G.; List, J.H.; Thieler, E.R. National Assessment of Shoreline Change: Historical Shoreline Change along the New England and Mid-Atlantic Coasts. In *U.S. Geological Survey Open File Report*; U.S. Geological Survey: Reston, VA, USA, 2011; Volume 57. [CrossRef]
2. Slott, J.M.; Murray, A.B.; Ashton, A.D. Large-scale responses of complex-shaped coastlines to local shoreline stabilization and climate change. *J. Geophys. Res. Earth Surf.* **2010**, *115*, 1–19. [CrossRef]
3. Stive, M.J.; Aarninkhof, S.G.; Hamm, L.; Hanson, H.; Larson, M.; Wijnberg, K.M.; Nicholls, R.J.; Capobianco, M. Variability of shore and shoreline evolution. *Coast. Eng.* **2002**, *47*, 211–235. [CrossRef]
4. Boak, E.H.; Turner, I.L. Shoreline Definition and Detection: A Review. *J. Coast. Res.* **2005**, *21*, 688–703. [CrossRef]
5. Gens, R. Remote sensing of coastlines: Detection, extraction and monitoring. *Int. J. Remote Sens.* **2010**, *31*, 1819–1836. [CrossRef]
6. Pardo-Pascual, J.E.; Almonacid-Caballer, J.; Ruiz, L.A.; Palomar-Vázquez, J. Automatic extraction of shorelines from Landsat TM and ETM+ multi-temporal images with subpixel precision. *Remote Sens. Environ.* **2012**, *123*, 1–11. [CrossRef]
7. Virdis, S.G.P.; Oggiano, G.; Disperati, L.; Beach, W.P.; Virdis, S.G.P.; Oggiano, G.; Disperati, L. A Geomatics Approach to Multitemporal Shoreline Analysis in Western Mediterranean: The Case of Platamona-Maritza Beach (Northwest Sardinia, Italy). *J. Coast. Res.* **2012**, *28*, 624.
8. Wu, W. Coastline evolution monitoring and estimation - A case study in the region of Nouakchott, Mauritania. *Int. J. Remote Sens.* **2007**, *28*, 5461–5484. [CrossRef]
9. Fletcher, C.H., III; Murray-Wallace, C. V Introduction: Coastal environmental change during sea-level highstands. *Sediment. Geol.* **2002**, *150*, 1–2. [CrossRef]
10. Salik, K.M.; Hashmi, M.Z.U.R.; Ishfaq, S.; Zahdi, W.U.Z. Environmental flow requirements and impacts of climate change-induced river flow changes on ecology of the Indus Delta, Pakistan. *Reg. Stud. Mar. Sci.* **2016**, *7*, 185–195. [CrossRef]
11. Nazeer, M.; Nichol, J.E. Improved water quality retrieval by identifying optically unique water classes. *J. Hydrol.* **2016**, *541*, 1119–1132. [CrossRef]
12. Siddiqui, M.N.; Maajid, S. Monitoring of geomorphological changes for planning reclamation work in coastal area of Karachi, Pakistan. *Adv. Space Res.* **2004**, *33*, 1200–1205. [CrossRef]
13. Del Río, L.; Gracia, F.J.; Benavente, J. Shoreline change patterns in sandy coasts. A case study in SW Spain. *Geomorphology* **2013**, *196*, 252–266. [CrossRef]
14. McGranahan, G.; Balk, D.; Anderson, B. The rising tide: Assessing the risks of climate change and human settlements in low elevation coastal zones. *Environ. Urban.* **2007**, *19*, 17–37. [CrossRef]
15. Ozturk, D.; Sesli, F.A. Shoreline change analysis of the Kizilirmak Lagoon Series. *Ocean Coast. Manag.* **2015**, *118*, 290–308. [CrossRef]
16. Waqas, M.; Nazeer, M.; Shahzad, M.; Zia, I. Spatial and Temporal Variability of Open-Ocean Barrier Islands along the Indus Delta Region. *Remote Sens.* **2019**, *11*, 437. [CrossRef]

17. Kittinger, J.N.; Ayers, A.L. Shoreline armoring, risk management, and coastal resilience under rising seas. *Coast. Manag.* **2010**, *38*, 634–653. [CrossRef]
18. SEDAC. *Percentage of Total Population Living in Coastal Areas*; Socioeconomic Data and Applications Center: Columbia, SC, USA, 2011.
19. Esteves, L.S.; Williams, J.J.; Nock, A.; Lymbery, G. Quantifying Shoreline Changes along the Sefton Coast (UK) and the Implications for Research-Informed Coastal Management. *J. Coast. Res.* **2009**, *56*, 602–606.
20. Greening, H.; Doering, P.; Corbett, C. Hurricane impacts on coastal ecosystems. *Estuaries Coasts* **2006**, *29*, 877–879. [CrossRef]
21. Knutson, T.R.; McBride, J.L.; Chan, J.; Emanuel, K.; Holland, G.; Landsea, C.; Held, I.; Kossin, J.P.; Srivastava, A.K.; Sugi, M. Tropical cyclones and climate change. *Nat. Geosci.* **2010**, *3*, 157–163. [CrossRef]
22. Sankar, R.D.; Murray, M.S.; Wells, P. Decadal scale patterns of shoreline variability in Paulatuk, N.W.T, Canada. *Polar Geogr.* **2019**, *42*, 196–213. [CrossRef]
23. Liu, H.; Jezek, K.C. Automated extraction of coastline from satellite imagery by integrating Canny edge detection and locally adaptive thresholding methods. *Int. J. Remote Sens.* **2004**, *25*, 937–958. [CrossRef]
24. Ford, M. Shoreline changes interpreted from multi-temporal aerial photographs and high resolution satellite images: Wotje Atoll, Marshall Islands. *Remote Sens. Environ.* **2013**, *135*, 130–140. [CrossRef]
25. Mujabar, P.S.; Chandrasekar, N. Shoreline change analysis along the coast between Kanyakumari and Tuticorin of India using remote sensing and GIS. *Arab. J. Geosci.* **2013**, *6*, 647–664. [CrossRef]
26. Fletcher, C.; Rooney, J.; Barbeeh Siang-Chyn Limt, M.; Richmond, B.; Palm Beach, W.; Fletcher, F. Mapping Shoreline Change Using Digital Orthophotogrammetry on Maui, Hawaii. *J. Coast. Res.* **2004**, *SI*, 106–124.
27. Maiti, S.; Bhattacharya, A.K. Shoreline change analysis and its application to prediction: A remote sensing and statistics based approach. *Mar. Geol.* **2009**, *257*, 11–23. [CrossRef]
28. Morton, R.A.; Sallenger, A.H., Jr. Morphological Impacts of Extreme Storms on Sandy Beaches and Barriers. *J. Coast. Res.* **2003**, *19*, 560–573.
29. Morton, R.A.; Miller, T.L.; Moore, L.J. National assessment of shoreline change: Part 1: Historical shoreline changes and associated coastal land loss along the US Gulf of Mexico. In *U.S. Geological Survey Open File Report 2004–1043*; U.S. Geological Survey: Reston, VA, USA, 2004; Volume 45. [CrossRef]
30. Marfai, M.A.; Almohammad, H.; Dey, S.; Susanto, B.; King, L. Coastal dynamic and shoreline mapping: Multi-sources spatial data analysis in Semarang Indonesia. *Environ. Monit. Assess.* **2008**, *142*, 297–308. [CrossRef]
31. Sekovski, I.; Stecchi, F.; Mancini, F.; Del Rio, L. Image classification methods applied to shoreline extraction on very high-resolution multispectral imagery. *Int. J. Remote Sens.* **2014**, *35*, 3556–3578. [CrossRef]
32. Ghoneim, E.; Mashaly, J.; Gamble, D.; Halls, J.; AbuBakr, M. Nile Delta exhibited a spatial reversal in the rates of shoreline retreat on the Rosetta promontory comparing pre- and post-beach protection. *Geomorphology* **2015**, *228*, 1–14. [CrossRef]
33. McFeeters, S.K. The use of the Normalized Difference Water Index (NDWI) in the delineation of open water features. *Int. J. Remote Sens.* **1996**, *17*, 1425–1432. [CrossRef]
34. Feyisa, G.L.; Meilby, H.; Fensholt, R.; Proud, S.R. Automated Water Extraction Index: A new technique for surface water mapping using Landsat imagery. *Remote Sens. Environ.* **2014**, *140*, 23–35. [CrossRef]
35. Xu, H. Modification of normalised difference water index (NDWI) to enhance open water features in remotely sensed imagery. *Int. J. Remote Sens.* **2006**, *27*, 3025–3033. [CrossRef]
36. Nassar, K.; Mahmod, W.E.; Fath, H.; Masria, A.; Nadaoka, K.; Negm, A. Shoreline change detection using DSAS technique: Case of North Sinai coast, Egypt. *Mar. Georesour. Geotechnol.* **2018**, 1–15. [CrossRef]
37. Cenci, L.; Disperati, L.; Persichillo, M.G.; Oliveira, E.R.; Alves, F.L.; Phillips, M. Integrating remote sensing and GIS techniques for monitoring and modeling shoreline evolution to support coastal risk management. *GISci. Remote Sens.* **2018**, *55*, 355–375. [CrossRef]
38. Bheeroo, R.A.; Chandrasekar, N.; Kaliraj, S.; Magesh, N.S. Shoreline change rate and erosion risk assessment along the Trou Aux Biches–Mont Choisy beach on the northwest coast of Mauritius using GIS-DSAS technique. *Environ. Earth Sci.* **2016**, *75*, 1–12. [CrossRef]
39. Kermani, S.; Boutiba, M.; Guendouz, M.; Guettouche, M.S.; Khelfani, D. Detection and analysis of shoreline changes using geospatial tools and automatic computation: Case of jijelian sandy coast (East Algeria). *Ocean Coast. Manag.* **2016**, *132*, 46–58. [CrossRef]

40. Kuleli, T.; Guneroglu, A.; Karsli, F.; Dihkan, M. Automatic detection of shoreline change on coastal Ramsar wetlands of Turkey. *Ocean Eng.* **2011**, *38*, 1141–1149. [CrossRef]
41. Salik, K.M.; Jahangir, S.; Zahdi, W.U.Z.; Hasson, S.U. Climate change vulnerability and adaptation options for the coastal communities of Pakistan. *Ocean Coast. Manag.* **2015**, *112*, 61–73. [CrossRef]
42. Ali Khan, T.M.; Razzaq, D.A.; Chaudhry, Q.U.Z.; Quadir, D.A.; Kabir, A.; Sarker, M.A. Sea level variations and geomorphological changes in the coastal belt of Pakistan. *Mar. Geod.* **2002**, *25*, 159–174. [CrossRef]
43. Shahzad, M.I.; Meraj, M.; Nazeer, M.; Zia, I.; Inam, A.; Mehmood, K.; Zafar, H. Empirical estimation of suspended solids concentration in the Indus Delta Region using Landsat-7 ETM+ imagery. *J. Environ. Manag.* **2018**, *209*, 254–261. [CrossRef]
44. Hay, C.C.; Morrow, E.; Kopp, R.E.; Mitrovica, J.X. Probabilistic reanalysis of twentieth-century sea-level rise. *Nature* **2015**, *517*, 481–484. [CrossRef]
45. Oppenheimer, M.; Glavovic, B.; Hinkel, J.; Van de Wal, R.; Magnan, A.K.; Abd-Elgawad, A.; Cai, R.; Cifuentes-Jara, M.; DeConto, R.M.; Ghosh, T.; et al. Sea Level Rise and Implications for Low Lying Islands, Coasts and Communities. In *IPCC Special Report on the Ocean and Cryosphere in a Changing Climate*; IPCC: Geneva, Switzerland, 2019; pp. 321–411.
46. Martínez, M.L.; Intralawan, A.; Vázquez, G.; Pérez-Maqueo, O.; Sutton, P.; Landgrave, R. The coasts of our world: Ecological, economic and social importance. *Ecol. Econ.* **2007**, *63*, 254–272. [CrossRef]
47. MFF Pakistan. *A Handbook on Pakistan's Coastal and Marine Resources*; MFF Pakistan: Karachi, Pakistan, 2016; ISBN 9789696430193.
48. Klemas, V. Remote Sensing Techniques for Studying Coastal Ecosystems: An Overview. *J. Coast. Res.* **2011**, *27*, 2–17.
49. Giosan, L.; Constantinescu, S.; Clift, P.D.; Tabrez, A.R.; Danish, M.; Inam, A. Recent morphodynamics of the Indus delta shore and shelf. *Cont. Shelf Res.* **2006**, *26*, 1668–1684. [CrossRef]
50. Tolman, H. User manual and system documentation of WAVEWATCH III TM version 3.14. *Tech. Note MMAB Contrib.* **2009**, *276*, 220.
51. Vakarelov, B.K.; Ainsworth, R.B.; MacEachern, J.A. Recognition of wave-dominated, tide-influenced shoreline systems in the rock record: Variations from a microtidal shoreline model. *Sediment. Geol.* **2012**, *279*, 23–41. [CrossRef]
52. Crowell, M.; Leatherman, S.P.; Buckley, M.K. Historical shoreline change: Error analysis and mapping accuracy. *J. Coast. Res.* **1991**, *7*, 839–852.
53. Bilal, M.; Nazeer, M.; Nichol, J.E.; Bleiweiss, M.P.; Qiu, Z.; Jäkel, E.; Campbell, J.R.; Atique, L.; Huang, X.; Lolli, S. A simplified and robust surface reflectance estimation method (SREM) for use over diverse land surfaces using multi-sensor data. *Remote Sens.* **2019**, *11*. [CrossRef]
54. Nazeer, M.; Nichol, J.E.; Yung, Y.-K. Evaluation of atmospheric correction models and Landsat surface reflectance product in an urban coastal environment. *Int. J. Remote Sens.* **2014**, *35*, 6271–6291. [CrossRef]
55. Chávez, P.S.J. Image-based atmospheric corrections - revisited and improved. *Photogramm. Eng. Remote Sens.* **1996**, *62*, 1025–1036.
56. Vos, K.; Harley, M.D.; Splinter, K.D.; Simmons, J.A.; Turner, I.L. Sub-annual to multi-decadal shoreline variability from publicly available satellite imagery. *Coast. Eng.* **2019**, *150*, 160–174. [CrossRef]
57. Chander, G.; Markham, B.L.; Helder, D.L. Summary of current radiometric calibration coefficients for Landsat MSS, TM, ETM+, and EO-1 ALI sensors. *Remote Sens. Environ.* **2009**, *113*, 893–903. [CrossRef]
58. Genz, A.S.; Fletcher, C.H.; Dunn, R.A.; Frazer, L.N.; Rooney, J.J. The Predictive Accuracy of Shoreline Change Rate Methods and Alongshore Beach Variation on Maui, Hawaii. *J. Coast. Res.* **2007**, *231*, 87–105. [CrossRef]
59. Li, X.; Zhou, Y.; Zhang, L.; Kuang, R. Shoreline change of Chongming Dongtan and response to river sediment load: A remote sensing assessment. *J. Hydrol.* **2014**, *511*, 432–442. [CrossRef]
60. Romine, B.M.; Fletcher, C.H.; Frazer, L.N.; Genz, A.S.; Barbee, M.M.; Lim, S.-C. Historical Shoreline Change, Southeast Oahu, Hawaii; Applying Polynomial Models to Calculate Shoreline Change Rates. *J. Coast. Res.* **2009**, *256*, 1236–1253. [CrossRef]
61. Sahoo, P.K.; Soltani, S.; Wong, A.K.C. A survey of thresholding techniques. *Comput. Vis. Graph. Image Process.* **1988**, *41*, 233–260. [CrossRef]
62. Pardo-Pascual, J.E.; Sánchez-García, E.; Almonacid-Caballer, J.; Palomar-Vázquez, J.M.; de los Santos, E.P.; Fernández-Sarría, A.; Balaguer-Beser, Á. Assessing the accuracy of automatically extracted shorelines on microtidal beaches from landsat 7, landsat 8 and sentinel-2 imagery. *Remote Sens.* **2018**, *10*, 1–20. [CrossRef]

63. Li, R.; Liu, J.K.; Felus, Y. Spatial modeling and analysis for shoreline change detection and coastal erosion monitoring. *Mar. Geod.* **2001**, *24*, 1–12. [CrossRef]
64. Thieler, E.R.; Himmelstoss, E.A.; Zichichi, J.L.; Ergul, A. *The Digital Shoreline Analysis System (DSAS) Version 4.0 -An ArcGIS Extension for Calculating Shoreline Change*; U.S. Geological Survey: Reston, VA, USA, 2009.
65. Ryabchuk, D.; Leont'yev, I.; Sergeev, A.; Nesterova, E.; Sukhacheva, L.; Zhamoida, V. The morphology of sand spits and the genesis of longshore sand waves on the coast of the eastern Gulf of Finland. *Baltica* **2011**, *24*, 13–24.
66. Rasul, G.; Mahmood, A.; Sadiq, A.; Khan, S.I. Vulnerability of the Indus Delta to Climate Change in Pakistan. *Pak. J. Meteorol.* **2012**, *89–107*.
67. Haider, K.W.; Rasul, G.; Afzaal, M. A study on tropical cyclones of the Arabian sea in June 2007 and their connection with sea surface temperature. *Pak. J. Meteorol.* **2008**, *4*, 37–48.
68. Wells, J.T.; Coleman, J.M. Deltaic morphology and sedimentology, with special reference to the Indus River Delta. *Mar. Geol. Oceanogr. Arab. Sea Coast. Pak.* **1985**, *424*, 85–100.
69. Zia, I.; Zafar, H.; Shahzad, M.I.; Meraj, M.; Kazmi, J.H. Assessment of sea water inundation along Daboo creek area in Indus Delta Region, Pakistan. *J. Ocean Univ. China* **2017**, *16*, 1055–1060. [CrossRef]
70. Beer, T. Beaufort wind scale. In *Encyclopedia of Earth Sciences Series*; Springer: Berlin/Heidelberg, Germany, 2013; pp. 42–45. ISBN 9780123822253.

remote sensing

MDPI

Technical Note

Simulation of a Wideband Radar Echo of a Target on a Dynamic Sea Surface

Wang-Qiang Jiang [1], Liu-Ying Wang [2], Xin-Zhuo Li [3], Gu Liu [2] and Min Zhang [1,*]

[1] School of Physics and Optoelectronic Engineering, Xidian University, Xi'an 710071, China; wqjiang@mail.xidian.edu.cn
[2] Xi'an Research Institute of High Technology, Xi'an 710025, China; wangliuying1971@163.com (L.-Y.W.); liugu5032@163.com (G.L.)
[3] China Academy of Space Technology, Xi'an Branch, Xi'an 710100, China; xzli_1@stu.xidian.edu.cn
* Correspondence: mzhang@mail.xidian.edu.cn; Tel.: +86-029-88202663

Abstract: Unlike a generally rough ground surface, the sea surface varies over time. To analyze the impact of the motion of sea waves on the synthetic aperture radar (SAR) image of a target, the wideband echo simulation method based on a frequency domain is used. For the wideband echo, the electromagnetic (EM) scattering properties of the main frequency components are analyzed by the simulation method. Based on the EM scattering properties, the echo can be accurately simulated by using the inverse fast Fourier transformation (IFFT). Combined with the flight path of the radar, the echo of each pulse can be simulated to obtain the SAR image. The correct evaluation of the EM scattering properties is indispensable to the acquisition of an accurate SAR image. For complex targets, such as ships, the multiple scattering effects have a significant influence on the EM scattering properties. Thus, a rectangular wave beam-based geometrical optics and physical optics (GO/PO) method is introduced to calculate the EM scattering properties, which is more efficient than the traditional GO/PO. The GO/PO method is suitable to simulate SAR images in which the EM scattering properties of each pulse need to be calculated. With these methods, the SAR images of the target on the sea surface are simulated. Based on the comparison of the SAR images between a static and dynamic sea surface, it is found that the region corresponding to the target is blurred and the texture of the dynamic sea is blurrier. The impact of multiple scattering and sea wave motion on target recognition are also analyzed with the SAR images that were generated under different conditions. Some strong scattering points appear when multiple scattering effects are considered. It is also found that the texture of the SAR images, corresponding to the sea surface, changes with the synthetic aperture time.

Keywords: synthetic aperture radar (SAR); target; sea surface; multiple scattering

Citation: Jiang, W.-Q.; Wang, L.-Y.; Li, X.-Z.; Liu, G.; Zhang, M. Simulation of a Wideband Radar Echo of a Target on a Dynamic Sea Surface. *Remote Sens.* **2021**, *13*, 3186. https://doi.org/10.3390/rs13163186

Academic Editors: Weicheng Wu and Yalan Liu

Received: 17 June 2021
Accepted: 9 August 2021
Published: 11 August 2021

1. Introduction

Target recognition based on synthetic aperture radar (SAR) images is a hot topic in the remote sensing field [1,2]. With the development of technology, the bandwidth is increased to improve the range resolution and more details can be obtained from the SAR image [3,4]. The strong scattering points in SAR images are usually used to detect the target. However, the background may affect the accuracy of the target detection. The objective of the paper is to efficiently provide reliable simulation data for ship target recognition in the dynamic sea. In general, the sea surface is stirred up when the wind grows fierce, and the strong scattering points of the ship may be submerged in the background noise caused by scattering from the sea surface. Thus, many researchers have worked against a complex and noisy background [5,6]. The traditional methods are usually based on a prior detection window, such as the constant false alarm rate methods [7,8]. Because the prior detection window needs to be initialized, the application of the traditional method is limited for a complex background. To detect a target against a complex background, the self-adaptive method based on the local variance weighted information entropy (VWIE) has been developed [9,10]. In recent years, a ship detection method based on convolutional neural networks (CNN) has been proposed [11,12]. The method is conventional and performs well. Thus, it is widely used in the field of target recognition. However, its precision relies on a large-scale, high-quality training dataset. As we all know, accurately measured data are difficult to obtain. However, it is much easier to acquire simulation data. To ensure the accuracy of the simulation data, the SAR imaging process follows the practical situation. Thus, the simulated sea surface has changed over time during the flight of the radar, and the multiple scattering effects have been taken into account for a complex target, such as a ship, to enhance the electromagnetic (EM) scattering echo. Moreover, it takes lots of time to calculate the EM scattering properties of all the pulses during the SAR imaging process. Thus, the geometrical optics and physical optics (GO/PO) method, which is efficient and accurate enough, is employed to simulate the SAR image of a target on a dynamic sea surface for a wideband signal.

In terms of the general time-domain simulation method of the SAR echo, the SAR scene is divided into many scattering units, and the total echo is the sum of the echo from each unit. In addition, the echo from each unit is approximated by the scattering properties of the center-frequency EM wave [13,14]. However, for the wideband signal, the differences in scattering properties between the main frequency components increase and the error of the traditional approximation becomes large. To accurately simulate the SAR echo, the method in the frequency domain is employed, which considers the scattering properties of all the frequency components. After the scattering properties of each frequency are obtained, the time domain scattering echo can be calculated by using the inverse fast Fourier transformation (IFFT). Then, the SAR image of the whole scene can be generated by using the simulated echo.

To calculate the scattering properties, frequency domain EM scattering algorithms can be used, such as method of moment (MoM), equivalent edge currents (EEC), finite element method (FEM), and physical optics (PO) [15]. However, as the scattering properties of all the frequency components need to be analyzed, the calculation load is quite heavy. To reduce the simulation time, a high-efficiency method would be a better choice. Here, the GO/PO hybrid method, which has high efficiency and considers the multiple scattering effects, is employed to simulate the scattering fields. Although the efficiency of the GO/PO method is high, it still takes a lot of time to search for the reflected rays. Moreover, the calculation is mainly focused on the ray tracing process. To accelerate the ray tracing process, the kd-tree method is usually used [16]. The method of hardware acceleration is frequently used too. For instance, the graphical electromagnetic computing method uses a graphics processing unit (GPU) to estimate the illuminated area [17]. In recent years, the parallel acceleration method has been very popular because of the significant improvement in the parallel computing capability of the GPU [18]. However, most of these methods trace the ray according to the facets, e.g., the traditional GO/PO method, which calculates the EM scattering properties by analyzing the EM scattering fields of the illuminated facets [17]. Thus, these methods require the density of rays to be higher than the density of the facets. Their efficiency depends on the number of facets and their accuracy is affected by the size of the facet; thus, the size of the facet should be small enough to ensure accuracy. Of course, this will increase the number of facets and calculation load. To further improve the efficiency, the rectangular wave beam-based GO/PO method can be used. It calculates the EM scattering properties by analyzing the EM scattering fields of the area illuminated by a small rectangular wave beam. Furthermore, it does not have a requirement for the facet size as the traditional GO/PO method does. Thus, the same level of efficiency and accuracy can be achieved with a lower number of pixels [19].

During the process of the SAR platform movement, the fluctuation of the dynamic sea surface changes. The longer the time, the more obvious the change. Unlike the static rough surface, the dynamic sea surface at a different time, corresponding to each pulse will change. Thus, the impact of the sea surface movement on SAR imaging will be discussed.

After the analysis, it was found that it is more difficult to recognize the shape of the ship on a dynamic sea surface than a static one. The longer the synthetic aperture time, the more blurred the shape of the ship. It means that the SAR images of a target ship on a dynamic sea are affected by the synthetic aperture time.

2. Materials and Methods

2.1. Scattering Echo Simulation

For many emission signals of radar, the energy concentrates on the frequency components near the center frequency f_0, with a range between $f_0 - f'$ and $f_0 + f'$, as shown in Figure 1a. In addition, the frequency components out of this range are so weak that the components can be ignored. Its spectrum, $F_e(f)$, can be written as follows [20]:

$$\begin{aligned} F_e(f) &= \sum_i \delta(f - f_i) A_{ei} \quad f_0 - f' \le f < f_0 + f' \\ \delta(f) &= \begin{cases} 1 & f = 0 \\ 0 & f \neq 0 \end{cases} \end{aligned}, \tag{1}$$

where A_{ei} is the complex amplitude corresponding to frequency f_i.

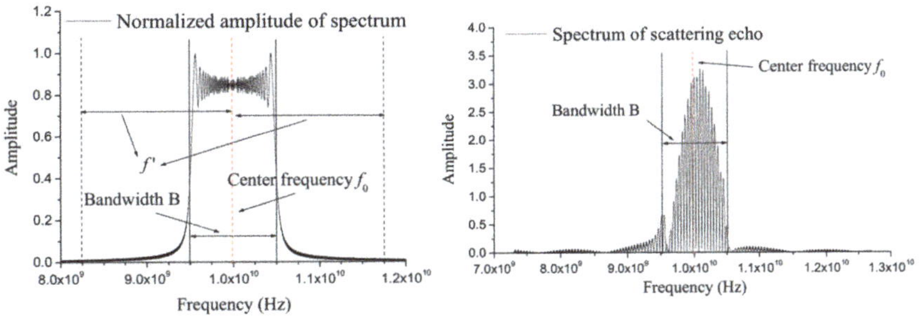

(a) **(b)**

Figure 1. Normalized spectrum of a signal: (**a**) emission signal; (**b**) scattering echo.

Figure 1a shows the normalized spectrum of the emission signal for a linear frequency-modulated (LFM) signal. As shown in Figure 1a, the energy mainly distributes near the center frequency, f_0, with bandwidth B. Figure 1b shows the normalized spectrum of the corresponding scattering echo. Its energy is also mainly distributed near the center frequency, f_0. Then, the frequency-domain form of the scattering echo signal can be written as follows [20]:

$$F_s(f) = \sum_i \delta(f - f_i) A_{ei} \sqrt{\sigma}_i \cdot \varphi(f_i), \tag{2}$$

where σ_i is the radar cross section (RCS) and $\varphi(f_i, t)$ is the phase delay corresponding to the frequency of f_i. Before the scattering echo processing, it should be demodulated to the baseband. Its corresponding spectrum can be written as follows [20]:

$$F_{sB}(f) = \sum_i \delta(f - f_i + f_0) A_{ei} \sqrt{\sigma}_i \cdot \varphi(f_i) \cdot e^{-i4\pi f_0 R_0 / C}, \tag{3}$$

where C is the light speed and R_0 is the distance between the target and the radar sensor.

With $F_{sB}(f)$, the demodulated scattering echo in the time domain is obtained by using IFFT:

$$s_B(t) = IFFT[F_{sB}(f)], \tag{4}$$

During irradiation, the radar emits a pulse periodically. The demodulated scattering echo of the *i*th pulse is marked as $s_B(t, \eta_i)$. Then, the SAR echo can be written as follows:

$$s_B(t, \eta) = \sum_i \delta(\eta - \eta_i) s_B(t, \eta_i), \tag{5}$$

where η represents the time when the radar emits the pulse, and η_i is the time when the radar emits the *i*th pulse.

With the SAR echo, the SAR image can be calculated using imaging algorithms, such as the Range–Doppler (RD) algorithm [21] or the Back Projection (BP) algorithm [22,23]. Here, the RD algorithm is applied to calculate the SAR image with the simulated echoes.

In order to verify the reliability of the simulation method, the radar moves along the same designed trajectory as the actual illuminating process. A scene of five cubes is presented to demonstrate that the simulated echo with different parameters is in accordance with the actual situation. For example, the simulated echo with a wider bandwidth can generate a SAR image with a higher range resolution. Figure 2 shows a schematic diagram of the scene. The airborne radar moves along the flight path with the height of H; see the red line in Figure 2. R is the distance between the target and radar. θ is the incident angle. The five cubes in Figure 2 have a volume of 1 m^3. The five cubes are placed on the XOY plane, as shown in Figure 2.

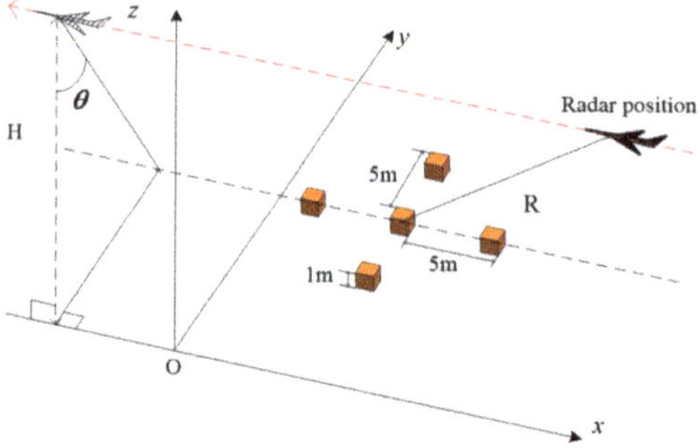

Figure 2. Schematic diagram of the scene.

While the radar moves along the flight path, it emits pulses periodically. The echo from the cubes corresponding to each pulse is simulated by the proposed method. With these echoes, the SAR images of the five cubes can be generated. Figure 3 gives the SAR images of the five cubes with different bandwidths B and azimuth resolution ρ_a. The antenna lengths of Figure 3a–d are 4 m, 2 m, 1 m, and 0.5 m, respectively. The corresponding bandwidths are 0.0375 GHz, 0.075 GHz, 0.15 GHz, and 0.3 GHz, respectively. The incident angle θ is 57°. The carrier frequency f_0 is 9.375 GHz. The range resolution is C/(2B). As shown in Figure 3, when the azimuth resolution and the range resolution are both low, it is difficult to detect the cubes, shown in Figure 3a, where the resolution is 4.0 m. When the resolution is improved, the cubes can be gradually identified in the SAR image, as shown in Figure 3b–d, with increasing bandwidth B.

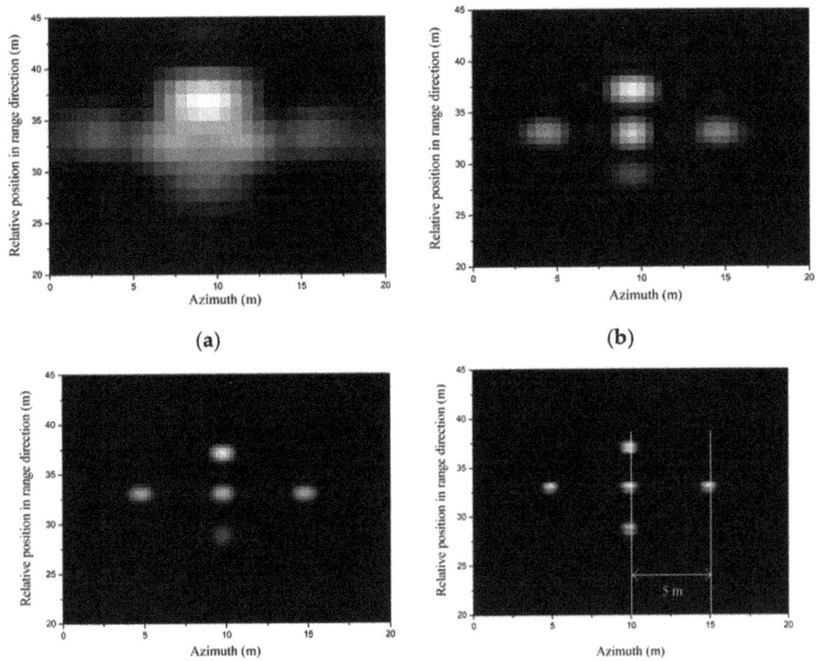

(a) (b) (c) (d)

Figure 3. SAR images of the cubes: (**a**) ρ_a = 4 m, B = 0.0375 GHz; (**b**) ρ_a = 2 m, B = 0.075 GHz; (**c**) ρ_a = 1 m, B = 0.15 GHz; (**d**) ρ_a = 0.5 m, B = 0.3 GHz.

2.2. Rectangular Wave Beam-Based GO/PO Method

According to Equation (3), the scattering echo can be calculated when obtaining the RCS σ_i of all the frequency components. Because the multiple scattering effect of the complex target is obvious, the GO/PO method is used to calculate the RCS from the target on the sea surface.

The GO/PO method is based on the PO method, and also considers the scattering fields of the facets illuminated by the reflected rays that are traced according to the GO method. For the traditional GO/PO method, the scattering property is the sum of the contribution of each illuminated facet. The total RCS σ can be obtained by the formula below.

$$\begin{aligned}\sqrt{\sigma} = \; & \sum\nolimits_0 j\frac{k}{\sqrt{\pi}}\int_{s_m}\hat{e}_r\cdot\left[\hat{s}\times\left(\hat{n}_m\times\hat{h}_i\right)\right]\exp(jk\vec{r}\cdot(\hat{i}-\hat{s}))ds_m+ \\ & \sum\nolimits_1 j\frac{k}{\sqrt{\pi}}\int_{s_m}\hat{e}_r\cdot\left[\hat{s}\times\left(\hat{n}_m\times\hat{h}_{ir1}\right)\right]\exp(jk\vec{r}\cdot(\hat{i}_{r1}-\hat{s}))ds_m+ \\ & \sum\nolimits_2 j\frac{k}{\sqrt{\pi}}\int_{s_m}\hat{e}_r\cdot\left[\hat{s}\times\left(\hat{n}_m\times\hat{h}_{ir2}\right)\right]\exp(jk\vec{r}\cdot(\hat{i}_{r2}-\hat{s}))ds_m+\cdots\end{aligned} \quad (6)$$

where k is the wave number and j is the imaginary unit; $\hat{i}$ is the direction of the incident wave; $\hat{i}_{r1}$ and $\hat{i}_{r2}$ are the directions of the first-order and second-order reflected waves, respectively; $\hat{h}_i$ is the polarization direction of the incident wave; $\hat{h}_{ir1}$ and $\hat{h}_{ir2}$ are the polarization direction of the first-order and second-order reflected waves, respectively; $\hat{s}$ is the scattering direction; $\hat{e}_r$ is the polarization direction of the receiver; $\hat{n}_m$ is the normal vector of the mth facet; and ds_m is the area of the mth facet.

To accurately determine which facets are illuminated requires a larger density of rays than that of facets. If the density of rays is low, some facets will be missed. The accuracy of the traditional GO/PO method depends on the size of the facets. To ensure accuracy requires that the size of facet be small enough. However, this will increase the number of facets and the density of the rays. Thus, the calculation load is large.

To improve the efficiency, it is an effective way to reduce the density of the rays, but the problem of the facets missing must be solved when the density of the rays is low. Here, the rectangular wave beam-based GO/PO method is used [19]. The improved method treats the ray as a rectangular wave beam, and analyzes the EM scattering properties of the area illuminated by the rectangular wave beam. Each rectangular wave beam corresponds to an illuminated area, and these areas just fill the whole illuminated area. For the improved GO/PO method, there is no illuminated area missed, with the same density of rays. Thus, it is more accurate than the traditional GO/PO method with the same density of rays. Since the density of the rays is lower, its computational load is lower, and the improved method is more efficient than the traditional method.

To calculate the EM scattering properties, the contribution of each rectangular wave beam should be considered. According to the ray direction, the illuminated area is mapped to a picture with pixel matrix (see Figure 4). Then the beam is divided into many small rectangular wave beams according to the pixel matrix. Each small rectangular wave beam corresponds to one pixel. For example, the nth rectangular wave beam corresponds to the nth pixel. Furthermore, the boundary of the cross-section corresponding to the nth rectangular wave beam is rectangular with two edge vectors of $\vec{L}_{xrn}$ and $\vec{L}_{yrn}$, as shown in Figure 4.

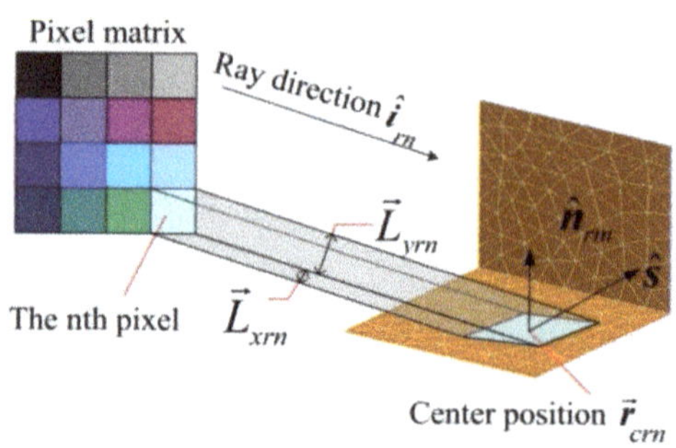

Figure 4. Schematic diagram of a small rectangular ray.

The nth rectangular wave beam illuminates an area. The RCS of this illuminated area is marked as σ_{prn}, which satisfies the equation below:

$$\sqrt{\sigma_{prn}} = j\frac{4k}{\sqrt{\pi}}\hat{e}_r \cdot \left[\hat{s} \times \left(\hat{n}_{rm} \times \hat{h}_{irn}\right)\right] \exp(jk\vec{r}_{crn} \cdot (\hat{i}_{rn} - \hat{s})) \cdot F(\hat{i}_{rn}, \hat{n}_{rm}, \vec{L}_{xrn}, \vec{L}_{yrn}), \quad (7)$$

where $\hat{n}_{rm}$ is the normal vector of the facet illuminated by the nth ray, and its number is marked as m; and $\vec{r}_{crn}$ is the center position of the area illuminated by the nth ray, with a ray direction of $\hat{i}_{rn}$. The expression of $F(\hat{i}_{rn}, \hat{n}_{rm}, \vec{L}_{xrn}, \vec{L}_{yrn})$ is shown in [19].

The total RCS is the sum of the contribution of all rays. In addition, multiple scattering effects are considered. Thus, the nth ray, which illuminates the facet, may be part of the incident wave or the reflected wave. To determine the distribution of the different kinds of waves, the RCS caused by the nth ray of the incident wave is marked as σ_{prn0}, and the corresponding number is marked as $n0$. Similarly, the RCS, which is caused by the nth ray of the first-order reflected wave, is marked as σ_{prn1}, and the corresponding number is marked as $n1$. The RCS, which is caused by the nth ray of the second-order reflected wave, is marked as σ_{prn2}, and the corresponding number is marked as $n2$. Thus, the total RCS σ of the scene satisfies the equation below:

$$\sqrt{\sigma} = \sum_{n0} \sqrt{\sigma_{prn0}} + \sum_{n1} \sqrt{\sigma_{prn1}} + \sum_{n2} \sqrt{\sigma_{prn2}} + \cdots, \tag{8}$$

where $\sum_{n0}$, $\sum_{n1}$, and $\sum_{n2}$ are operated on the small rectangular wave beams of the incident wave as well as the first time and the second time reflected ray, respectively.

2.3. Scene of Ship and Sea

Generally, a rough surface is simulated by a superposition of harmonic waves. Thus, the surface can be generated by the given spectrum. For a dynamic sea surface, the height $h(r,t)$ of the sea surface at time t can be expressed as follows:

$$h(r,t) = \mathrm{Re}\left\{ \sum_{k} A(k, v_w) \cdot \exp(iwt) \cdot \exp(ik \cdot r) \right\}, \tag{9}$$

where $A(k)$ is the Elfouhaily sea spectrum used to simulate the dynamic sea surface with a wind speed of v_w [24]. In addition, there is an obvious orientation for the dynamic sea surface along the direction of the wind. Then, the angular spreading function $\Phi(k, \varphi')$ [25] is added to modify Equation (8):

$$h(r,t) = \mathrm{Re}\left\{ \sum_{k} A(k, v_w) \cdot \exp(iwt) \cdot \Phi(k, \varphi') \exp(ik \cdot r) \right\}, \tag{10}$$

where φ' is the angle between the wave vector k and the direction of the wind.

Figure 5 shows the sea surface simulated with a wind speed of 5 m/s and a wind direction of 45°. As shown in Figure 5, the direction of the sea surface is obvious along the wind direction. The size of the sea surface is 150×150 m^2, which is large enough to put a general ship on it. As shown in Figure 6, the ship model is 120 m long, 20 m wide, and 25 m high. Generally, the scattering properties of a ship change obviously with the azimuth. Here, two typical orientations are considered. One orientation is that the ship's bow is perpendicular to the moving direction of the airborne radar (see Figure 6a). The other is that the ship's bow is parallel to the moving direction of the airborne radar (see Figure 6b).

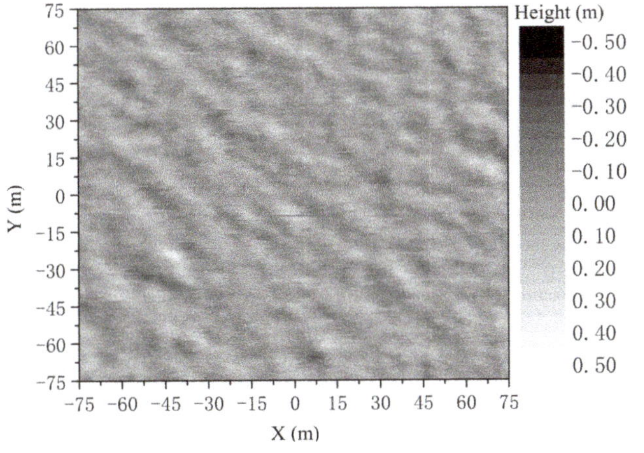

Figure 5. Sea surface with a wind speed of 5 m/s.

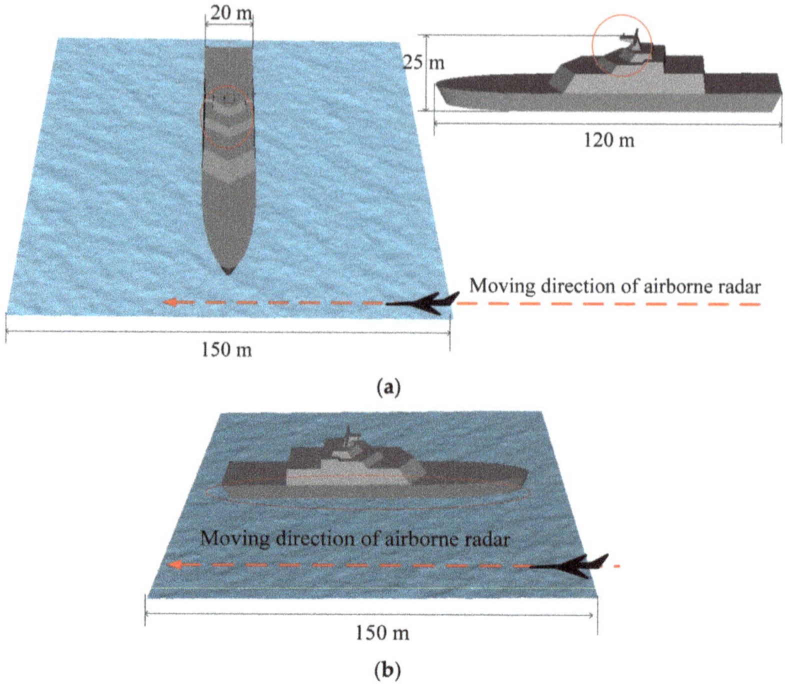

Figure 6. Model of a ship on a dynamic sea surface with different orientations: (**a**) perpendicular to the moving direction of the airborne radar; (**b**) parallel to the moving direction of the airborne radar.

The incident angle θ is 45°. The carrier frequency is 9 GHz. The bandwidth is 0.15 GHz and its corresponding range resolution is 1 m. The azimuth is along the direction of motion of the airborne radar. The range direction is perpendicular to the azimuth. The antenna length is 2 m and its corresponding azimuth resolution, ρ_a, is 1 m.

3. Results

3.1. Results of the SAR Simulation

Figure 7a shows the SAR image of the scene corresponding to Figure 6a, when the sea surface is static. The velocity of the airborne radar, V_{radar}, is 300 m/s and the synthetic aperture time, T_R, is about 0.62 s. Figure 7b shows the SAR image where the multiple scattering effects are considered. Comparing to Figure 7a, some regions in Figure 7b become brighter where multiple scattering is considered. These brighter regions correspond to the place marked with the red elliptical frame in Figure 6a and make the ship more obvious in the SAR image. Thus, it is necessary to analyze multiple scattering effects, which have an important impact on target recognition. Because the deck of the ship is flat and the EM wave is reflected to the mirror direction, its back scattering is weaker and its corresponding regions in the SAR image are obviously darker than the other regions. The shape of the darker regions is also close to the shape of the ship, which can serve as another feature for detecting the ship from the sea background.

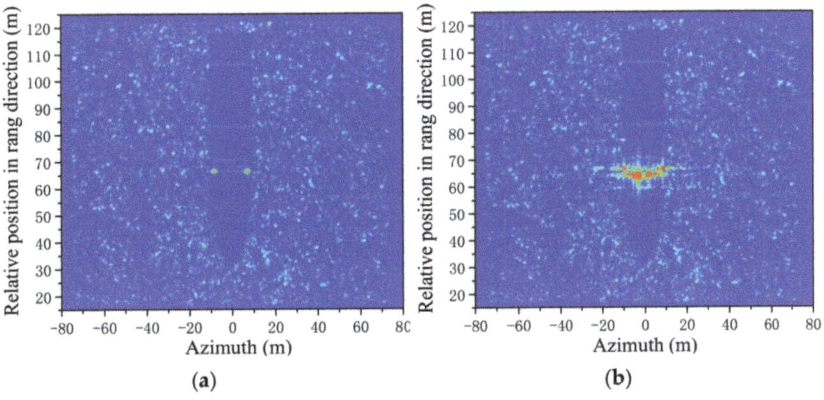

Figure 7. SAR images of the ship on a static sea surface: (**a**) single scattering; (**b**) multiple scattering.

Figure 8 shows the SAR images with the same conditions as in Figure 7, but with the sea surface being dynamic. The synthetic aperture time, T_R, is 0.62 s. Compared to the static one shown in Figure 7, the boundary of the region for the deck is not as clear as the static one. Thus, it is not as easy to detect the shape of the ship as with the static one.

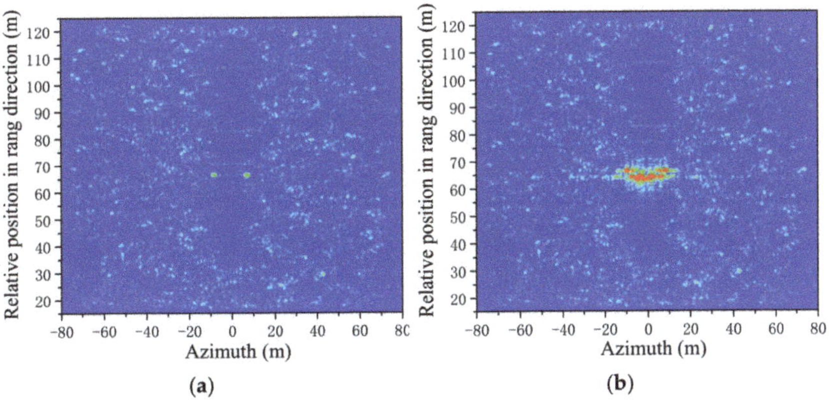

Figure 8. SAR images of a ship on a dynamic sea surface (0.62 s): (**a**) single scattering; (**b**) multiple scattering.

Figure 9 shows the SAR image of the ship on a dynamic sea surface with a longer synthetic aperture time $T_R = 1.86$ s. Because the synthetic aperture time gets longer, the sea surface changes much more and the SAR image of the dynamic sea is blurrier than the one shown in Figure 7. As shown in Figure 7, the dark region corresponding to the ship looks close to the ship's shape. In this case, it is a feasible method to recognize the ship by analyzing the shape of the dark region. However, this method fails when the dark region is disturbed by the dynamic sea surface. As shown in Figure 9, the area corresponding to the dark region in Figure 7 has a similar texture to the sea surface. In this situation, the strong scattering points are reliable features to recognize the ship. The strong scattering points are caused by the multiple scattering effects, which are considered in the GO/PO method. Comparing Figure 9a,b, there are some scattering points much stronger than the background. These strong scattering points are caused by the multiple scattering effects of the structure marked with the red elliptical frame in Figure 6a. Thus, it is necessary to use the GO/PO method to calculate the multiple scattering effects. Furthermore, the bright region caused by the multiple scattering plays an import role in target recognition, especially for a long synthetic aperture time.

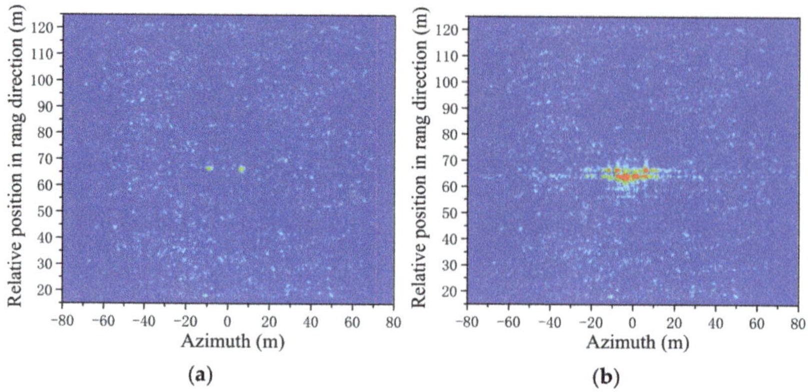

Figure 9. SAR images of a ship on a dynamic sea surface (1.86 s): (**a**) single scattering; (**b**) multiple scattering.

3.2. *Efficiency of the Rectangular Wave Beam-Based GO/PO Method*

Figure 10a,b shows the SAR images of the scene corresponding to Figure 6b, with the sea surface being dynamic. The SAR imaging conditions are the same as in Figure 8. Comparing these two SAR images, the region below the ship gets brighter when multiple scattering is considered. This is because that the side of the ship and the sea surface form an angular structure marked with the red elliptical frame in Figure 6b. It means that the multiple scattering effects between the target and the background may enhance the scattering echo under special conditions when forming some special structures.

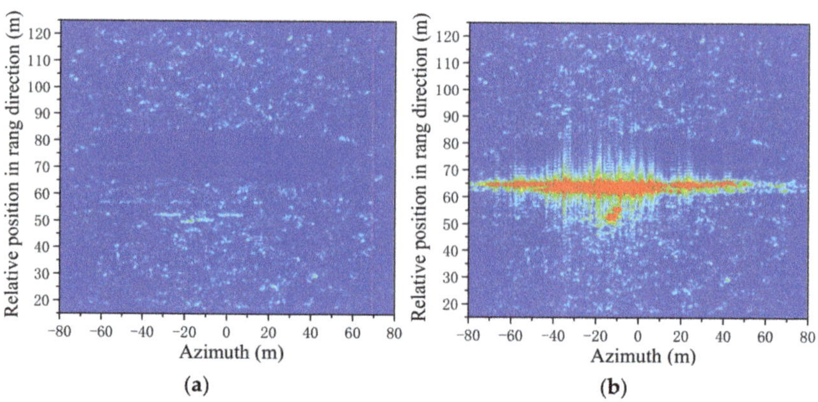

Figure 10. *Cont.*

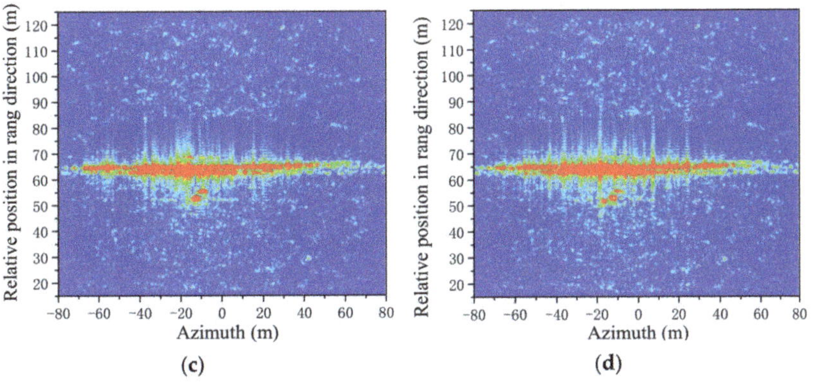

Figure 10. SAR images with different sizes of the pixel matrices: (**a**) single scattering (128×128); (**b**) multiple scattering (128×128); (**c**) multiple scattering (256×256); (**d**) multiple scattering (512×512).

Figure 10 shows the SAR images with different sizes of the pixel matrix. It takes more time to calculate the scattering echoes from the ship when the size of the pixel matrix is larger. Table 1 gives the time to calculate the scattering echo with different matrix sizes. There are 256 frequency components calculated. The time required drastically increases when the size of the matrix is larger. Fortunately, the proposed method is accurate enough with a small size pixel matrix. As shown in Figure 10b,d, the SAR image obtained with the pixel matrix size of 128 × 128 is almost the same as the SAR image with a size of 512 × 512. The proposed rectangular wave beam-based GO/PO method has the ability to improve the efficiency with low errors when the pixel matrix size is small.

Table 1. Simulation time of the EM scattering field for one pulse (256 frequency components) *.

Pixel Matrix Size	Time Needed with Proposed Method (s)
128×128	16.364
256 × 256	65.224
512×512	260.925

* Calculated by computer with Intel Core™ i7-6700K CPU and NVIDIA GeForce GTX 1080 display card.

4. Conclusions

The scattering echo simulation method in the frequency domain was used to calculate the echo from a ship on a dynamic sea surface. The method introduced considers the EM scattering properties of all the frequencies. Thus, the method is able to simulate the scattering echo for a wideband signal. For a complex ship with many corners, the rectangular wave beam-based GO/PO method can accurately calculate the EM scattering properties when considering multiple scattering effects. The improved GO/PO method is efficient and accurate when the size of the pixel matrix is small. Combined with the SAR imaging process, the rectangular wave beam-based GO/PO method was used to calculate the EM scattering properties of all the frequency components when the radar emits a wideband pulse. With these EM scattering properties, the scattering echo of each pulse can be calculated by using the introduced echo simulation method. The sea surface is dynamic when the radar emits a pulse periodically. The SAR image results show that the static sea surface is quite different from the dynamic sea surface, which gets blurrier as the synthetic aperture time increases. In addition, multiple scattering, which increases the back scattering, has an important impact on the SAR image of a complex target. Therefore, the proposed method can simulate the SAR image of a target on a sea surface. Lastly, the method can improve the simulation efficiency with good accuracy by using a small pixel matrix size.

Author Contributions: Conceptualization, W.-Q.J. and M.Z.; methodology, W.-Q.J.; software, X.-Z.L.; validation, W.-Q.J., and M.Z.; formal analysis, M.Z.; investigation, W.-Q.J.; resources, L.-Y.W.; data curation, G.L.; writing—original draft preparation, W.-Q.J.; writing—review and editing, M.Z.; visualization, W.-Q.J.; supervision, M.Z.; project administration, M.Z.; funding acquisition, W.-Q.J. and M.Z. All authors have read and agreed to the published version of the manuscript.

Funding: This work was supported in part by the National Natural Science Foundation of China under Grant Nos. 61771355 and 41701386, and the National Key Laboratory Foundation under Grant No. 614241103010417/6142A01180206/JCKYS2019204020.

Institutional Review Board Statement: Not applicable.

Informed Consent Statement: Not applicable.

Data Availability Statement: Not applicable.

Conflicts of Interest: The authors declare no conflict of interest.

References

1. Dong, G.; Kuang, G. Classification on the Monogenic Scale Space: Application to Target Recognition in SAR Image. *IEEE Trans. Image Process.* **2015**, *24*, 2527–2539. [CrossRef] [PubMed]
2. Zhao, P.; Liu, K.; Zou, H.; Zhen, X. Multi-Stream Convolutional Neural Network for SAR Automatic Target Recognition. *Remote Sens.* **2018**, *10*, 1473. [CrossRef]
3. Kajbaf, H.; Case, J.T.; Yang, Z.; Zheng, Y.R. Compressed Sensing for SAR-based Wideband Three-dimensional Microwave Imaging System Using Non-uniform Fast Fourier Transform. *IET Radar Sonar Navig.* **2013**, *7*, 658–670. [CrossRef]
4. Peng, X.; Wang, Y.; Hong, W.; Tan, W.; Wu, Y. Autonomous Navigation Airborne Forward-Looking SAR High Precision Imaging with Combination of Pseudo-Polar Formatting and Overlapped Sub-Aperture Algorithm. *Remote Sens.* **2013**, *5*, 6063–6078. [CrossRef]
5. Gao, G.; Liu, L.; Zhao, L.; Shi, G.; Kuang, G. An adaptive and fast CFAR algorithm based on automatic censoring for target detection in high-resolution SAR images. *IEEE Trans. Geosci. Remote Sens.* **2009**, *47*, 1685–1697. [CrossRef]
6. Pourmottaghi, A.; Gazor, S. A CFAR detector in a nonhomogenous Weibull clutter. *IEEE Trans. Aerosp. Electron. Syst.* **2012**, *48*, 1747–1758. [CrossRef]
7. Hou, B.; Chen, X.; Jiao, L. Multilayer CFAR detection of ship targets in very high resolution SAR images. *IEEE Geosci. Remote Sens. Lett.* **2015**, *12*, 811–815.
8. Leng, X.; Ji, K.; Yang, K.; Zou, H.X. A bilateral CFAR algorithm for ship detection in SAR images. *IEEE Geosci. Remote Sens. Lett.* **2015**, *12*, 1536–1540. [CrossRef]
9. Wang, X.; Chen, C. Adaptive ship detection in SAR images using variance WIE-based method. *Signal, Image Video Process* **2016**, *10*, 1219–1224. [CrossRef]
10. Wang, X.; Chen, C. Ship Detection for Complex Background SAR Images Based on a Multiscale Variance Weighted Image Entropy Method. *IEEE Geosci. Remote Sens. Lett.* **2017**, *14*, 184–187. [CrossRef]
11. Geng, X.; Shi, L.; Yang, J.; Li, P.; Zhao, L.; Sun, W.; Zhao, J. Ship Detection and Feature Visualization Analysis Based on Lightweight CNN in VH and VV Polarization Images. *Remote Sens.* **2021**, *13*, 1184. [CrossRef]
12. You, J.; Hu, Z.; Peng, C.; Wang, Z. Generation and Annotation of Simulation-Real Ship Images for Convolutional Neural Networks Training and Testing. *Appl. Sci.* **2021**, *11*, 5931. [CrossRef]
13. Zhao, Y.W.; Zhou, P.; Zhang, X.; Zhang, M. Application of improved equivalent edge currents in synthetic aperture radar imaging. *J. Syst. Eng. Electron.* **2010**, *21*, 556–571. [CrossRef]
14. Chen, H.; Zhang, M.; Zhao, Y.W.; Luo, W. An Efficient Slope-deterministic Facet Model for SAR Imagery Simulation of Marine Scene. *IEEE Trans. Antennas Propag.* **2010**, *58*, 3751–3756. [CrossRef]
15. Ghavidel, A.; Camps, A. Time-domain Statistics of the Electromagnetic Bias in GNSS-Reflectometry. *Remote Sens.* **2013**, *7*, 11151–11162. [CrossRef]
16. Tao, Y.B.; Lin, H.; Bao, H.J. Kd-tree Based Fast Ray Tracing for RCS Prediction. *Prog. Electromagn. Res.* **2008**, *81*, 329–341. [CrossRef]
17. Rius, J.M.; Ferrando, M.; Jofre, L. GRECO: Graphical Electromagnetic Computing for RCS Prediction in Real Time. *IEEE Antennas Propag. Mag.* **1993**, *35*, 7–17. [CrossRef]
18. Wei, P.B.; Zhang, M.; Niu, W.; Jiang, W.Q. GPU-Based Combination of GO and PO for Electromagnetic Scattering of Satellite. *IEEE Trans. Antennas Propag.* **2012**, *60*, 5278–5285. [CrossRef]
19. Jiang, W.Q.; Zhang, M.; Zhao, Y.; Nie, D.; Jiao, Y.C. Rectangular Wave Beam Based GO/PO Method for RCS Simulation of Complex Target. *Prog. Electromagn. Res. M* **2017**, *53*, 53–65. [CrossRef]

20. Jiang, W.Q.; Zhang, M.; Nie, D.; Jiao, Y.C. Improved GO/PO method and its application to wideband SAR image of conducting objects over rough surface. *Waves Random Complex Media* **2018**, *28*, 310–325. [CrossRef]
21. Neo, Y.L.; Wong, F.H.; Cumming, I.G. Processing of Azimuth-Invariant Bistatic SAR Data Using the Range Doppler Algorithm. *IEEE Trans. Geosci. Remote Sens.* **2008**, *46*, 14–21. [CrossRef]
22. Li, Z.; Wang, J.; Liu, Q.H. Frequency-domain backprojection algorithm for synthetic aperture radar imaging. *IEEE Geosci. Remote Sens. Lett.* **2015**, *12*, 905–909.
23. Tong, X.; Bao, M.; Sun, G.; Han, L.; Zhang, Y.; Xing, M. Refocusing of Moving Ships in Squint SAR Images Based on Spectrum Orthogonalization. *Remote Sens.* **2021**, *13*, 2807. [CrossRef]
24. Elfouhaily, T.; Chapron, B.; Katarsos, K. A unified directional spectrum for long and short wind-driven waves. *J. Geophys. Res.* **1997**, *102*, 15781–15796. [CrossRef]
25. Longuet-Higgins, M.S.; Cartwright, D.E.; Smith, N.D. Observations of the directional spectrum of sea waves using the motions of a floating buoy. In *Ocean Wave Spectra*; Prentice-Hall: Hoboken, NJ, USA, 1963; pp. 111–132. [CrossRef]

remote sensing

MDPI

Article

Towards Sustainable Management of Mussel Farming through High-Resolution Images and Open Source Software—The Taranto Case Study

Carmine Massarelli [1,*], Ciro Galeone [2], Ilaria Savino [1], Claudia Campanale [1] and Vito Felice Uricchio [1]

[1] Water Research Institute—Italian National Research Council (CNR-IRSA), Via F. De Blasio 5Zona Industriale, 70132 Bari, Italy; ilaria.savino@ba.irsa.cnr.it (I.S.); claudia.campanale@ba.irsa.cnr.it (C.C.); vito.uricchio@ba.irsa.cnr.it (V.F.U.)

[2] Environmental Protection Agency of Apulia (ARPA Puglia), Corso Trieste 27, 70126 Bari, Italy; c.galeone@arpa.puglia.it

* Correspondence: carmine.massarelli@ba.irsa.cnr.it

Citation: Massarelli, C.; Galeone, C.; Savino, I.; Campanale, C.; Uricchio, V.F. Towards Sustainable Management of Mussel Farming through High-Resolution Images and Open Source Software—The Taranto Case Study. *Remote Sens.* **2021**, *13*, 2985. https://doi.org/10.3390/rs13152985

Academic Editor: Weicheng Wu

Received: 21 June 2021
Accepted: 27 July 2021
Published: 29 July 2021

Abstract: This research activity, conducted in collaboration with the Aero-Naval Operations Department of the Guardia di Finanza of Bari as part of the Special Commissioner for urgent measures of reclamation, environmental improvements and redevelopment of Taranto's measurement, is based on the use of a high-resolution airborne sensor, mounted on board a helicopter to identify and map all in operation and abandoned mussel farming in the first and second inlet of Mar Piccolo. In addition, factors able to compromise the environmental status of the Mar Piccolo ecosystem were also evaluated. The methodological workflow developed lets extract significant individual frames from the captured video tracks, improves images by applying five image processing algorithms, georeferences the individual frames based on flight data, and implements the processed data in a thematic Geographical Information System. All mussel farms, in operation and derelict, all partially submerged and/or water-coated invisible to navigation poles and other elements such as illegal fishing nets and marine litter on the seabed up to about 2 m deep, have been identified and mapped. The creation of an instant, high-precision cartographic representation made it possible to identify the anthropogenic pressures on the Mar Piccolo of Taranto and the necessary actions for better management of the area.

Keywords: mussel farming; high-resolution image; transitional water management; environmental pollution; open source software

1. Introduction

The recent development of remote sensing techniques has contributed significantly to improving the ability to monitor coastal areas more efficiently, providing cost-fitting spatially distributed data [1,2].

Many different observing sensors are used to measure the qualitative parameters of water bodies, i.e. suspended sediments, dissolved organic matter (DOM), chlorophyll-a, and pollutants, map coastal areas and monitor anthropogenic pressures [3].

They are divided into two main categories based on the platforms used to locate them. Airborne sensors are assembled on a platform within the Earth's atmosphere (i.e. helicopter, aircraft, etc.), and spaceborne sensors are mounted on spacecraft or satellites outside of the Earth's atmosphere [4].

Airborne sensors have greater spectral and spatial resolutions compared to spaceborne sensors [5,6].

Among the most commonly used sensors, there are digital cameras, Light Detection and Ranging (LiDAR), Synthetic-aperture radar (SAR), MultiSpectral and Iperspectral Scanner [7]. However, each platform has advantages and disadvantages necessary to know to choose the more appropriate instruments to achieve work objectives (Supplementary Material, Tables S1–S3) [7–10].

Among the remote-sensing applications interesting is mapping mussels aquaculture farms using high-resolution images acquired from different sources [11].

The primary role of mussel aquaculture is to provide food, but it also has an important ecological role through carbon fixation and, therefore, contributes to improving coastal ecosystems' capacity to absorb atmospheric CO_2 from anthropic sources [12,13].

However, the massive expansion of mussel aquaculture started in the 1990s continued to proliferate rapidly [12]. This significant development took place in establishing several illegals farms due to inadequate control and management. Consequently, mussel aquaculture became a considerable pressure for natural resources by producing large quantities of decomposing biomass that leads to eutrophication of the waters and enhancing potentially harmful algal blooms with effects on local and migratory wildlife and the aesthetic value of the coastal area [14].

Traditional mapping aquaculture farms are currently being developed on-site, including Global Positioning Systems (GPS) [2,15]. However, these methods appear expensive for the cost of field survey, time-consuming and potentially dangerous for the physical restrictions that make navigation difficult. In this context, mussel farms, distributed over vast areas, are difficult to detect and monitor with traditional monitoring devices.

1.1. Framework and Criticalities of the Study Area

The experimental site is located in the *Southern Italy—Apulia Region*; it is the *Mar Piccolo* of *Taranto*, in the northern part of the town (Figure 1). This transitional water system is characterised by an enclosed basin that shows lagoon characteristics due to numerous submarine freshwater springs, knows as "citri" [16–18]. These submarine springs, supplied by an underlying karst aquifer, recharge the *Mar Piccolo* with freshwater, which influences the equilibrium of its ecosystem and provides it with the features of a transition environment [19,20].

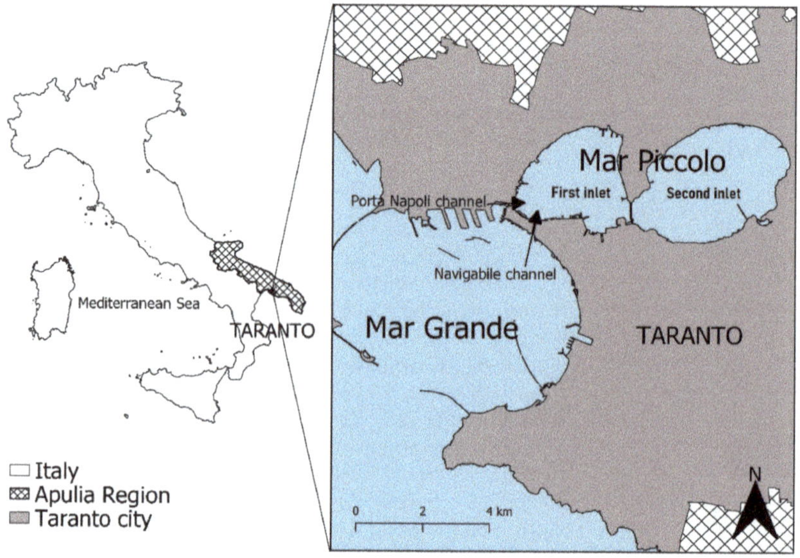

Figure 1. Study area.

The total surface area of the basin (20.72 km^2), divided into two inlets, called first (west) and second (east) inlet, which have a maximum depth of 18 m and 29 m [21] at the *citri* and 8 m and 13 m as average values, respectively. The *Mar Piccolo* is connected with the *Mar Grande* through two channels, the *Navigabile* and the *Porta Napoli* channels (Figure 1). The tidal range does not exceed 30–40 cm [22]. The circulation of sea waters in the *Mar Piccolo* is influenced not only by the semi-enclosed characteristics but also by human infrastructure [23]. Due to its peculiar characteristics, the *Mar Piccolo* is one of the major mussel farming sites in the *Mediterranean Sea*, with a maximum production capacity of about 60.000 t/y [24,25].

Mussel farming is one of the oldest practices in aquaculture of the *Taranto* area [24], particularly in the *Mar Piccolo*. It is dated back to the end of the 17th century and was carried out on fixed structures that have remained almost similar to this day. Techniques are based on the empirical knowledge of mussel farmers handed down from father to son for centuries. It assumes high socio-economic importance in Taranto, and cooperatives or individual firms traditionally practise it often family-run without a real management collaboration plan. The entire sector comprises 37 individual cooperatives, employing about 900 farmers (300 full time and 600 part-time) [24]. However, it is not easy to quantify the annual turnover, which is in the order of M€ [26].

The current legal framework, through the Ordinance of the *Taranto* Port Authority n°107/2005 (subsequently amended and supplemented by Ordinance n°222/2009 [27] regulates the mooring, transit and anchorage of nautical vehicles and people in the first and second inlets of the *Mar Piccolo*, outlining the profile of mussel plants in the two water basins. In addition, different regulations apply to the two inlets: for the first inlet, the Ordinance of the President of the Regional Council n°532 of 13 September 2018 [28] in which it is forbidden for mussels to exceed 28 mm in this sub-basin due to accumulation of water contaminants; for the second inlet, a fixed monitoring station was identified for the health surveillance of mussel farming areas under EC regulation n°854/04 [29].

If, on the one hand, mussel farming represents a characteristic of the *Taranto* area and products are particularly appreciated, on the other hand, this practice contributes to increasing pollution. In particular, mussel farming causes the accumulation of marine litter, especially plastic waste *ghost nets* [30] in the *Mar Piccolo* basin.

In recent decades serious organisational and management deficiencies have led to a deterioration in the quality of the mussels produced. In July 2011, following the detection of dioxins and polychlorinated biphenyl dioxins-like (PCB—DL) in the mussels of the first inlet, its collection and handling was banned [26].

In addition, the lack of maintenance has favoured over the years the accumulation of ancient structures to the point that abandoned farming areas become a source of danger to shipping or are illegally occupied without authorisation. Hence, through accurate spatial planning, regulating mussel production is essential to ensure sustainable use of marine resources [31].

Due to this poor management of permits, licenses and concessions, illegal phenomena, industrial pollution, and climate change affected mussel quality and productivity [32–34]. This area is included within the Italian Sites of National Interest (SIN) of *Taranto*, which needs urgently environmental remediation [35]. More recently, the *Mar Piccolo* has been the subject of several studies to identify strategies for the site's reclamation [36,37].

As generally happens in this type of activity, data coming from mussels cultivation are highly variable, aggregated and poorly documented. However, valid and reliable data is here reported: the production of mussels in 2010 is close to 40,000 tons/year, in the *Mar Piccolo*, of which about 13,000 t/year in the first inlet [24,26].

1.2. Mussel Farms of Mar Piccolo

The Mar Piccolo of Taranto is the *Mytilus galloprovincialis* L. historical breeding site, representing a significant socio-economic activity.

A high density of mussel farms characterises the study site. Among these are recognisable traditional plants called *quadri* (Figure 2), built with chestnut poles and synthetic strings [38], and those created using plastic materials [39], to form the latest *long lines* (Figure 3) [24].

Figure 2. Traditional mussel farms called "quadri" in the first inlet of Mar Piccolo (Puglia Region). May 2015.

Figure 3. Floating systems of mussel farms called "long lines" where the plastic buoys are clearly visible (in the background, a former shipyard is visible) in the first inlet of *Mar Piccolo* (*Puglia Region*). May 2018.

Mussel crops, called *squares*, are formed by a variable number (4–6) of parallel rows of chestnut wood poles for the tip at the bottom of the sea. Under the sea's surface, the poles are connected by ropes of synthetic material, called *libàni*.

Each module consists of a rope, generally of polypropylene, called *ventìa*, whose thickness varies between 40 and 50 mm and whose length is not less than 5 m. The floats placed at the two ends of the *ventìa* were anchored to the bottom using dead bodies in concrete, with an anchoring function, with polyethene or polypropylene cables, whose length varies according to the bathymetry of the site. The *ventìa* has the function of supporting the *reste*, tubular stockings in which the mussels are grafted and supporting growth.

In the last decades, the availability of floats in plastic material has allowed the construction of floating systems, called *long lines*, which have spread, progressively replacing the systems with poles. These floating suspension farms are typical of inland seas and lagoons, such as the *Mar Piccolo.*

1.3. Objectives of the Work

The activity reported in the present work was born as part of the actions undertaken by *"Special Commissioner for urgent measures of reclamation, environmental improvements and redevelopment of Taranto"*, in cooperation with National Research Council, University of Bari Aldo Moro and Apulian Aero Naval Regional Command of the Finance Police. It is titled: *"Interventions for the mitigation of impacts on the Mar Piccolo"*.

The general objective of the present study is to develop a methodology that combines High-Resolution airborne images with open-source software for the elaboration of data aimed to map aquaculture mussel farms in the "Mar Piccolo" of Taranto city (South Italy).

Moreover, the specific aims of the study referred to the study case of Taranto were:

- realise a census of all the mussels farms in the first and second inlet of the *Mar Piccolo* providing accurate information on the number of units, the precise location through georeferentiation, the area, perimeter and density of mussel farms;
- identify and georeferencing abandoned cultivation fields for which navigation can be dangerous;
- identify illegal installations and provide details necessary to local port authorities to define the appropriate management interventions of the area
- detect any unauthorised anthropogenic action and environmental pressures that may alter the profile of the marine ecosystem in terms of overfishing
- plan strategies for the surveillance, remediation and restoration of the *Mar Piccolo* area

These problems and criticalities have been showing up lately as over time no remediation operations have ever been carried out on the systems that have gone into disrepair and now creates severe problems for the *Mar Piccolo* ecosystem and navigation, as they are real hidden dangers (Figure 4).

Figure 4. Ancient mussel farms abandoned and never restored (the red arrows indicate the barely visible poles on the water). First inlet of Mar Piccolo (Puglia Region). May 2018.

2. Materials and Methods

2.1. Data Acquisition

2.1.1. Sensor Characteristics

The Leonardo Electro-Optic Surveillance System (LEOSS) is the latest multi-sensor, high accuracy, four axes gyro-stabilised for aerial surveillance applications, designed to be compliant with demanding vibration profiles. The system combines high-performance sensors capable of meeting the requirements of advanced fixed and rotary-wing aerial platforms. The Optronic Turret Unit (OTU) is the central unit of the system that, through the IR (Infrared) and CCD (Change Couple Device) sensors, produces the visible and infrared vision of the framed scene. Among the electro-optical sensors of the system, there are a High Definition Color Camera, TVC-HD; an IR SD (Standard Definition) camera and a Short-wave Infrared (SWIR) camera.

The TVC-HD Camera is a High Definition Serial Digital Interface whit CMOS technology. It produces an image of 1920 × 1080 pixels and has a 40X optical zoom up to NFOV 0.5° any more than a 4X digital zoom.

The IR SD Camera works in the 3–5 micron IR spectrum. It produces an image of 640 × 512 pixels and has optics SD and 2× optical zoom whit a Narrow Field of View (NFOV) of 0.6° × 0.48°.

The SWIR camera, on the other hand, offers the possibility to observe scenarios with low visibility (haze, fog, fumes); it is set to work in the 0.9–1.7 micron spectrum and produces an image of 640 × 512 pixels. The optic has a High Field of View (HFOV) from 27° to 0.7° and a 40× optical zoom integrated whit 4× digital zoom.

The sensor is mounted on an AgustaWestland AW-109N Nexus helicopter.

2.1.2. Planning of Flight

The flights were carried out in eight days of activity from 17 to 25 June 2018. In total, eighty-three video tracks for over 34GB of archive data have been acquired.

The optical data for each swipe consist of video files with an average duration of 3 minutes and 30 seconds, with a frame rate of 25 frames/second and resolution of 1920 × 1080 pixels in *AVI* format, for a final resolution on the ground of 5 cm.

All the flight paths used were georeferenced to understand the totality of the activity carried out in the overflights of the *Mar Piccolo*. The trajectories were processed by combining the information provided by the GNSS receiver and the inertial platform on board the helicopter with the information provided by the permanent GNSS stations of Ginosa, Valenzano and Fasano. The precision processing of the trajectories is necessary to correctly identify the position and orientation of sensors during the flight. The files generated by the onboard navigation system are called logbooks; they have been exported in *CSV* exchange format and contain the following information: date, time, latitude, longitude, altitude of helicopter and the observed object, roll, pitch, yaw. Figure 5 shows all the flight trajectories (set of all georeferenced points) performed obtained from the logbooks of the onboard instrumentation. All data were planimetrically reported in the WGS 84/UTM zone 33N reference system, EPSG 32633.

Figure 5. All flight paths from the logbook.

2.1.3. Field Inspections

To verify the correctness of the coordinates of some acquired details and when the ground control points (GCPs) were not sufficient, additional inspections were carried out aimed at thickening the network of GCPs with Leica CS25 differential GPS.

2.2. Image Processing

To pursue our mapping objectives, they are generally used push-broom sensors (also known as an along-track scanner) installed on aircraft. However, given the necessary presence of the *Guardia di Finanza* to monitor other aspects with a high degree of confidentiality, to optimise resources, we found it helpful to develop an ad-hoc operational workflow based on interoperability tools. It is based on the steps reported in Figure 6. As shown in Figure 6, for video files, substantial processing (six distinct phases) is required to georeferences the individual frames through the data of the flight logbooks. Only after georeferenciation, it is possible to obtain the information indispensable to create the thematic maps helpful in achieving our objectives.

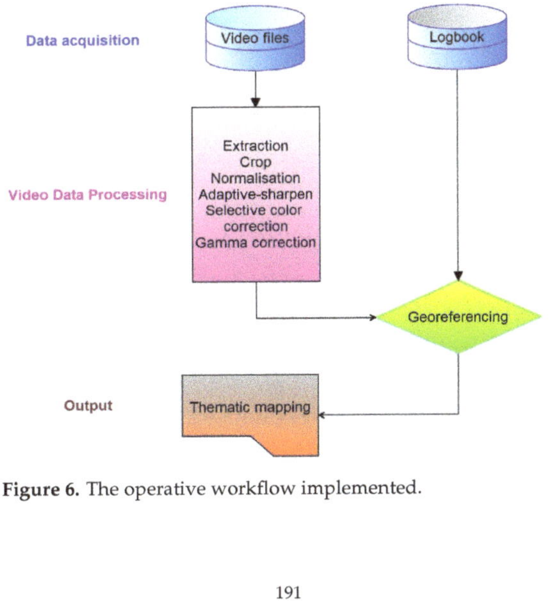

Figure 6. The operative workflow implemented.

Differently from many satellite images and that produced by well-known airborne sensors for which many algorithms and software for the image processing are available, our novel LEOSS sensor miss of any correlated tools. Therefore, our efforts aimed also to develop a new methodology fitted for LEOSS sensors that fulfil our application purposes.

Right from the design of the activity, it was known that to adapt the specific engineering of the sensor and the acquisition methods via helicopter to our specific mapping targets, several changes in the acquisition-processing phase were necessary to reach the set goal. In addition, we always considered the possibility to introduce several new useful functions according to other needs that could emerge during the operations.

Driven by these needs, we opted for the choice of Open Source software focusing exclusively on the high-level programming language python, considered very robust and highly versatile and with a consistent number of python bindings [40], that are packages and extensions tools for programming and manipulating the Geospatial Data Abstraction Library (GDAL) [41].

The CNR-IRSA working group has therefore developed a series of scripts capable of extracting a considerable amount of information from the video data acquired at the same time as the aircraft's logbooks, preserving the considerable degree of detail (as said resolution of 5 cm) allowing to obtain an instant representation of the investigated area focused mainly on the mapping of mussel farms present.

2.2.1. Video Data Processing

For our purpose, first, we used FFmpeg [42], a leading multimedia framework, able to decode, encode, transcode, mux, demux, stream, filter and play the video under the GNU Lesser General Public License [43] under an Ubuntu server [44]. So, the first step consists of a frame extraction technique that implements video content by selecting a set of summary keyframes to represent video sequences [45].

Secondly, a crop filter was applied [46] to remove undesired logos such as compass and navigation data. The number of pixels removed is always the same because these are portions of the video which are fixed and which report navigation data. In this way, each frame is cleared of the raw data frame of the video file in which the superimposed information relating to the aeronautical acquisition software appears.

After these first two phases, we included the use of another free software delivered as a ready-to-run binary distribution or as source code that can be used, copied, modified, and distributed in both open and proprietary applications [47] under a derived Apache 2.0 license [48].

The third stage of video processing consists of a normalisation operation increasing the contrast in an image by stretching the range of intensity values. Normalisation transforms any n-dimensional grayscale image (in the case of an RBG, every single channel) into a new image according to the formula with the desired range 0–255 to obtain more contrast through the dynamic range expansion techniques [49]. The linear normalisation of a greyscale digital image is performed according to the formula:

$$I_N = (I - \mathrm{Min})\,\frac{\mathrm{newMax} - \mathrm{newMin}}{\mathrm{Max} - \mathrm{Min}} + \mathrm{newMin} \qquad (1)$$

with:

I_N new pixel value;
I old pixel value;
Min and Max are the minimum and maximum pixel value measured;
newMin and newMax are the new minimum and maximum desired pixel values.

An adaptive-sharpen filter was then applied to the processed frame aimed at improving the contours by borrowing the technique from the processing of medical images that are known to have a blur with a special tuning weight that controls the amount of the produced sharpness helped to increase the acutance of edges and improve the overall sharpness significantly in the processed images [50].

Subsequently, to correct the intrinsic characteristics related to the acquisition in the optical field of the sensor with an IR cut filter for low light visibility (not removed due to other mission purposes), selective colour correction had to be applied. The result is that the red band was shifted towards the thermal infrared and selective colour correction had to be applied by selectively correcting the green colour by an empirical value of 20% to counterbalance the shift towards red.

This step is highly effective for improving colours for coastal environments and improving the contrast between semi-submerged and submerged objects near the coast in the water matrix.

A gamma value correction was applied, which indicates the slope of the logarithmic curve representative of the relationship between signal and response according to (2) (specifically, it represents the derivative of the relationship between input and output in a logarithmic space).

$$\gamma = \frac{\mathrm{d}\ \log(\mathrm{Vout})}{\mathrm{d}\ \log(\mathrm{Vin})} \tag{2}$$

with:

γ gamma value;
d log(Vout) logarithmic value of the digital signal in output;
d log(Vin) logarithmic value of the digital signal in input.

To keep the data in a domain as uniform as possible from the perceptual point of view, the gamma value of the images in post-processing was empirical. For the case study, we based on the acquisition of 10 keyframes representing different conditions, set at +0.9, which was also valuable for enhancing the contrast. Indeed, if the gamma value increases, the resulting curve will be convex, and the resulting image will be more bright while shrinking that value will darken the image, and the curve will be concave [51]. In Figure 7. we report three of the ten keyframes used.

An example of a result relating to the overall improvement of the image for each frame is shown in Figure 8. As it is possible to observe, the improvement has also significantly reduced the slight blur of the starting image because the sensor was unable to focus well due to the speed of the aircraft and the focal values. The reader can also notice how the green colour looks (much closer to reality) in the final result (Figure 8).

2.2.2. Frames Georeferencing

At the end of the image enhancement phase, we inserted the georeferencing procedure prepared with the data coming from the logbooks of the onboard instruments. The first critical aspect of this phase is linked to the fact that there was only one point for georeferencing the helicopter position from which we had to obtain the coordinates of the point in the centre of the frame with geometric calculations.

The second critical aspect is related to the frame rotation value derived from the sensor positioning angle that was not recorded in the flight logbooks. Therefore, every effort was made to derive this data from the helicopter position values assuming that the helicopter and sensor moved in unison.

So, we used the following Free and Open Source libraries named Geospatial Data Abstraction Library (GDAL) [41] and the following python bindings [40]. The main python modules implemented are *osgeo, gdal, osr, math, utm*.

The georeferencing of the single video frames in geographic raster takes place through the creation of a geotransform array that gdal uses to describe a raster position that needs the following input information: upper left position in x and y coordinates, pixel scale for x and y directions, values of x and y inclinations and finally the epsg code of the projection.

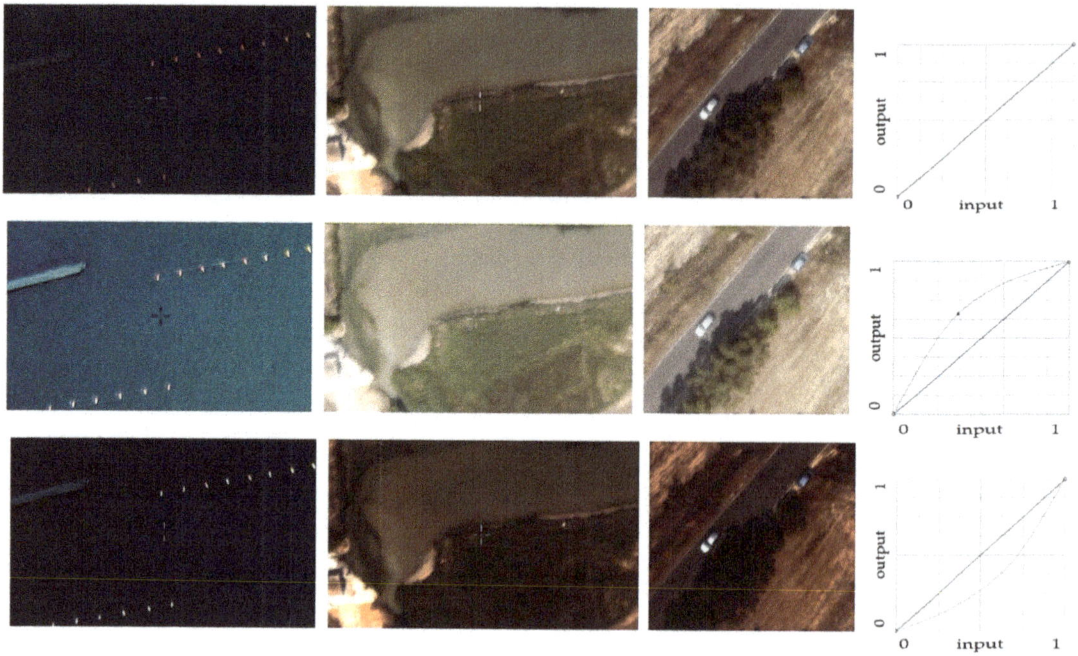

Figure 7. The figure represents how to change three of the ten keyframes used to empirically testing how to change three of the ten keyframes used to set empirically the gamma value based on acquired video files. On the right, the slope of the curve representing the relationship (in logarithmic terms) between signal and response.

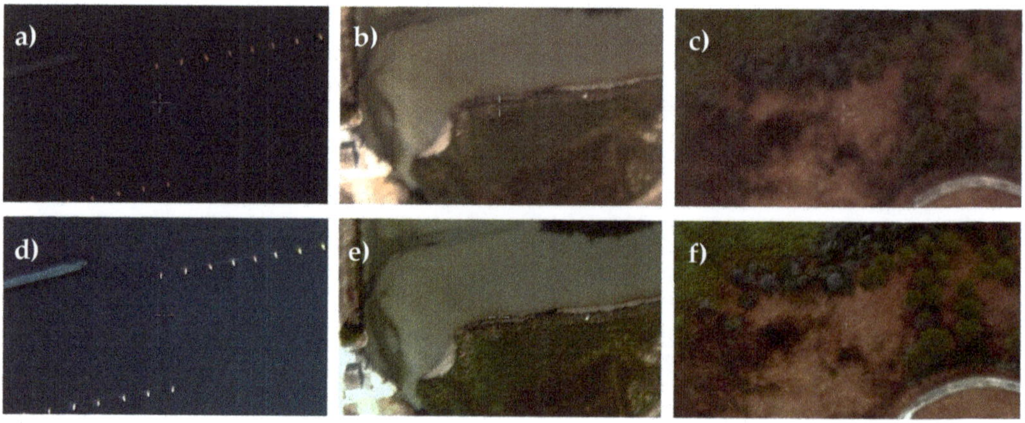

Figure 8. Image enhancement in three different environments: (**a–c**) the acquired image, (**d–f**) the same reworked image.

Since there was often wind and consequently the rotation value increased, to perform a more precise georeferencing we preferred using an Affine Transformation through the implementation of the *affine* package and calculating the necessary input parameters through the *gdalinfo* and *pil* packages, helpful retrieving geometric information from the images. Data from logbooks provided information on the position in the spatial and temporal moment of the acquisition.

In essence, an Affine Transformation represents and gives life to a relation between two images of affine self-similarity to the point of giving life to fractals [52]. Affine transformation can be expressed in matrix multiplication (linear transformation) followed by a vector addition (translation). We can use an Affine Transformation [53] to express:

1. rotations (linear transformation), if we think to the sensor;
2. translations (vector addition) if we think to the shift due to wind shifts;
3. scale operations (linear transformation), introduced because the wide-angle lens created a difference in surfaces re-covered by the pixels, i.e. the central pixels cover less surface than the lateral ones.

In summary, the following parameters were calculated, and some necessary correction factors were implemented (the onboard instrumentation was still in the testing phase, and there was no detailed information on the position of the sensor concerning that of the helicopter) to obtain the correct centimetre georeferencing of the frames can be automatic:

- total framed width as a function of the shooting height;
- average coverage area of the pixel recorded on the monitor;
- calculation of the upper left corner upwards in degrees for final translation;
- semi-diagonal frame calculation;
- calculation of the number of base map pixels involved;
- calculation of the scale parameter between the two maps;
- correction factors for the above reasons.

Once imported into the GIS, each *frame* was rechecked using different information *layers* relating to the most recent years to verify the correct georeferencing and carrying out, where necessary, a further orthorectification. All the images produced were subsequently used to create a mosaic, overlapping them from east to west and creating a unique raster for each captured video. In this way, it was possible to analyse the entire area of interest and create a shapefile as a union framework (Figure 9a) from a mosaic of more frames (Figure 9b), appreciable when superimposed on a recent base map, aerial image of Puglia Region dated 2016 [54].

In addition, operating in this way, it is possible to see in Figure 9c a result of the superimposition of the georeferenced frames and their relative variability.

This variability is due to the wind, above for the rotation and with the flight paths that instead seem straight in the very short instant of time.

Unfortunately, despite the introduction of some correction factors and even if the script created extracts the individual "tiles" from the acquired video and places them in a reference system thanks to the data contained in the flight route logbooks, sometimes due to synchrony discrepancies due to instrumental problems, the activity required further manual corrections for single frames, this correction was made in a Geographical Information System (GIS) environment with a tool to perform an interactive raster georeferencing [55]. However, the number of frames thus rectified amounts to less than 5%.

More complicated and precise georeferencing algorithms could not be applied; for example, polynomial algorithms need at least n. 3 points.

Only the framework is shown as the data are reserved for ongoing investigations aimed at verifying any spaces illegally occupied and being judicial evidence. In the event of publication the charges, for formal defects provided for by the Italian legislation, would lapse and therefore cannot be made public.

Through the processing carried out, it was possible to investigate the *Mar Piccolo* (I and II inlet) entire basin for all the areas affected by the phenomena under investigation. Whereas fish farms were not found, the frames were discarded to lighten the files produced to make them easier to use on each machine.

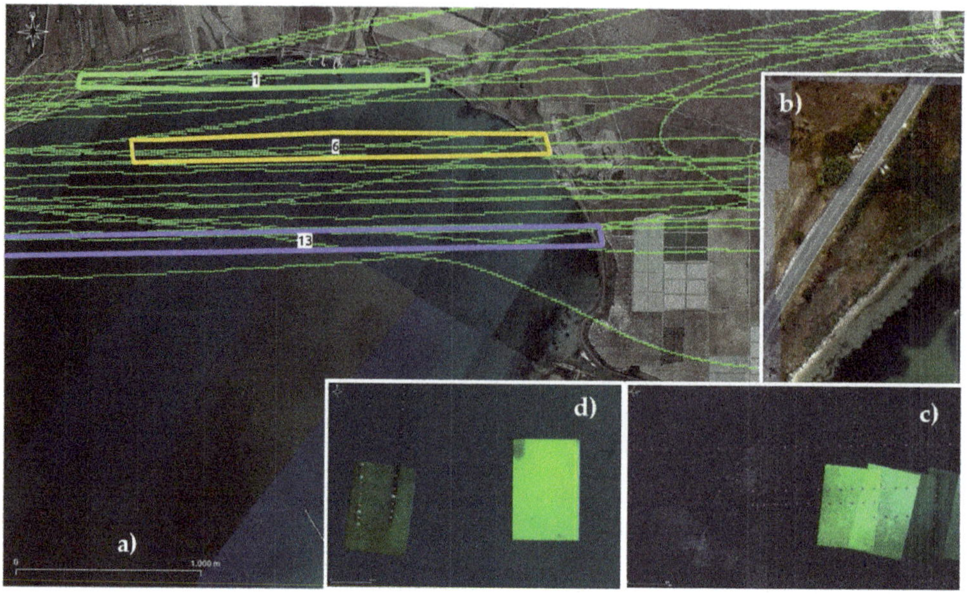

Figure 9. (**a**) phase of the creation of the union framework. Detail of some video tracks during processing, a detail in which the set of significant individual frames is superimposed for each flight path. (**b**) Mosaic of two frames of the investigated study area. The perfect overlap of them is shown with the example of a street; (**c**) Overlapping of the frames after the georeferencing phase with the flight route data identified with dots; (**d**) Example of how a frame, not affected by the presence of anthropogenic objects, is coloured (right).

This activity was relatively immediate to carry out because, with the implemented procedure in the absence of objects of anthropogenic origin (i.e. buoy) in the sensor's field of view, the single frame is coloured in intense green, very easy to find and put aside (Figure 9d).

We did not consider it helpful to use many "machine hours" to create a mosaic of raster to reduce the visibility of the edges of the frames as this lengthy procedure would not have added anything useful but could only be applied for aesthetic purposes.

3. Results

For the first time, a complete mapping of the mussel farming plants present, active and abandoned, in the first and second inlet of the *Mar Piccolo* of *Taranto* was carried out. This achieved result represents a significant milestone to understand the dynamics taking place in the territory

useful to implement initiatives for the sustainable management of this delicate transition ecosystem.

In total, over 5000 frames were georeferenced for a total of over 126 GB of files in Tagged Image File Format (TIFF) format have been produced relating to ultra-detailed mapping (resolution < 5 cm) of the over 80 videos acquired by the *Guardia di Finanza*. Images acquired and processed provided the mapping of each cultivation module of the mussel farm.

Pole mussel farms, floating buoys and longlines were easily visualised because their identification on the image is related to the sensor's spatial resolution. The socks, generally made of polypropylene, have a length ranging from 5 to 6 m, are located at a distance of about 1.5 m from each other and have meshes whose size depends on the size of the mussels. All these details were not possible to map with the aerial images of the Puglia region in 2016 as they are subject to the influence of waves that confuse the interpretation of the images, while this does not happen with the LEOSS sensor (Figure 10a).

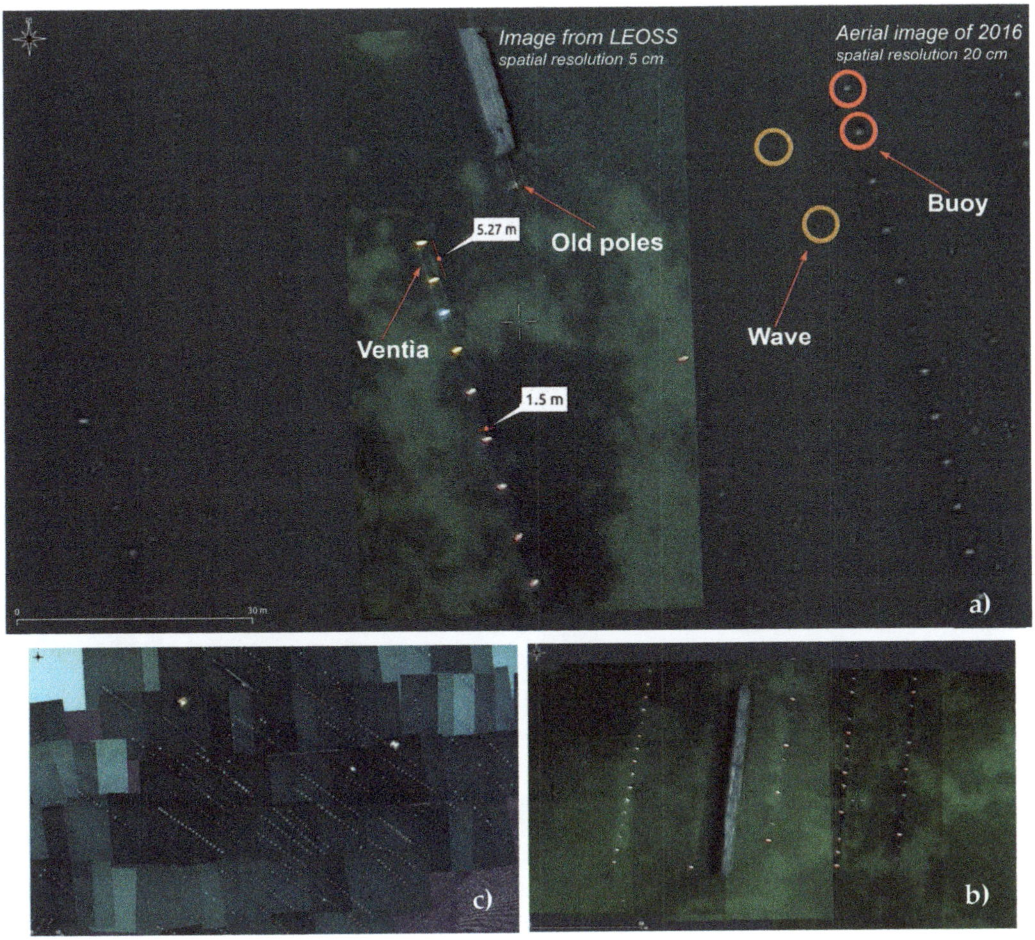

Figure 10. (**a**) Example of details identifiable with LEOSS unlike the aerial images of Puglia of 2016; (**b**) Detail of the part of the overall mosaic of all mussel farms; (**c**) Magnification of the high degree of detail provided.

Aiming to detect any unauthorised anthropogenic actions that may alter the ecosystem profile, efforts were made to provide new, highly detailed information.

The *long lines* identified in each module is located at a distance very variable in the interval of 10–20 m one to each other (Figure 10b,c).

In addition, through the on-screen interpretation, with GIS Desktop software, it is possible, for example, to identify variations in the perimeter of mussel farming fields. These variations, in removal and addition, have been identified in a percentage >20% as a total number.

Furthermore, old poles mussel farms in wood and galvanised steel alloys (semi-submerged and submerged), colour (red, blue, yellow, white) of floating polyethene buoys of varying sizes (generally 1 m in length) were also identified (Figure 11a,b).

Figure 11. (**a**) Invisible and submerged poles (red arrows), (First Inlet of the Mar Piccolo, Puglia Region); (**b**) Abandoned mussel fields. First Inlet of the Mar Piccolo near Galeso river mouth, Puglia Region (**c**) Identification of an abandoned mussel farm. (**c**) Aerial image of Puglia in 2016; (**d**) LEOSS recent image on the right after the processing here reported. We could call it the "luminol" effect.

Thanks to work carried out, it is also possible to identify and recognise those fields once used and now abandoned (Figure 11c,d) that are not visible with the aerial images available [54] and not even with the well-known Google services.

On the one hand, in the first inlet, the estimated area dedicated to mussel farming is 2.8 km^2 (on a total area of 8.3 km^2), representing 34% coverage. The 13% of the area dedicated to mussel cultivation (0.38 km^2) is affected by ancient abandoned structures. These structures are mainly located near the Galeso submarine water spring and at the mouth of the homonymous river.

On the other hand, in the second inlet, the estimated area dedicated to mussel aquaculture activities is 6.54 km^2 (on a total area of 12.4 km^2), representing 53% coverage. Differently from the first inlet, in the second one, just 2% of the area dedicated to mussel farms (0.15 km^2) is affected by the presence of abandoned structures.

This difference is probably due to the presence of several shipyards active in the first inlet until a few years ago, which prevented the installation of the mussel facilities to ensure the navigation of ships.

In addition, lately, water pollution phenomena due to PCBs and dioxins, have led to a further abandonment of mussel crops by fishermen in the first inlet.

In the second inlet, it was possible to increase the density of farms because no active shipyards and no pollution phenomena subsist.

The high quality of the images processed, according to the procedures used, also made it possible to identify other partially submerged structures, bulky waste (wreck) and/or near "on the surface of the water" (tire) Figure 12. Precision mapping activities are still ongoing.

Figure 12. Some waste and wrecks identified by the mapping activities (the photos are related to post-mapping inspections. First and second inlet of the Mar Piccolo (Puglia Region).

4. Discussion

The video data available were processed for two purposes:

- Mapping of mussel crops (Figure 10) for a comparison with the authorised areas in the entire study area;
- Identify semi-submerged poles, mussel cemeteries and other significant aspects of the environmental degradation of the area and ecosystem (Figures 11 and 12).

Based on these objectives, the applied methodology has led to remarkable results (Figure 10) that can implement the cognitive status of the area of interest due to the higher resolution acquired than that obtained to date [54].

Given the importance of the ongoing investigations, using images certified by the *Guardia di Finanza,* it is possible to capture a picture of the state of the art of the entire ecosystem at a given moment. Otherwise, using images from free providers, we would notice that we have no absolute certainty of the acquisition date of the various strips and that the dates of contiguous strips often have different dates.

Therefore it would not be possible to have a representative image of the entire area at a given time. In addition, Italian law provides that evidence must be collected by state police bodies to initiate judicial activities.

In previous studies, mussel farms were often mapped and monitored using images from Landsat, QuickBird, SPOT, RADARSAT and ENVISAT satellites [11,56]. However, the limited spatial resolution of satellite images could not provide sufficient information to map in detail mussel aquaculture systems.

Among the different satellites images, the maximum spatial resolution achievable is not confrontable with the visible range of LEOSS.

Comparing aerial images with a 20 cm spatial resolution with our images acquired by LEOSS sensor with a higher resolution (5 cm) (Figure 10), it is possible to observe an evident improvement of the level of detail of our images, allowing us to distinguish the mapped buoys farm's typology and colour, the presence of *ventia* and pole farm's shape. Moreover, it should be considered that detailed aerial images are not freely accessible, and their retrieval is not so easy even for port authorities.

Analysing the overall results, is it possible to observe a very disordered situation in the *Mar Piccolo* of *Taranto* (Figure 13), resulting from clear overexploitation of the area also due to the presence of old mussel aquaculture systems. Moreover, different types of waste, submerged and semi-submerged wrecks and abandoned structures, were also observed (Figures 11 and 12). All these aspects translate into a real obstacle course that makes navigation extraordinarily complex and dangerous for boats and local authorities.

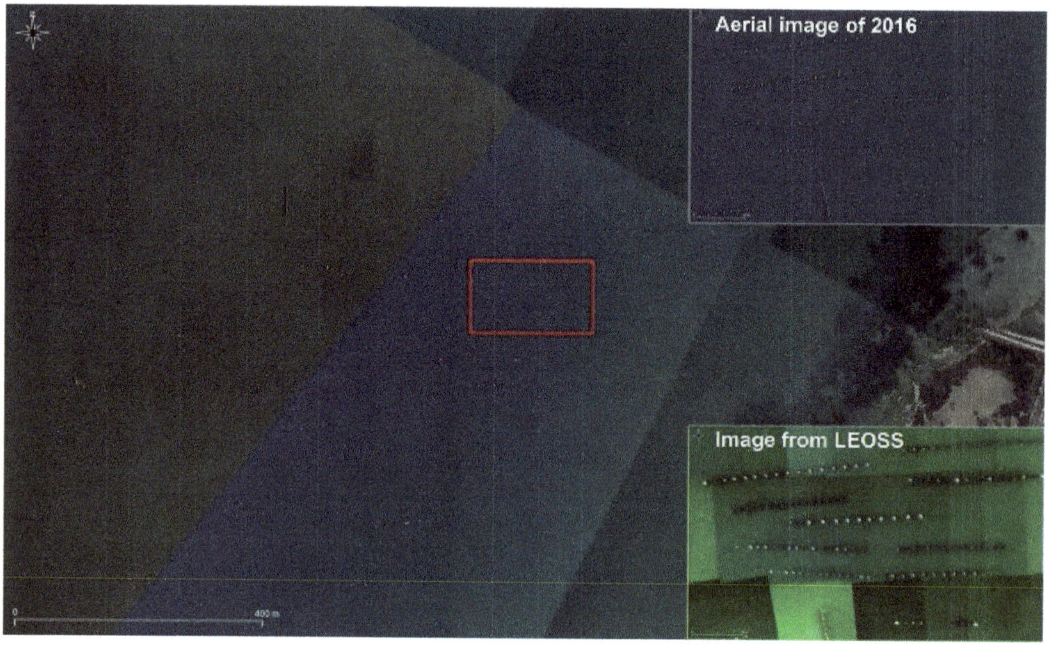

Figure 13. Example of disorder due to overexploitation.

Proper management of the areas of the *Mar Piccolo* granted in concession for the development of mussel farming activities should take into account the load-bearing capacity of the system to guarantee both the environmental and economic sustainability of the production site.

However, in the absence of specific studies in this sense, it has not been possible to give more precise indications on the maximum load for each plant (e.g. the number of pergolas per farm room) relying on the direct experience of mussel farmers.

In this context, based on our data, manufacturers will have indications on the optimal load to be respected to have a quality product.

The work carried out has provided important information, which, however, cannot be fully carried out because of the ongoing investigations by the competent authorities. Many mussel farms had no authorisation at the time of video acquisition. In addition, this study made it possible to identify the distance between the different plants and their location in the two inlets. This aspect is important to quantify the production capacity of the considering marine area also considering the benefits to the community in terms of ecosystem services [12].

According to [56], the spatial planning of mussel fields must occur in a specific site way analysing the different variables that characterise the area of interest. Therefore, according to current regulations, a specification for the use and management of the areas granted for mussel farming activities is required considering the studies and the surveys carried out [25,57].

The single area covered by the state concession, i.e. the water mirror intended for mussel farming, is assigned by procedure and public evidence.

The spaces that can be granted in the *Mar Piccolo* of *Taranto* are located in the First inlet and the Second inlet only within the areas delimited for this purpose by the Ordinance of the Port Authority 107/2005 [58]. According to the provisions of the abovementioned Ordinance, the individual fields suitable for the necessary installation have been identified within each area consented to mussel farming.

This assessment took into account the bathymetric development of the seabed, the presence along the coast of canals and rivers [59], potential landing points, the necessary safety distances for the navigation of nautical means and their passage through mussel fields. The individual modules that can be obtained for mussel cultivation, each rectangular, have an extension of 2 Ha (125 m × 160 m). For the First Inlet, 63 areas are planned, for a total area of 126 Ha. In the Second Inlet, 238 areas of 2 hectares each are designed, for a total area of 476 Ha. Therefore, the total number of areas in the *Mar Piccolo* is equal to 126 + 476 = 602 Ha. The current spatial planning has a minimum distance of 20 m between adjacent rows of areas and 5 m from offshore to coast. A minimum distance of 50 m has been provided among the sectors identified, suitably increased at potential landing areas. Dealers must ensure this minimum distance. Each area granted is characterised by a unique code consisting of the sector name and an identification number. At each of the four vertices of the area in use, the dealer must place a yellow bi-conic buoy. Among these, the one at the S-E summit of the concession must be equipped with a flag and must have applied an identification plate of the concession. The plant structure consists of a series of parallel modules, organised according to the "monoventìa" or "biventìa" system, i.e. with one or two main ropes suspended on the same series of floats.

Each module consists of a polypropylene rope, usually called "ventìa", whose thickness varies between 40 and 50 mm and the length of which is not less than 6 metres. The long line consists of a number between 10 and 20 "ventìe" [60]. Floats are PE-HD buoys for mussel breeding in various colours and generally varying in size (generally 1 m in length). Floats at both ends of the "ventìa" are anchored to the bottom by dead concrete bodies, having an anchor function, with polyethene or polypropylene cables, the length of which varies with the bathymetry of the site. The "ventìa" has the function of supporting the "reste", tubular socks that act as a support for the mussel growth [25]. The socks, generally made of polypropylene, have a length ranging from 4 to 6 m, are located at about 40–50 cm from each other. The mesh size depends on the size of the mussels. The distance between the floats carrying the parallel "ventìe" is between 10 and 20 m (Figure 14).

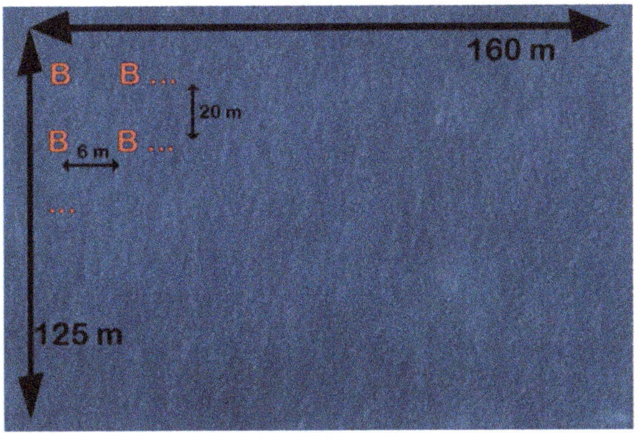

Figure 14. Scheme of the mussel farm module according to the proposed specification. B = Buoy.

Each concession must adopt floats of the same colour and size (unless otherwise indicated for any housing areas), taking care to adopt colours other than adjacent concessions, to distinguish the ownership of the area. Finally, particular attention should be paid to waste that can create problems for the marine ecosystem [30,61]. In this regard, the municipality of *Taranto* has already signed an agreement protocol for the use of bioplastics [62].

Possible alternatives to current mussel farming systems could be identified by multipurpose structures such as shellfish towers that can be used for mussels and other possible candidates as seaweed, lobsters, sponges or tunicates [63,64]. However, these structures are currently in the prototype phase, and they need deep waters, present only in limited areas of the *Mar Piccolo*, to be installed. This is a big deal because the sediment of the *Mar Piccolo* is susceptible to remoulding and resuspension due to its very low consistency [36]. Therefore this solution may not be optimal for the reference area.

The possibility of installing mussel farms hooked to the supporting structures of offshore wind farms appears to be very advantageous [65], but even in this case, we must consider some aspects. Regarding the positioning of mussel systems in the Puglia region, the legislation provides that mussel farms must limit the impact on the environment through mitigation actions locating them far from protected areas. In addition, the solution proposed by [65] has to take into account the health and hygiene, landscape constraints and cultural and economic aspects (cost of authorisations) that subsist in the area.

Theoretically, it would not be possible to build mussel farms in the first inlet due to the proximity to the SIN (currently we go in derogation of the regulations).

Moreover, the role of mussels in water filtering and remediation [66] remains essential as an ecosystem service for the environment [67].

Lately, mussel farming practices in the first inlet are in decline due to heavy pollution, which forces farmers to pay higher costs, often to dispose of polluted and unsaleable crops. Therefore the point on which to focus attention is to remove the pollution sources and reclaim the area. In addition to this delicate aspect, the right approach is to make the practice more sustainable by reusing waste products such as mussel shells and recovering plastic materials, as highlighted by the many project proposals present in the Italian Recovery and Resilience Plan (NRRP) [68] which are based on studies on hotspots as evaluation more critical of the life cycle of mussel farms [69].

5. Conclusions

The specific activity carried out supports the interventions to be implemented for the reduction and/or interruption of contamination flows in the *Mar Piccolo* as well as the exploitation of marine resources.

The territorial analysis carried out led to the achievement of an instant and current cartographic representation of the state of the places in the entire basin of the *Mar Piccolo.*

The digital cartography produced provides a knowledge database thanks to which those structures that impact the ecosystem of interest can be identified with high precision and a large scale of detail.

The results achieved by these activities have led to the definition of detail of all the mussel farms in the *Mar Piccolo* sea and the areas affected by illegal mussel cultivation activities. Therefore, it was possible to map those areas where it was necessary to remove abusive structures and sanction fishers and lay the groundwork to develop a sustainable management model in the two inlets of the *Mar Piccolo* concerning mussel farming. In addition, the foundations have been laid to implement a specification of use for these activities, which has become an integral part of the Municipal Coastal Area Management Plan [70].

The level of detail reached with our methodology (*airborne sensors and open-source software*) is considerably higher than satellite imagery, allowing us to frame in great detail the state of mussel farming and propose guidelines for crops for the first time and allowing the identification of submerged objects up to 1m deep.

Moreover, through the operational workflow developed, it was also possible to identify different types of marine litter (e.g., out-of-use tires), illegal activities along the banks, uncontrolled mini landfills and illegal fishing and submerged objects otherwise not visible (e.g., "*ghost nets*", wrecked boats and sunken mussel seed capture installations).

The adaptability of open source software to particular and complex contexts has been widely demonstrated as they can also integrate well with hardware components whose characteristics are in a certain way little known and sometimes secret.

Finally, during the processing of the images, it was also possible to identify algal communities; these would not have been identified with sensors operating in the same i.e., the visible range. This reveals new research perspectives for developing new sensors dedicated to the detailed mapping of these biological characteristics of transitional waters and shallow water bodies.

Other developments that can be hypothesised and on which appropriate investigations are being made are:

- identification of any illegal activities as the presence of the fishing net Figure 15;
- detailed mapping of all waste (Figure 16, Figure 17);
- algal blooms and species of high ecological interest (phanerogams, biocenosis) that have been registered in the study area on the days of image acquisition;
- identification of unfavourable locations for mussel farm sites due to turbidity or potential development of harmful algal blooms (we could perform this also using historical series of satellite maps);
- identification and mapping of marine biocoenosis and phanerogams of high ecological interest

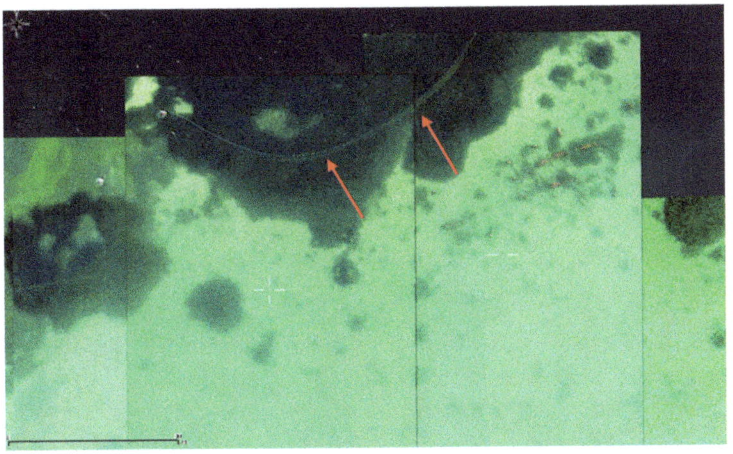

Figure 15. Illegal fishing net.

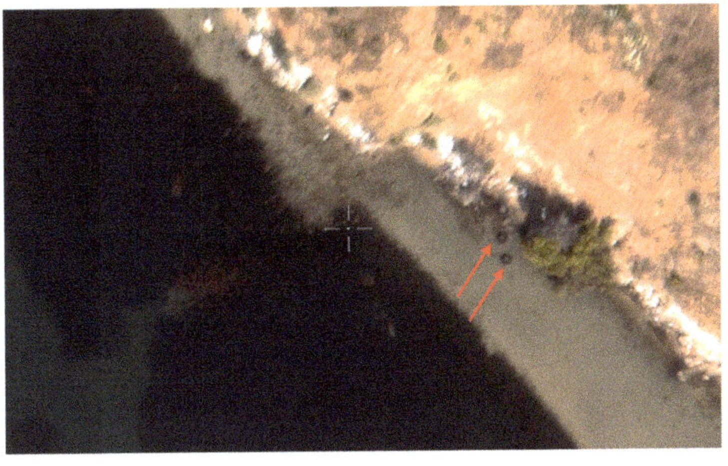

Figure 16. Detail of two tires (red arrows) on the beach.

Figure 17. Inspection in which we verified the presence of waste on the banks. First inlet of the Mar Piccolo (Puglia Region). September 2019.

Although our most ambitious goal would be to create other aircraft features, such as real-time counting and mapping, immediate change detection through implementing the harris algorithm [71].

Finally, it is hoped that this case study will become best practices in the combined use of open source technologies in previously precluded fields, especially for a valuable purpose to the community.

Supplementary Materials: The following are available online at https://www.mdpi.com/article/10.3390/rs13152985/s1.

Author Contributions: Conceptualization, C.M. and V.F.U.; methodology, C.M. and C.G.; software, C.M.; validation, C.M., C.G. and I.S.; formal analysis, C.M., C.G. and I.S.; investigation, C.M., C.G., I.S. and C.C.; resources, C.M. and V.F.U.; data curation, C.M. and C.G.; writing—original draft preparation, C.M, C.G., I.S. and C.C.; writing—review and editing, C.M., C.G., I.S. and C.C.; visualization, C.M., C.G., I.S. and C.C.; supervision, C.M. and V.F.U.; project administration, C.M. and V.F.U.; funding acquisition, V.F.U. All authors have read and agreed to the published version of the manuscript.

Funding: Not applicable.

Data Availability Statement: The datasets generated during the current study are not publicly available and are used through a specific clearance issued by the Special Commissioner for urgent measures of reclamation, environmental improvements and redevelopment of *Taranto*.

Acknowledgments: This study was promoted, supervised and funded by the Special Commissioner for urgent measures of reclamation, environmental improvements and redevelopment of *Taranto*, V. Corbelli, in the framework of the Collaboration Agreement (ex article 15 of law 241/90). The authors gratefully thank ROAN of *Bari* for the active collaboration.

Conflicts of Interest: The authors declare no conflict of interest.

References

1. Huang, W.; Fu, B. Remote sensing for coastal area management in China. *Coast. Manag.* **2002**, *30*, 271–276. [CrossRef]
2. Mumby, P.J.; Green, E.P.; Edwards, A.J.; Clark, C.D. The cost-effectiveness of remote sensing for tropical coastal resources assessment and management. *J. Environ. Manag.* **1999**, *55*, 157–166. [CrossRef]

3. McCarthy, M.J.; Colna, K.E.; El-Mezayen, M.M.; Laureano-Rosario, A.E.; Méndez-Lázaro, P.; Otis, D.B.; Toro-Farmer, G.; Vega-Rodriguez, M.; Muller-Karger, F.E. Satellite Remote Sensing for Coastal Management: A Review of Successful Applications. *Environ. Manag.* **2017**, *60*, 323–339. [CrossRef] [PubMed]
4. Gholizadeh, M.H.; Melesse, A.M.; Reddi, L. A comprehensive review on water quality parameters estimation using remote sensing techniques. *Sensors* **2016**, *16*, 1298. [CrossRef]
5. Transon, J.; d'Andrimont, R.; Maugnard, A.; Defourny, P. Survey of hyperspectral Earth Observation applications from space in the Sentinel-2 context. *Remote Sens.* **2018**, *10*, 157. [CrossRef]
6. Lu, B.; Dao, P.D.; Liu, J.; He, Y.; Shang, J. Recent advances of hyperspectral imaging technology and applications in agriculture. *Remote Sens.* **2020**, *12*, 2659. [CrossRef]
7. Fu, W.; Ma, J.; Chen, P.; Chen, F. Remote Sensing Satellites for Digital Earth. In *Manual of Digital Earth*; Springer: Singapore, 2020.
8. Gholizadeh, A.; Kopačková, V. Detecting vegetation stress as a soil contamination proxy: A review of optical proximal and remote sensing techniques. *Int. J. Environ. Sci. Technol.* **2019**, *16*. [CrossRef]
9. Chang, J.; Clay, E.D. Matching ReMatching Remote Sensing to Problems-Remote Sensing to Problems November 2016. In *iGrow Corn: Best Management Practices*; Clay, E.D., Clay, S.A., Carlson, C.G., Byamukama, E., Eds.; South Dakota State University: Brookings, South Dakota, 2016.
10. Haji Gholizadeh, M.; Melesse, A.M.; Reddi, L. Spaceborne and airborne sensors in water quality assessment. *Int. J. Remote Sens.* **2016**, *37*, 3143–3180. [CrossRef]
11. Grant, J.; Bacher, C.; Ferreira, J.G.; Groom, S.; Morales, J.; Rodriguez-Benito, C.; Saitoh, S.-I.; Sathyendranath, S.; Stuart, V. *Remote Sensing Applications in Marine Aquaculture-Remote Sensing in Fisheries and Aquaculture—Chapter 6 Remote Sensing Applications in Marine Aquaculture*; Forget, M.-H., Stuart, V., Platt, T., Eds.; IOCCG: Dartmouth, NS, Canada, 2009.
12. Suplicy, F.M. A review of the multiple benefits of mussel farming. *Rev. Aquac.* **2020**, *12*, 204–223. [CrossRef]
13. Ottinger, M.; Clauss, K.; Kuenzer, C. Aquaculture: Relevance, distribution, impacts and spatial assessments-A review. *Ocean. Coast. Manag.* **2016**, *119*, 244–266. [CrossRef]
14. Nikolaidis, G.; Koukaras, K.; Aligizaki, K.; Heracleous, A.; Kalopesa, E.; Moschandreou, K.; Tsolaki, E.; Mantoudis, A. Harmful microalgal episodes in Greek coastal waters. *J. Biol. Res.* **2005**, *3*, 77–85.
15. Green, E.P.; Mumby, P.J.; Edwards, A.J.; Clark, C.D. A review of remote sensing for the assessment and management of tropical coastal resources. *Coast. Manag.* **1996**, *24*, 1–40. [CrossRef]
16. Cerruti, A. Le sorgenti sottomarine (Citri) del Mar Grande e Mar Piccolo di Taranto. *Ann. Ist. Super. Nav. Di Napoli* **1938**, *7*, 171–196.
17. Spizzico, M.; Tinelli, R. Hydrogeology of Galese spring, Mar Piccolo of Taranto (South Italy). In Proceedings of the 9th Salt Water Intrusion Meet, Delft, The Netherlands, 12–16 May 1986. Water Manag. Group Delft Univ. Technol.: Delft, The Netherlands, 1986; pp. 85–97.
18. Massarelli, C.; Matarrese, R.; Uricchio, V.F.; Vurro, M. GRASS GIS processing to detect thermal anomalies with TABI sensor. *Conf. FOSS4G Eur. Como* **2015**, *1*, 477–482.
19. Zuffianò, L.E.; Basso, A.; Casarano, D.; Dragone, V.; Limoni, P.P.; Romanazzi, A.; Santaloia, F.; Polemio, M. Integrated Environmental Characterization of the Contaminated Marine Coastal Area of Taranto, Ionian Sea (Southern Italy)-The RITMARE Project. *Environ. Sci. Pollut. Res.* **2016**, 12491–12494. [CrossRef]
20. Valenzano, E.; D'Onghia, M.; De Giosa, F.; Demonte, P. Morfologia Delle Sorgenti Sottomarine Dell'area di Taranto (Mar Ionio). *Mem. Descr. Carta Geol. d'It.* **2020**, *105*, 65–69.
21. ARPA-Puglia. *Approfondimento Tecnico-Scientifico Sulle Interazioni tra il Sistema Ambientale ed i Flussi di Contaminanti da Fonti Primarie e Secondarie. Elaborazione di un Modello Concettuale Sito-Specifico*; Ecomondo—ReclaimEXPO: Rimini, Italy, 2014.
22. Cardellicchio, N.; Buccolieri, A.; Giandomenico, S.; Lopez, L.; Pizzulli, F.; Spada, L. Organic pollutants (PAHs, PCBs) in sediments from the Mar Piccolo in Taranto (Ionian Sea, Southern Italy). *Mar. Pollut. Bull.* **2007**, *55*, 451–458. [CrossRef]
23. Parenzan, P. Parenzan Il Mar Piccolo e il Mar Grande di Taranto. *Thalass. Salentina* **1969**, *3*. [CrossRef]
24. Caroppo, C.; Giordano, L.; Palmieri, N.; Bellio, G.; Bisci, A.P.; Portacci, G.; Sclafani, P.; Hopkins, T.S. Progress toward sustainable mussel aquaculture in Mar Piccolo, Italy. *Ecol. Soc.* **2012**, *17*. [CrossRef]
25. Giordano, L.; Portacci, G.; Caroppo, C. Multidisciplinary tools for sustainable management of an ecosystem service: The case study of mussel farming in the Mar Piccolo of Taranto (Mediterranean, Ionian Sea). *Ocean. Coast. Manag.* **2019**, *176*, 11–23. [CrossRef]
26. Portacci, G. *Contabilizzazione Dei Servizi Ecosistemici: Il Caso Della Molluschicoltura Environmental-Accounting of the Ecosystem Services*; 48° Congresso della Societá Italiana di Biologia Marina: Rome, Italy, 2017.
27. Guardia Costiera Taranto. Capitaneria di Porto di Taranto. 2005. Available online: https://www.guardiacostiera.gov.it/Taranto/Pages/ordinanze.aspx (accessed on 20 June 2021).
28. Regione Pugia. Ordinanza del Presidente Della Giunta Regionale, n. 532: Misure Sanitarie Straordinarie di Controllo del Rischio Per Diossina e PCB Nelle Produzioni di Mitili di Taranto. BURP: Regione Pugia, Italy, 2018.
29. European Parliament and Council of Europe. Corrigendum to Regulation (EC) No 854/2004 of the European Parliament and of the Council of 29 April 2004 Laying Down Specific Rules for the Organisation of Official Controls on Products of Animal Origin Intended for Human Consumption. Available online: https://eur-lex.europa.eu/LexUriServ/LexUriServ.do?uri=OJ:L:2004:226:0083:0127:EN:PDF (accessed on 20 June 2021).

30. Spirkovski, Z.; Ilik-Boeva, D.; Ritterbusch, D.; Peveling, R.; Pietrock, M. Ghost net removal in ancient Lake Ohrid: A pilot study. *Fish. Res.* **2019**, *211*, 46–50. [CrossRef]
31. Tecniche, A.N.D.I.; Giuseppe, O.A.; Lobasso, V.; Melucci, R.; Fabrizio Manzulli, I. DI NTA PCC. 2020. Available online: https://www.comune.taranto.it/attachments/article/3378/RELAZIONE%20%20PCC%20TARANTO_DIC_2020.pdf (accessed on 3 June 2021).
32. Mali, M.; Dell'Anna, M.M.; Notarnicola, M.; Damiani, L.; Mastrorilli, P. Combining chemometric tools for assessing hazard sources and factors acting simultaneously in contaminated areas. Case study: "Mar Piccolo" Taranto (South Italy). *Chemosphere* **2017**, *184*, 784–794. [CrossRef] [PubMed]
33. Di Leo, A.; Cardellicchio, N.; Giandomenico, S.; Spada, L. Mercury and methylmercury contamination in Mytilus galloprovincialis from Taranto Gulf (Ionian Sea, Southern Italy): Risk evaluation for consumers. *Food Chem. Toxicol.* **2010**, *48*, 3131–3136. [CrossRef]
34. Cardellicchio, N.; Buccolieri, A.; Di Leo, A.; Giandomenico, S.; Spada, L. Levels of metals in reared mussels from Taranto Gulf (Ionian Sea, Southern Italy). *Food Chem.* **2008**, *107*, 890–896. [CrossRef]
35. Republic Italian National Law n. 426/1998. In *Nuovi Interventi in Campo Ambientale*; Gazzetta Ufficiale n. 291del 14 dicembre 1998; 1998. Available online: https://www.mite.gov.it/sites/default/files/archivio/normativa/legge_09121998_426.pdf (accessed on 3 June 2021).
36. Cotecchia, F.; Vitone, C.; Sollecito, F.; Mali, M.; Miccoli, D.; Petti, R.; Milella, D.; Ruggieri, G.; Bottiglieri, O.; Santaloia, F.; et al. A geo-chemo-mechanical study of a highly polluted marine system (Taranto, Italy) for the enhancement of the conceptual site model. *Sci. Rep.* **2021**, *11*. [CrossRef] [PubMed]
37. Labianca, C.; De Gisi, S.; Todaro, F.; Notarnicola, M. DPSIR Model Applied to the Remediation of Contaminated Sites. A Case Study: Mar Piccolo Taranto. *Appl. Sci.* **2020**, *10*, 5080. [CrossRef]
38. Garen, P.; Robert, S.; Bougrier, S. Comparison of growth of mussel, Mytilus edulis, on longline, pole and bottom culture sites in the Pertuis Breton, France. *Aquaculture* **2004**, *232*, 511–524. [CrossRef]
39. Konstantinou, Z.I.; Kombiadou, K. Rethinking suspended mussel-farming modelling: Combining hydrodynamic and bio-economic models to support integrated aquaculture management. *Aquaculture* **2020**, *523*, 735179. [CrossRef]
40. Welcome to the Python GDAL/OGR Cookbook!—Python GDAL/OGR Cookbook 1.0 Documentation. Available online: http://pcjericks.github.io/py-gdalogr-cookbook/index.html# (accessed on 3 June 2021).
41. GDAL—GDAL Documentation. Available online: https://gdal.org/index.html (accessed on 3 June 2021).
42. FFmpeg License and Legal Considerations. Available online: https://ffmpeg.org/legal.html (accessed on 3 June 2021).
43. GNU licenses. Available online: https://www.gnu.org/licenses/lgpl-3.0.txt (accessed on 1 June 2021).
44. Ubuntu Server-for Scale Out Workloads | Ubuntu. Available online: https://ubuntu.com/server (accessed on 3 June 2021).
45. Shi, Y.; Yang, H.; Gong, M.; Liu, X.; Xia, Y. A Fast and Robust Key Frame Extraction Method for Video Copyright Protection. *J. Electr. Comput. Eng.* **2017**, *2017*. [CrossRef]
46. FFmpeg Filters Documentation. Available online: https://ffmpeg.org/ffmpeg-filters.html#crop (accessed on 8 June 2021).
47. The ImageMagick Development Team. Available online: https://imagemagick.org (accessed on 8 June 2021).
48. Apache Licenses. Available online: https://www.apache.org/licenses/ (accessed on 3 June 2021).
49. Gonzalez, R.C.; Woods, R.E. *Digital Image Processing*; Pearson: London, UK, 2008; ISBN 978-0-13-168728-8.
50. Al-Ameen, Z. Sharpness Improvement for Medical Images Using a New Nimble Filter. *3D Res.* **2018**, *9*, 1–12. [CrossRef]
51. Bull, D.R. Digital Picture Formats and Representations. In *Communicating Pictures*; Elsevier: Amsterdam, The Netherlands, 2014; pp. 99–132.
52. Barnsley, M.F. *Fractals Everywhere*; Elsevier: Amsterdam, The Netherlands, 1993; ISBN 978-0-12-079061-6.
53. Szeliski, R. *Computer Vision*; Texts in Computer Science; Springer: London, UK, 2011; ISBN 978-1-84882-934-3.
54. Apulia Region Territorial Information System. Available online: http://www.sit.puglia.it/portal/portale_cartografie_tecniche_tematiche/WMS (accessed on 3 June 2021).
55. GitHub-Gvellut/FreehandRasterGeoreferencer: QGIS Plugin for the Interactive Georeferencing of Rasters. Available online: https://github.com/gvellut/FreehandRasterGeoreferencer (accessed on 3 June 2021).
56. Alexandridis, T.K.; Topaloglou, C.A.; Lazaridou, E.; Zalidis, G.C. The performance of satellite images in mapping aquacultures. *Ocean Coast. Manag.* **2008**, *51*, 638–644. [CrossRef]
57. Corbelli, V.; Capasso, G.; Velardo, R. Talassografico "A. Cerruti". In *AUSL Taranto-Dipartimento di Prevenzione Disciplinare d'uso e Gestione Delle Aree Concesse per Attività di Mitilicoltura nel Mar Piccolo e Mar Grande di Taranto*; BURP: Puglia Region, Italy, 2019.
58. Guardia Costiera Taranto. *Orders nn. 222/2009 and 107/2005*; Guardia Costiera Taranto: Taranto, Italy, 2009.
59. Konstantinou, Z.I.; Kombiadou, K.; Krestenitis, Y.N. Effective mussel-farming governance in Greece: Testing the guidelines through models, to evaluate sustainable management alternatives. *Ocean Coast. Manag.* **2015**, *118*, 247–258. [CrossRef]
60. Regione autonoma della Sardegna. *Acquacoltura in Sardegna. Tradizioni, Innovazione e ambiente*; Laore Sardegna: Cagliari, Italy, 2016.
61. Da Ros, L. *LIFE-GHOST Project. Final Report*; ISMAR Istituto di Scienze Marine: Venice, Italy, 2016.
62. Retine in Bioplastica per la Mitilicoltura, Firmato un Protocollo D'intesa tra Comune di Taranto e Novamont. Available online: https://www.comune.taranto.it/index.php/elenco-news/3121-retine-in-bioplastica-per-la-mitilicoltura-firmato-un-protocollo-d-intesa-tra-comune-di-taranto-e-novamont (accessed on 14 June 2021).

63. Heasman, K.G.; Scott, N.; Smeaton, M.; Goseberg, N.; Hildebrandt, A.; Vitasovich, P.; Elliot, A.; Mandeno, M.; Buck, B.H. New system design for the cultivation of extractive species at exposed sites-Part 1: System design, deployment and first response to high-energy environments. *Appl. Ocean Res.* **2021**, *110*, 102603. [CrossRef]
64. Landmann, J.; Fröhling, L.; Gieschen, R.; Buck, B.H.; Heasman, K.; Scott, N.; Smeaton, M.; Goseberg, N.; Hildebrandt, A. New system design for the cultivation of extractive species at exposed sites-Part 2: Experimental modelling in waves and currents. *Appl. Ocean Res.* **2021**, *113*, 102749. [CrossRef]
65. Dalton, G.; Bardócz, T.; Blanch, M.; Campbell, D.; Johnson, K.; Lawrence, G.; Lilas, T.; Friis-Madsen, E.; Neumann, F.; Nikitas, N.; et al. Feasibility of investment in Blue Growth multiple-use of space and multi-use platform projects; results of a novel assessment approach and case studies. *Renew. Sustain. Energy Rev.* **2019**, *107*, 338–359. [CrossRef]
66. Montaudouin, X.; de Sandra, E. (Eds.) Shellfish aquaculture and the environment. *Environ. Sci. Pollut. Res.* **2014**, *21*, 7781–7783. [CrossRef]
67. Jansen, H.M.; Van Den Burg, S.; Bolman, B.; Jak, R.G.; Kamermans, P.; Poelman, M.; Stuiver, M. The feasibility of offshore aquaculture and its potential for multi-use in the North Sea. *Aquac. Int.* **2016**, *24*, 735–756. [CrossRef]
68. The National Recovery and Resilience Plan (NRRP)-Ministry of Economy and Finance. Available online: https://www.mef.gov.it/en/focus/The-National-Recovery-and-Resilience-Plan-NRRP/ (accessed on 22 July 2021).
69. Tamburini, E.; Turolla, E.; Fano, E.A.; Castaldelli, G. Sustainability of Mussel (Mytilus Galloprovincialis) Farming in the Po River Delta, Northern Italy, Based on a Life Cycle Assessment Approach. *Sustainability* **2020**, *12*, 3814. [CrossRef]
70. Comune di Taranto Avviso Pubblico Adozione Piano Comunale Delle Coste. Available online: https://www.comune.taranto.it/index.php/41-avvisi/3378-avviso-adozione-piano-comunale-delle-coste (accessed on 19 June 2021).
71. Tadonki, C.; Lacassagne, L.; Dadi, E.; El Daoudi, M. Accelerator-based implementation of the Harris algorithm. In *Lecture Notes in Computer Science (including subseries Lecture Notes in Artificial Intelligence and Lecture Notes in Bioinformatics)*; Springer: Berlin, Heidelberg, 2012; Volume 7340, pp. 485–492.

MDPI
St. Alban-Anlage 66
4052 Basel
Switzerland
Tel. +41 61 683 77 34
Fax +41 61 302 89 18
www.mdpi.com

Remote Sensing Editorial Office
E-mail: remotesensing@mdpi.com
www.mdpi.com/journal/remotesensing

www.ingramcontent.com/pod-product-compliance
Lightning Source LLC
Chambersburg PA
CBHW061441210926
51850CB00008B/1575
* 9 7 8 3 0 3 6 5 5 7 5 5 7 *